ADVANCES IN RENEWABLE ENERGY AND SUSTAINABLE DEVELOPMENT

Advances in Renewable Energy and Sustainable Development focuses on cutting-edge research areas including renewable energy and sustainable development. As a leader in the global megatrend of science and technology innovation, China has been creating an increasingly open environment for science and technology innovation, increasing the depth and breadth of academic cooperation, and building an innovation community that benefits all people. These efforts make a new contribution to globalization and the building of a community for a shared future. The proceedings feature the most cutting-edge research directions and achievements related to Renewable Energy and Sustainable Development.

Subjects in the proceedings include:

- Hydraulic Engineering
- Environmental Science and Environmental Engineering
- Energy Engineering and Energy Technologies
- Green Manufacturing
- Energy Policy and Economics
- Energy Security and Clean Use
- Geothermal Energy

PROCEEDINGS OF THE INTERNATIONAL CONFERENCE ON RENEWABLE ENERGY AND SUSTAINABLE DEVELOPMENT (IRESD 2022), NANNING, CHINA, 20–22 MAY 2022

Advances in Renewable Energy and Sustainable Development

Edited by

Zhoufu Liang
Nanning Normal University, China

Rafiziana Md. Kasmani
School of Chemical and Energy Engineering, Universiti Teknologi Malaysia, Malaysia

CRC Press is an imprint of the
Taylor & Francis Group, an **informa** business

A BALKEMA BOOK

First published 2023
by CRC Press/Balkema

4 Park Square, Milton Park / Abingdon, Oxon OX14 4RN / UK

e-mail: enquiries@taylorandfrancis.com

www.routledge.com – www.taylorandfrancis.com

CRC Press/Balkema is an imprint of the Taylor & Francis Group, an informa business

© 2023 selection and editorial matter, Zhoufu Liang and Rafiziana Md. Kasmani; individual chapters, the contributors

The right of Zongming Li and Mohd Johari Mohd Yusof to be identified as the authors of the editorial material, and of the authors for their individual chapters, has been asserted in accordance with sections 77 and 78 of the Copyright, Designs and Patents Act 1988.

The right of Zhoufu Liang and Rafiziana Md. Kasmani to be identified as the authors of the editorial material, and of the authors for their individual chapters, has been asserted in accordance with sections 77 and 78 of the Copyright, Designs and Patents Act 1988.

All rights reserved. No part of this book may be reprinted or reproduced or utilised in any form or by any electronic, mechanical, or other means, now known or hereafter invented, including photocopying and recording, or in any information storage or retrieval system, without permission in writing from the publishers.

Although all care is taken to ensure integrity and the quality of this publication and the information herein, no responsibility is assumed by the publishers nor the author for any damage to the property or persons as a result of operation or use of this publication and/or the information contained herein.

Library of Congress Cataloging-in-Publication Data
A catalog record has been requested for this book

ISBN: 978-1-032-39407-7 (hbk)
ISBN: 978-1-032-39408-4 (pbk)
ISBN: 978-1-003-34964-8 (ebk)

DOI: 10.1201/9781003349648

Typeset in Times New Roman
by MPS Limited, Chennai, India

Advances in Renewable Energy and Sustainable
Development – Liang & Kasmani (Eds)
© 2023 Copyright the Editor(s), ISBN: 978-1-032-39407-7

Table of contents

Preface	ix
Committee members	xi

Renewable energy application and energy chemical preparation

Current status of development and utilization of dry hot rock and its key technology prospect in China *Xiaolan Lv, Zhiyong Zhang, Lin Zhang & Beibei Yang*	3
Influence of different composite insulation methods on steam long-distance pipeline *Yanming Zhang, Xin Liu, Qun Liu, Xiufang Huang, Duxing Wei & Peng Guan*	12
Preparation and adsorption properties of CaO/modified biochar *Xiaowen Zhang, Lianghua Chen, Guofeng Yu & Yu Ding*	19
Structural design and performance analysis of a two-rotor vertical axis wind turbine applied to a new wind-wave coupled power generator *Guorui Zhu, Jiefei Gao, Yuxuan Liu, Yuankang Xin & Zixuan Li*	28
Design of small domestic sewage treatment device suitable for homestay *Jialin Cao, Hao Yang, Chengming Shi, Feihang Li & Chenghe Sun*	35
Study on cascade effect of combined sediment retention of gully check dam system in Loess small watershed *Weiying Sun & Pan Zhang*	40
The evolution of renewable energy in the United States in a warming climate *Hangzhou Wang*	45
Optimization control strategy of waste heat utilization in data center with renewable energy *Liangliang Zhu, Pengpai Feng, Kui Wang, Tianheng Chen, Yi Ding & Ye Li*	52
Allocation optimization of combined cooling and heating system based on high-voltage and high-temperature composite phase change heat storage device *Meixiu Ma, Wei Kang, Jibiao Hou, Lixiao Liang, Zhanfeng Deng,* *Na Zhang & Mengdong Chen*	59
Adsorption removal of Cu in wastewater with waste motherwort by citric acid modification *Chinghua Liao, Haiqi Lin, Chien Hung, Qiyong Li, Shengchung Chen & Chihung Wu*	67
Preparation of hydrophobically modified PVA/sponge and its properties for oil/water separation *Mouyuan Yang, Qin Liu & Junkai Gao*	73
Effects of wood vinegar on the growth of oil sunflower (*Helianthus annuus* L.) seedlings in a salt-affected soil of Yellow River Delta, China *Shuai Wu, Yanfei Yuan & Qiang Liu*	79
Quantitation and extraction of flavonoids from Okra flowers *Ziping Zhu, Linhua Zhao, Zhenda Xie & Na Li*	84

Detection and analysis of six toxic elements in *Codonopsis lanceolata* in
different areas of Jilin Province in China 89
Fenglin Li, Zhimin Liu, Li Li, Zhongkui Lu & Wuyang Hua

Different approaches to animal testing from scientific and ethical perspectives 95
Ziqi Liu

Screening of fungicides for mulberry sclerotinia in the field 102
*Honglin Mou, Minghai Zhang, Jiequn Ren, Li Chen, Lixin Tan, Huaxian Yu,
Mingjian Guo, Zhimin Fan, Zhangyun Zheng & Yi Yang*

Effects of selenium application concentration on the content of secondary
metabolites in dandelion 113
Zhiguo Zhao, Qi Lu, Xianglong Meng & Xiangjun Lin

Improvement of EFB palm fiber pelletizing with castor meal 121
Jer-Yuan Shiu, Xiang Wang & Chih-Hung Wu

Research on the sublimation of distiller's grain quantum technology peptide
sublimation and the application of livestock feed technology and
market management and sales 130
Huai Lu

Effects of rumen fluid treatment on nutritional components and feeding value of
sweet sorghum 137
Baoyu Yang, Liangyao Bai, Feng Chen, Jiao Wang, Kai Zhang & Sujiang Zhang

Evaluation of the performance characteristics of colored melting ice-snow
asphalt concrete 145
Weidong Ji, Benju Zhang, Zhitao Zhang, Guangyu Men & Yuchuan Feng

Green energy security and urban sustainable development

Research on site selection of prefabricated component factory based on
P-median model 155
Ziji Liu

Traditional ecological wisdom of Guilin Longji terraced field landscape from the
perspective of water adaptability 160
Luyao Hu

Permanent basic farmland based on minimum cumulative resistance model 166
Lu Huang

Causes and countermeasures of poor habitat status of endangered birds 172
Mingrui Li

Evaluation of construction waste recycling schemes under the life cycle:
A proposed weight-TOPSIS approach improved by GRA 179
Runfei Chen, Jiawei Xu, Yudong Xie & Shaokun Zhang

Research on noise reduction of driving vehicles in Sponge City 190
*Zhou-Fu Liang, Xin-yu Li, Jie Pan, Zhi-ling Yang, Si-yu Tang,
Xiu-qin Yang & Zhan-hui Chen*

Construction and reflection of a new electricity market mechanism 195
*Yuhui Xing, Maolin Zhang, Qinggui Chen, Ling Chen, Meihan Jin,
Xuan Yang, Yiguang Zhou & Haoyue Wu*

A differential electricity tariff for high-energy-consuming industry based on
comprehensive energy consumption indicators 205
Zhixun Wang, Rengcun Fang, Xianguo Fan, Tingting Hou, Wei Liao,
Lanfei He & Chao Luo

Analysis of requirements and key problems of national carbon market construction 212
Xiaoxuan Zhang, Zheng Zhao, Yu Wang & Su Yang

Design of a market trading mechanism for energy storage in the context of a
new power system 217
Yuhui Xing, Zhelin Yang, Maolin Zhang, Qinggui Chen, Ling Chen,
Meihan Jin, Xuan Yang & Yiguang Zhou

The carbon market research hotspot and trend in China – Based on citespace
visualization analysis 230
Yujie Wu, Yue Cao & Yinying Duan

CiteSpace-based visualization analysis on the hotspots and trends in the
field of gas power generation 237
Bing He & Ting Ni

Study on prevention strategy of coal mine flood accident 243
Botao Bi & Hong Zhang

Carbon neutral development path under digitization—Based on MATLAB
three-party evolutionary game simulation 250
Kun Xie & Jingli Wu

Can ESG investment improve corporate green innovation performance?
Evidence from China 256
Hui Lyu, Yan Sun, Ruili Zhou & Yue Chu

Integrating technologies into urgent wildlife release 263
Junlin Shao

Spatial planning of eco-industrial park based on reverse logistics 268
Qing Yang, Chen Wang, Chenyang Cai & Xingxing Liu

Possibility analysis of three Gorges Dam to realize sustainable urban
development goals—Taking Yichang City as an example 275
Hsiao-Hsien Lin, Su-Fang Zhang, Penghui Liu, Heyong Wei, Mei-Ling Chan,
Chin-Hsien Hsu & Po Hsuan Wu

Author index 281

Advances in Renewable Energy and Sustainable
Development – Liang & Kasmani (Eds)
© 2023 Copyright the Editor(s), ISBN: 978-1-032-39407-7

Preface

Nowadays, mass gatherings are not permitted by the government. It is uncertain when the COVID-19 pandemic will end, so it remains unclear how long the meeting needed to be postponed, while many scholars and researchers wanted to attend this long-waited conference and have academic exchanges with their peers. Therefore, in order to actively respond to the call of the government, and meet author's request, the IRESD 2022, which was planned to be held in Nanning, China from May 20-22, 2022, was changed to be held online through Zoom software. This approach not only reduces people gathering, but also meets their communication needs.

The conference aimed to broaden the international scientific and technological academic exchange channels, build a platform for shared academic resources, thus promoting scientific and technological innovation on a global scale and enhancing academic cooperation in China and abroad in this field. Frontier information exchange in different fields is also encouraged. We connected the most advanced academic resources at home and abroad, transform research results into industrial solutions, and brings together talents, technologies and capitals to help development. Experts and scholars from universities and scientific research institutions in China and abroad, members of business community, and other prominent personalities are invited to attend the conference.

During the conference, the conference model was divided into three sessions, including oral presentations, keynote speeches, and online Q&A discussion. In the first part, some scholars, whose submissions were selected as the excellent papers, were given about 5-10 minutes to perform their oral presentations one by one. Then in the second part, keynote speakers were each allocated 30-45 minutes to hold their speeches. There were over 60 participants attending the meeting.

In the second part, we invited three professors as our keynote speakers. The first keynote speaker, Prof. Liang Zhoufu, Nanning Normal University, China. His research interests cover civil engineering (sponge city), disaster prevention engineering, soil and water conservation engineering, disaster prevention engineering, environmental engineering and engineering management. And then we had Assoc. Prof. Ayyoob Sharifi, Hiroshima University, Japan. He performed a speech: *Co-benefits and trade-offs between climate change mitigation and adaptation measures and policies in cities.* Lastly, we invited Prof. Nur Islami, Universitas Riau, Pekanbaru, Indonesia as our finale keynote speaker. He delivered a wonderful speech: *Renewable Energy Exploration. A case study in the Riau Indonesia.* Their insightful speeches had triggered heated discussion in the third session of the conference. Every participant praised this conference for disseminating useful and insightful knowledge.

The proceedings of IRESD 2022 span over 3 topical tracks, that include: Renewable Energy, Sustainable Energy and Environment, Power System Technology and other related fields. All the papers have been through rigorous review and process to meet the requirements of international publication standards.

We would like to acknowledge all of those who have supported IRESD 2022. Each individual and institutional help was very important for the success of this conference. Especially we would like to thank the organizing committee for their valuable advices in the organization and helpful peer review of the papers. We hope that IRESD 2022 will be a forum for excellent discussions that will put forward new ideas and promote collaborative researches. We are sure that the proceedings will serve as an important research source of references and the knowledge, which will lead to not only scientific and engineering progress but also other new products and processes.

The Committee of IRESD 2022

Advances in Renewable Energy and Sustainable Development – Liang & Kasmani (Eds)
© 2023 Copyright the Editor(s), ISBN: 978-1-032-39407-7

Committee members

Conference Chairman
Prof. Liang Zhoufu, Nanning Normal University, China

Publication Chair
Prof. Liang Zhoufu, Nanning Normal University, China

Local Organizing Chairs
Prof. Liang Zhoufu, Nanning Normal University, China

Program Committees Chair
Prof. Liang Zhoufu, Nanning Normal University, China
Prof. Sung Ching Chen, Nanning Normal University, China

Organizing Committees
Prof. Zhou-Fu Liang, Nanning Normal University, China
Prof. Qu Bingpeng, Nanning Normal University, China
Prof. Su Chih-Yi, Guilin University of Electronic Technology, China
Prof. Sheng-Chung Chen, Hubei Polytechnic University, China
Assoc. Prof. Xinyu Wang, Nanning Normal University, China
Assoc. Prof. LingLing Qin, Nanning Normal University, China
Assoc. Prof. Liping Mo, Nanning Normal University, China
Assoc. Prof. Chi-En Hung, Huaiyin Normal University, China
Dr. Hsiao Hsien Lin, Jiaying University, China
Dr. Ching-Hua Liao, Sanming University, China

Technical Program Committees
Prof. Chih-Hung Wu, Sanming University, China
Prof. Liang Zhoufu, Nanning Normal University, China
Assoc. Prof. Xinyu Wang, Nanning Normal University, China
Assoc. Prof. LingLing Qin, Nanning Normal University, China
Assoc. Prof. Liping Mo, Nanning Normal University, China
Assoc. Prof. Xiaojuan Li, Fujian Agriculture and Forestry University, China
Assoc. Prof. Lien-Chieh Lee, Hubei Polytechnic University, China
Assoc. Prof. Srinivasa R Popuri, The University of the West Indies, Cave Hill Campus, Barbados
Dr. Penghui Liu, Jiaying University, China

*Renewable energy application and
energy chemical preparation*

Advances in Renewable Energy and Sustainable Development – Liang & Kasmani (Eds)
© 2023 Copyright the Author(s), ISBN: 978-1-032-39407-7

Current status of development and utilization of dry hot rock and its key technology prospect in China

Xiaolan Lv
Development Research Center of China Geological Survey, Beijing, China

Zhiyong Zhang*
Beijing Stimlab Oil & Gas Technology Co, Ltd, Beijing, China

Lin Zhang
Haikou Marine Geological Survey Center of China Geological Survey, Haikou, Hainan Province
Oil and Gas Resources Survey Center of China Geological Survey, Beijing, China

Beibei Yang
Development Research Center of China Geological Survey, Beijing, China

ABSTRACT: Most of the proven geothermal resources in the world are dry hot rock geothermal resources, which have good prospects for development and utilization. In this paper, the development and utilization practice of foreign dry hot rock resources is combed, the research and practice of dry hot rock exploration and development in China are introduced, and the main technologies of geothermal exploration and utilization of dry hot rock are analyzed. Finally, three conclusions are drawn. First, although the development and utilization of dry hot rock abroad are still in the experimental stage, it shows good technical feasibility. Second, there is a great potential for dry hot rock geothermal resources in China, and the existence of dry hot rock has been confirmed by exploration in Gonghe Basin, Nothern Hainan Island (Qiongbei), and other places. Third, the technologies of well completion, fracturing, heat exchange, and power generation for dry hot rock are in the experimental stage, which still have many difficulties to solve.

1 INTRODUCTION

The 14th Five-Year Plan is a key period to lay a good foundation for peaking carbon dioxide emissions and achieving carbon neutrality. Geothermal energy, one of the five non-carbon-based energy sources, has unique advantages compared with other renewable energy sources in the energy revolution. Nowadays, most proven geothermal resources in the world are dry hot rock (hereafter called *DHR*) resources. There is no or only a small amount of fluid in the DHR, with high temperature and great potential for development and utilization. The development and utilization of DHR in the world show good technical feasibility; the existence of DHR in China has been confirmed, and the technologics for DHR exploration are progressing. This paper will sort out DHR development and utilization practice in foreign countries, introduce the research and practice of DHR in China, and analyze the main technologies of DHR exploration and utilization. It is necessary to research the development and utilization of DHR and its prospects to adapt to the progress of technology and the reduction of exploration and development costs in the near future.

---

*Corresponding Author: lffyzzy@163.com

DOI 10.1201/9781003349648-1

2 REPRESENTATIVE PROJECTS OF DHR DEVELOPMENT ABROAD

The development of DHR began in the United States in the 1970s. The development means are mainly enhanced geothermal systems (hereafter called *EGS*). Many countries in the world have carried out relevant research; however, the countries and representative projects carrying out on-site drilling, hydraulic fracturing test, and power generation development are mainly concentrated in the United States and Europe (Fu et al. 2018; He et al. 2021; Jing et al. 2018; Li et al. 2019; Xie et al. 2020; Xu et al. 2016; Zhang et al. 2019; Zhou et al. 2018). Through more than 40 years of exploration, 41 DHR development projects have been implemented in the world, 25 belong to the traditional DHR development projects, 16 belong to EGS (Mao et al. 2019), and some projects have entered the power generation operation stage.

2.1 *United States*

Since 1973, the industrial test of DHR development was carried out in Fenton hill. It was the first EGS project in the world to establish an industrial scale in the deep underground. The field test was suspended in 2000. Although it is not a commercial-scale DHR development project, it has obtained a large number of research results. The desert peak EGS project, which was implemented in 2013, was the first DHR power generation project in the United States. Through the Frontier Observatory for Research in Geothermal Energy (FORGE) (2018-2023), DHR drilling, reservoir reconstruction, and flow testing will be carried out at the Utah Milford site.

2.2 *France*

In 1977, the research on shallow enhanced geothermal system/DHR was started. Since 1985, Germany, France, and Britain have jointly carried out the most successful EGS project in the world in Soultz. In 2008, geothermal power generation was carried out. In 2011, Organic Rankine Cycle (hereafter called *ORC*) was used for power generation. It is the first megawatt (1.5MW) geothermal power generation project in the world. In 2013, DHR was successfully used to realize continuous power generation.

2.3 *Britain*

Since 1977, the EGS project has been implemented in the Cornwall area, with the main purpose of developing relevant equipment and technology for deep geothermal exploitation projects. In 1980, the test of technology for developing equipment was carried out. In 1989, Britain joined the Soultz enhanced geothermal system/DHR project in Soultz Sous Forets, the border between Germany and France, and officially formed the Britain (Germany, France, and Britain) industrial consortium.

2.4 *Germany*

Since 2003, the DHR power generation test has been carried out in the Alsace area, which was put into operation at the end of 2007. In 2009 and 2012, duplex ORC units were successfully used in Landau and Inhim areas, with an installed capacity of 8mW. At present, the commercial operation has been fully realized.

2.5 *Japan*

In 1984, the first EGS demonstration research project in Japan was carried out in Hijiori, and four geothermal wells (SKG-2, HDR-1, HDR-2, HDR-3) with a depth of 2,000–2,205m were drilled. Since 2000, the project has been in open cycle test. SKG-2 and HDR-1 are injection wells with injection temperatures of 36°C, and HDR-2 and HDR-3 are production wells with flow production temperatures of 163°C and 172°C, respectively. In 2002, 130 kW duplex mass cycle power generation was tested and stopped due to the reduction of heat storage temperature.

2.6 *Australia*

The feasibility study of DHR was carried out in the 1990s. In 2006, the geoscience Bureau carried out enhanced geothermal system/DHR research in the form of the geothermal project team. Since 2003, the world's largest EGS project has been carried out in the Cooper Basin. The well depth is 4,000m, and the reservoir temperature is as high as 250°C. In 2008, a good hydraulic connection was achieved by hydraulic fracturing and fractured thermal reservoir, water circulation, and power generation tests were carried out, and 1MW DHR geothermal test power generation was successfully realized in 2013.

3 DOMESTIC RESEARCH AND PRACTICE ON EXPLORATION AND DEVELOPMENT OF DHR

3.1 *Geothermal resource reserves and distribution*

Geothermal resources are mainly divided into shallow geothermal energy, hydrothermal geothermal resources, and DHR. The recoverable reserves of geothermal resources in China are mainly DHR. The regional distribution of DHR in China (Zhang et al. 2021) is shown in Figure 1. The development prospect is broad, but the development is difficult.

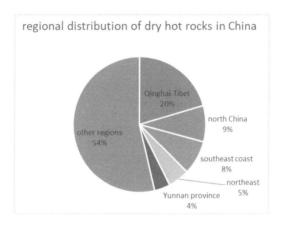

Figure 1. Regional distribution of DHR in China.

3.1.1 *Reserves and distribution of shallow geothermal energy resources*
In China, shallow geothermal energy resources less than 200 m are equivalent to 9.5 billion tons of standard coal. The annual exploitable amount of shallow geothermal energy resources in 336 cities above the prefecture level is equivalent to 700 million tons of standard coal. The areas more suitable for the application of groundwater ground source heat pump systems are mainly distributed in the eastern plain basin and areas with good water abundance (China Geological Survey: Investigation report on geothermal resources in China 2016). Buried pipe ground source heat pump system generally has good suitability.

3.1.2 *Reserves and distribution of hydrothermal geothermal resources*
In China, the amount of hydrothermal geothermal resources with a depth of 4,000 m is equivalent to 1.25 trillion tons of standard coal, and the annual recoverable amount is equivalent to 1.865 billion tons of standard coal. Medium and low-temperature geothermal resources are mainly distributed in large and medium-sized sedimentary basins such as North China Plain, Hehuai Plain, Subei Plain, Songliao Basin, Xialiaohe Plain, and Fenwei Basin. They are areas with the greatest development potential of geothermal resources. High-temperature geothermal resources are mainly distributed

in the intensive hydrothermal activity zone of southern Tibet - western Sichuan - western Yunnan (China Geological Survey: Investigation report on geothermal resources in China 2016).

3.1.3 *DHR reserves*

In China, DHR resources have great potential and broad development prospects. According to the preliminary calculation, the prospective resources of DHR with a buried depth of 3,000-10,000 km are equivalent to 856 trillion tons of standard coal, and 2% is equivalent to 17.12 trillion tons of standard coal according to the international DHR standard (China Geological Survey.: Investigation report on geothermal resources in China 2016). DHR resource is the most potential strategic alternative energy, but it is difficult to develop.

3.2 *Investigation and evaluation of DHR resources and exploration of typical areas*

The exploration and development of DHR in China are still in the stage of exploration and practice, and the EGS demonstration research site has not been built.

3.2.1 *Organized by China Geological Survey*

During the 12th Five Year Plan period, China Geological Survey surveyed shallow geothermal energy resources in 336 cities above the prefecture-level, launched the survey of DHR, basically found out the occurrence, distribution, development, and utilization of geothermal resources in China, preliminarily evaluated the potential of geothermal resources in China and delineated the priority development target area of DHR (China Geological Survey: Investigation report on geothermal resources in China 2016; Fu et al. 2018; He et al. 2021; Jing et al. 2018; Mao et al. 2019; Xie et al. 2020; Xu et al. 2016; Zhang et al. 2021; Zhou et al. 2018). In 2013, the DHR exploration project was carried out in Gonghe Basin in Qinghai Province. In 2014, the first DHR well (DR3 well) in China was drilled, and 168°C high-temperature DHR was drilled at a depth of 2735 m, which is the first proven DHR resource in China; In 2017, 236°C DHR mass was drilled at 3,705 m deep in well (GR1 well) in the north of the basin. The successful exploration achieved a major breakthrough in China's DHR exploration and delineated the distribution area of DHR mass with an area of 3000 square kilometers.

3.2.2 *Organization by oil companies*

After years of technical research, CNPC and Sinopec have mastered key technologies such as oil field geothermal resource exploration, resource evaluation, geothermal energy development and utilization, and engineering construction. In particular, significant progress has been made in technologies such as reinjection in sandstone formation, the transformation of abandoned wells into geothermal wells, utilization of waste heat from produced water, gas heat pump, high-temperature geothermal drilling, and completion (Wang et al. 2020). Sinopec attaches great importance to the development prospect of DHR. It has set up relevant research projects since 2013, with a cumulative investment of more than 31 million yuan. It has tracked the progress of DHR research and field tests at home and abroad and initially formed a number of characteristic technologies, which have provided strong technical support for the development scheme design of DHR in Gonghe Basin in Qinghai Province].

3.2.3 *Organization by geothermal enterprise*

In 2018, Hengtai Aipu (Hainan) clean Resources Development Co., Ltd. invested in drilling a DHR exploration well in the north of Chengmai County and drilled a high-temperature DHR mass (unsteady temperature measurement) exceeding 185°C at 4,387m, which is the first DHR drilling in Hainan Province (Jing et al. 2018; Li et al. 2019).

3.3 *Research results of DHR 863 Project*

In 2012, the "Research on key technologies for thermal energy development and comprehensive utilization of DHR" led by the "863" plan project of the Ministry of Science and Technology opened special research on 1,000 hot rocks in China, which has since opened the prelude to the DHR drilling project. After three years of tackling key scientific and technological problems, the project has carried out research on positioning technology and engineering testing technology of the DHR target area, completed the target area of 1km × 1km geophysical exploration and interpretation, established the experimental simulation platform for DHR development, and developed the geothermal resource display and evaluation system software. The project has developed a multi-field coupled heat transfer simulation program for the underground system of DHR, proposed three schemes of the composite power generation system, completed the design of a demonstration project, and developed three anti-corrosion and anti-scaling processes and coating materials to simulate the application conditions of DHR, which provides a theoretical basis and key technical support for the development and utilization of DHR resources in China (Li et al. 2019; Xu et al. 2016).

3.4 *Research on the development and utilization of DHR in China*

Since 2010, the research on the development and utilization of DHR in China has entered a small peak. The development phase focuses on resource exploration, drilling, fracturing, and other related technology research. The utilization phase is mainly focused on policy, power generation principles, and basic technology research, which have achieved certain results. Until April 2022, the data from CNKI shows that since 2013, 618 articles of "dry hot rock" have been searched in the keyword, 341 articles of "development and utilization of dry hot rock" have been searched in the full text, and 440 articles of "dry hot rock development technology" have been searched in the full text. The number of documents in each year is shown in Figure 2.

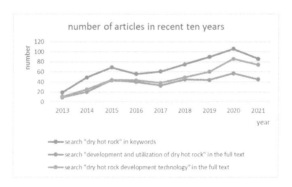

Figure 2. Number of documents on development and utilization of DHR in recent ten years.

There is little fluid in the DHR, and it is deeply buried, so it is necessary to exchange heat with an ultra-long buried pipe. Therefore, the development and utilization of DHR mainly include five stages: drilling, well completion, fracturing, heat extraction, and utilization. Technologies such as dry hot rock drilling and completion, dry hot rock fracturing, dry hot rock heat exchange, and dry hot rock power generation are required (Cao & Gai 2021; Chen et al. 2019; China Geological Survey.: Investigation report on geothermal resources in China 2016; Dong & Tian 2016; Fu et al. 2018; He 2018; He et al. 2021; Jing et al. 2018; Li et al. 2019; Mao et al. 2019; Wang et al. 2022; Xie et al. 2020; Xu et al. 2016; Ye 2018; Zhang et al. 2021; Zhao et al. 2020, 2018). According to a CNKI search, until April 2022, among 618 articles since 2013, the number of "drilling," "well completion," "fracturing," "heat exchange," and "power generation" in the full text is 229, 66, 223, 126 and 266 respectively.

4 DEVELOPMENT AND UTILIZATION OF TECHNOLOGY OF DHR

4.1 *Drilling and completion technology*

The technologies of DHR drilling and completion mainly include optimization design of geothermal well drilling and completion engineering, new downhole tools for geothermal drilling, new measurement while drilling and completion testing tools, geothermal well control safety technology, geothermal well complex accident treatment and workover technology, geothermal well corrosion and protection, geothermal reservoir protection technology, etc. (Cao & Gai 2021; Chen et al. 2019; Dong & Tian 2016; He 2018; Wang et al. 2022; Ye 2018; Zhao et al. 2020) The United States is the most advanced country in high-temperature geothermal well drilling technology, and Japan and the United Kingdom are also relatively mature. CNPC has carried out high-temperature drilling services and special technology research and development in Kenya for many years. The high-temperature geothermal drilling technology is generally equivalent to the advanced international level. Most of the rock lithology of the DHR reservoir is an igneous rock, and it is generally hard, resulting in poor drill ability and low rock breaking efficiency of the reservoir. Complex formation conditions and high reservoir temperature lead to frequent downhole accidents during drilling, and most drilling tools can't normally work under high temperatures.

There are four main difficulties: low rock breaking efficiency, easy change of drilling fluid performance, frequent downhole accidents, and insufficient high-temperature resistance of drilling tools.

There are four main research directions: research on rock mechanical behavior and failure mechanism of DHR reservoir under high temperature and complex environment, research on wellbore instability prediction model of DHR reservoir, research and development of high temperature resistant drilling fluid materials, and research on high-temperature resistant tools of DHR reservoir.

4.2 *Fracturing technology*

DHR fracturing technology mainly includes hydraulic fracturing and acid fracturing. Hydraulic fracturing is the most commonly used method. Key technologies include high-temperature fracturing tools and materials, numerical simulation of hydraulic fracturing of DHR reservoir, optimization of reservoir reconstruction scheme, and construction design (Cao & Gai, 2021; Ye 2018; Zhao et al. 2020). After more than 40 years of global exploration, the fracturing technology is still not mature enough, and the limited experiments are still focused on hydraulic fracturing. Currently, the technology used for HDR fracturing in China adopts the oil fracturing standard, mainly aimed at the fracturing standard and equipment for a sedimentary rock formation. For the drilling and fracturing of igneous or metamorphic rocks, large-scale fracturing tests have not been carried out, relevant standards have not been formed, and fracturing machines matching with HDR formation have not been developed.

There are three main difficulties: extremely high rock fracture pressure, serious fracturing fluid filtration, and difficult prediction of hydraulic fracture propagation.

There are five main research directions: research on fracture characteristics of DHR fracturing, development of ultra-high temperature layered/segmented fracturing tools and materials, development of new technology and software for fracture network fracturing in high-temperature hard formation, research on fracture real-time monitoring technology and evaluation method, and numerical simulation of hydraulic fracturing in DHR reservoir.

4.3 *Heat exchange and power generation technology*

After the fracturing of rock mass, it is also necessary to carry out a convective circulation test to evaluate the heat exchange and heat production capacity and efficiency (Dong, Tian 2016; Jing et al. 2018; Wang et al. 2022). In the convective circulation test, low-temperature water is injected into the injection well, and the water production capacity is tested in the production well. It is necessary to achieve circulation stability and produce high-temperature fluid for development and utilization (Dong & Tian 2016).

The heating principle of DHR is similar to the geothermal energy utilization technology of medium and deep buried pipes. The medium flows in the coaxial casing to exchange heat with the underground DHR mass without pumping groundwater (Wang et al. 2022).

The medium used for traditional EGS power generation in DHR is water. The heat recovery treatment method of the power generation process is shown in Figure 3(Wang et al. 2022). The low-temperature water is input into the thermal reservoir through the injection well. When passing through the high-temperature rock mass, the water is heated. Under the critical temperature state, it is recycled through the geothermal production well in the form of a mixture of high-temperature water and gas for power generation. After power generation, the cooled geothermal water is discharged into the injection well again, recycled, and recycled again (He 2018; Si et al. 2018; Wang et al. 2022).

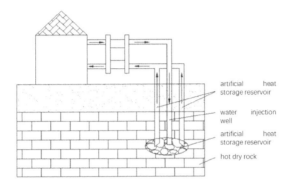

Figure 3. Schematic diagram of EGS power generation in DHR.

The main process of the new EGS power generation mode with CO_2 as the medium includes the transformation of heat storage structure, structural heat recovery system, and structural circulating heat recovery system (He 2018; Zhang et al. 2021), as shown in Figure 4. When CO_2 is cooled through the condensation system, it exchanges heat with the high-temperature rock mass of the underground reservoir, resulting in CO_2 at high temperatures. When the high-temperature gas passes through the power generation system, it can be used for power generation, heating, etc. Compared with the traditional EGS power generation mode, the power generation mode with CO_2 as the medium has a faster heat transfer rate, higher environmental protection, and economy.

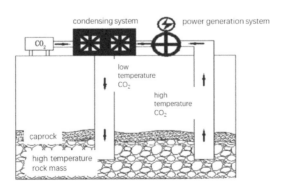

Figure 4. Schematic diagram of CO2 extraction geothermal power generation.

There are four main difficulties: serious water loss, fast energy attenuation, susceptibility to earthquake, poor stability of tubular pump, rod pump, and electric submersible pump used in high-temperature deep wells.

5 CONCLUSION

This paper adopts the comparative method to research the development and utilization of DHR resources in China. The main conclusions can be summarized as follows: (1) Although the development and utilization of DHR worldwide are still in the experimental stage, it shows good technical feasibility. (2) There is a great potential for DHR in China. The exploration in Gonghe Basin, Northern Hainan Island (Qiongbei), and other places has confirmed the existence of DHR, and no substantive EGS fracturing test has been carried out. (3) Well completion, fracturing, heat exchange, and power generation technologies for the development and utilization of DHR are in the experimental stage, and there are still many difficulties to be solved. In terms of future work, we should strengthen our country's geothermal exploration and development. The demonstration projects of geothermal heating and power generation in Gonghe Basin and Tibet should be well established. Relevant standards and preferential policies are regulated by combining domestic and foreign geothermal exploration and development technology to enhance the development and utilization of DHR geothermal in China.

ACKNOWLEDGMENTS

This paper is financially supported by the Special Project of Geological Mineral Resources and Environment Investigation (No.: DD20190470 and No.: DD20221826) and the National Science and technology major special project (No.: 2019 YFB1504103).

REFERENCES

Cao Jifei, Gai Yu. (2021) Applicability analysis of high-temperature oil and gas engineering technology in the field of dry and hot rock. In: *Western Exploration Engineering*. 9, pp. 42–44.

Chen Zuo, Xu Guoqing, Jiang Manqi. (2019) Present situation and development suggestions of dry hot rock fracturing technology at home and abroad. In: *Oil drilling technology*. Vol. 47, 6, pp. 1–8.

China Geological Survey.: Investigation report on geothermal resources in China, 2016.

Dong Ying, Tian Tingshan. (2016) *Theory and practice of dry hot rock power generation technology*, Beijing: Geological Publishing House.

Fu Yarong, Li Minglei, Wang Shuyi, et al. (2018) Current situation and prospect of exploration and development of dry hot rock. In: *Petroleum drilling and production technology*. Vol. 40, 4, pp. 526–540.

He Kai.: Prospect of carbon dioxide development of dry hot rock technology. In: *Modern chemical industry*. 6, 2018, pp. 56–58.

He Miao, Gong Wuzhen, Xu Mingbiao, et al. (2021) Research status and Prospect Analysis of dry hot rock development technology. In: *Renewable energy*. Vol. 39, 11, pp. 1447–1454.

Jing Tieya, Zhao Wentao, Gao Shiwang, et al. (2018) Geothermal development practice and technical feasibility study of dry hot rock. In: *Sino foreign energy*. Vol. 23, 11, pp. 17–22.

Li Ruixia, Huang Jin, Zhang Ying, et al. (2019) Analysis on development and utilization status and development trend of dry hot rock. In: *Contemporary petroleum and petrochemical*. Vol. 27, 3, pp. 47–52.

Mao Xiang, Guo Dianbin, Luo Lu, et al. (2019) Development progress and geological background analysis of dry hot rock geothermal resources in the world. In: *Geological review*. Vol. 65, 6, pp. 1462–1472.

Si Yang, Zhang Xuelin, Mei Shengwei, et al. (2018) Power generation technology of dry hot rock and its application in Gonghe dry hot rock, Qinghai. In: *Global energy Internet*. Vol. 1, 3, pp. 322–329.

Wang Shejiao, Chen Qinglai, Yan Jiahong. (2020) Development trend of geothermal energy industry and technology and suggestions for oil companies. In: *Petroleum Science and Technology Forum*. Vol. 39, 3, pp. 9–16.

Wang Wenzhong, Shao Dongyun, Cheng Xinke, et al. (2022) Development and utilization of shallow and middle deep geothermal energy in China. In: *Hydropower and new energy*. Vol. 36, 3, pp. 21–25.

Xie Wenping, Lu Rui, Zhang Shengsheng, et al. (2020) Exploration progress and development technology of dry hot rock in Gonghe Basin, Qinghai. In: *Oil drilling technology*. Vol. 48, 3, pp. 77–84.

Xu Tianfu, Yuan Yilong, Jiang Zhenjiao, et al. (2016) Dry hot rock resources and enhanced geothermal Engineering: international experience and China's prospect. In: *Journal of Jilin University* (Earth Science Edition). Vol. 46, 4, pp. 1139–1152.

Ye Zhaojun. (2018) Research on high temperature and high-pressure cementing technology. In: *China Science and Technology Expo*. 25, pp. 1009.

Zhang Jie, Zhao Meng, Niu Shiwei. (2021)Key technology progress and development trend of EGS in dry hot rock. In: *District heating*. 2, pp. 79–84.

Zhang Senqi, Wen Dongguang, Xu Tianfu, et al. (2019) Comparison between the "geothermal energy frontier Observatory research program" of dry hot rock in the United States and the exploration status of typical EGS sites in China and the United States. In: *Geoscience frontier*. Vol. 26, 2, pp. 322–333.

Zhao Xu, Yang Yan, Liu Yuhong, et al. (2020) Current situation and technical development trend of the global geothermal industry. In: *World petroleum industry*. Vol. 27, 2, pp. 53–57.

Zhou Zhou, Jin Yan, Lu Yunhu, et al. (2018) Technical problems and suggestions on drilling and hydraulic fracturing engineering of dry hot rock geothermal reservoir. In: *Chinese science*. Vol. 48, 12, pp. 1–5.

Advances in Renewable Energy and Sustainable
Development – Liang & Kasmani (Eds)
© 2023 Copyright the Author(s), ISBN: 978-1-032-39407-7

Influence of different composite insulation methods on steam long-distance pipeline

Yanming Zhang*, Xin Liu & Qun Liu
Datang Northeast Electric Power Research Institute, Changchun, China

Xiufang Huang & Duxing Wei
Datang Guizhou Faer Power Co., Ltd Liupanshui, China

Peng Guan
Datang Changchun Second Power Co., Ltd. Changchun, China

ABSTRACT: For long-distance steam transmission pipelines, the quality of the heat preservation method directly affects the quality of steam reaching end users. The main purpose of the research is to determine a better composite thermal insulation method for long-distance steam pipelines. Through numerical calculation software, the control variable method is used to compare different composite thermal insulation methods. Finally, it is determined that the double-layer composite thermal insulation structure is calcium silicate and perlite from the inside to the outside. From the inside to the outside, calcium silicate, perlite, and foam glass are the best.

1   INTRODUCTION

With the gradual expansion of the scale of the city and the continuous increase of large-scale food, pharmaceutical, chemical, and textile enterprises, the demand for steam is also increasing each year. Most of these large enterprises are located in the outer edge areas far from the city center or in newly developed economic development zones. The distance between the enterprises and the heat source is relatively long. How to meet the steam demand of these enterprises needs to be solved urgently (Bao 2018; Cai et al. 2011; Liu et al. 1997; Xue et al. 2018; Zhou 2016; Zhang 2017). Using numerical calculation software, the control variable method is used to compare and analyze different composite thermal insulation structures. The steam outlet parameters and the total investment cost of the two-layer and three-layer composite thermal insulation structures are mainly studied under different composite methods, and the optimal analysis of the three-layer composite thermal insulation mechanism is carried out.

2   DOUBLE-LAYER COMPOSITE THERMAL INSULATION STRUCTURE

Calcium silicate and perlite are selected as thermal insulation materials to study the double-layer composite thermal insulation structure. The total thickness of the two thermal insulation materials is taken as 140mm, the thickness of each layer of thermal insulation materials is 70mm, and the heating distance is 4km. In order to comply with the regulations in the *Central Heating Design Manual*, when DN$\geq$200, the maximum flow rate of steam in the steam heating pipeline is 60m/s, and when the initial steam temperature is 300°C, the steam mass flow rate is taken as 20t/h, and

---

*Corresponding Author: 871894273@qq.com

12   DOI 10.1201/9781003349648-2

when the initial steam temperature is 400°C, the steam mass flow rate is taken as 19t/h. The first composite thermal insulation material is calcium silicate and perlite from inside to outside. The second type of composite thermal insulation material is perlite and calcium silicate from inside to outside. The thermal conductivity of calcium silicate at 300°C is 0.083W/(m·K), and the thermal conductivity of perlite at 300°C is 0.104W/(m·K). The thermal conductivity of calcium silicate at 400°C is 0.097W/(m·K), and the thermal conductivity of perlite at 400°C is 0.122W/(m·K). For four kinds of steam heating pipes with different initial parameters and different nominal diameters, the thermal insulation effects of two different composite thermal insulation structures were simulated, respectively. It can be seen from Table 1 to Table 4 that: the thermal insulation effect of the first composite method is better than that of the second composite method. And the prices of the two insulation materials are as follows: the average price of calcium silicate is 750yuan/cubic meter, the average price of perlite is 550 yuan/cubic meter, and the total investment of the thermal insulation material of the first composite method is also lower than that of the second composite method. To sum up, the first composite thermal insulation method is better than the second composite thermal insulation method.

Table 1. Initial parameters 300°C 1.6 MPa DN250 composite insulation results.

Composite way	1	2
Output temperature (°C)	238.2	235.6
Outlet pressure (MPa)	1.399	1.399
Total cost (Ten thousand yuan)	45.97	48.44

Table 2. Initial parameters 300°C 2.5MPa DN300 composite insulation results.

Composite way	1	2
Output temperature (°C)	240.6	238.4
Outlet pressure (MPa)	2.451	2.452
Total cost (Ten thousand yuan)	51.92	54.38

Table 3. Initial parameters 400°C 2.5MPa DN350 composite insulation results.

Composite way	1	2
Output temperature (°C)	293.9	290.4
Outlet pressure (MPa)	2.475	2.476
Total cost (Ten thousand yuan)	57.86	60.32

Table 4. Initial parameters 400°C 4MPa DN400 composite insulation results.

Composite way	1	2
Output temperature (°C)	294.8	291.8
Outlet pressure (MPa)	3.991	3.991
Total cost (Ten thousand yuan)	63.46	65.92

3 THREE-LAYER COMPOSITE THERMAL INSULATION STRUCTURE

The foam glass, calcium silicate and perlite are selected as the thermal insulation materials to study the composite thermal insulation structure. The thermal conductivity of the three thermal insulation materials is shown in Table 5 and Table 6. The total thickness of the three thermal insulation materials is taken as 140mm, and the thickness of each layer of thermal insulation materials is 46.67mm, the heating distance is 4km, the steam mass flow rate is 20t/h at 300°C, and the steam mass flow rate is 19t/h at 400°C.

Table 5. Thermal conductivity of three thermal insulation materials at 300°C.

Insulation materials	Calcium silicate	Perlite	Foam glass
Thermal Conductivity W/(m·K)	0.083	0.104	0.115

Table 6. Thermal conductivity of three thermal insulation materials at 400°C.

Insulation materials	Calcium silicate	Perlite	Foam glass
Thermal Conductivity W/(m·K)	0.097	0.122	0.154

As shown in Figure 1, the three thermal insulation materials are divided into six cases for compounding. The first composite thermal insulation material is calcium silicate, perlite, and foam glass from inside to outside. The second type of composite thermal insulation material is calcium silicate, foam glass and perlite from inside to outside. The third composite thermal insulation material is perlite, calcium silicate and foam glass from inside to outside. The fourth type of composite thermal insulation material is perlite, foam glass and calcium silicate from inside to outside. The fifth composite thermal insulation material is foam glass, calcium silicate and perlite from inside to outside. The sixth type of composite thermal insulation material is foam glass, perlite and calcium silicate from inside to outside.

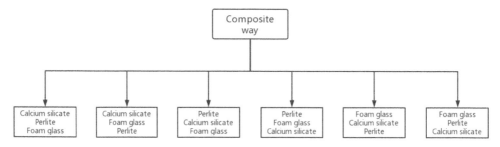

Figure 1. Six composite methods.

Six different composite thermal insulation structures were simulated for four kinds of steam heating pipes with different initial parameters and different diameters. The simulation results are shown in Table 7 to Table 10, and it can be seen from the simulation results that the composite thermal insulation structures with thermal insulation effects from high to low are as follows: The first, the third, the second, the fourth, the fifth, the sixth. And when the initial steam supply temperature is higher, and the pipe diameter is larger, the difference in thermal insulation performance of the composite thermal insulation structure is greater. The best thermal insulation effect should also consider the economy and calculate the total investment cost of six different composite methods of thermal insulation materials. The average price of calcium silicate is 750yuan/cubic meter, the average price of perlite is 550 yuan/cubic meter, and the average price of foam glass is 500 yuan/cubic meter. The calculation results are shown in Table 7 to Table 10. The calculation results show that the total investment of thermal insulation materials from high to low is as follows: sixth, fourth, fifth, third, second, and first. It can be seen that the first composite method with the best thermal insulation effect also has the least total investment cost, and the first composite method considering the economy and thermal insulation effect is the best.

Table 7. Initial parameters 300°C 1.6MPa DN250 composite insulation results.

Composite way	Output temperature (°C)	Outlet pressure (MPa)	Total investment (ten thousand yuan)
First	241.1	1.398	42.21
Second	239.8	1.398	42.48
Third	239.9	1.398	43.30
Fourth	237.1	1.399	44.67
Fifth	235.9	1.399	43.85
Sixth	234.9	1.400	44.94

Table 8. Initial parameters 300°C 2.5MPa DN300 composite insulation results.

Composite way	Output temperature (°C)	Outlet pressure (MPa)	Total investment (ten thousand yuan)
First	243.4	2.451	47.69
Second	242.0	2.451	47.97
Third	242.4	2.451	48.79
Fourth	239.6	2.452	50.15
Fifth	238.5	2.452	49.33
Sixth	237.6	2.452	50.43

Table 9. Initial parameters 400°C 2.5MPa DN350 composite insulation results.

Composite way	Output temperature (°C)	Outlet pressure (MPa)	Total investment (ten thousand yuan)
First	297.2	2.475	53.18
Second	293.8	2.475	53.45
Third	295.7	2.475	54.27
Fourth	289.7	2.476	55.64
Fifth	287.2	2.476	54.82
Sixth	285.6	2.476	55.91

4 OPTIMIZATION OF THREE-LAYER COMPOSITE THERMAL INSULATION STRUCTURE

The thermal insulation structure of the steam heating pipeline is taken as a composite thermal insulation structure composed of three thermal insulation materials, calcium silicate, perlite and

Table 10. Initial parameters 400°C 4MPa DN400 composite insulation results.

Composite way	Output temperature (°C)	Outlet pressure (MPa)	Total investment (ten thousand yuan)
First	297.9	3.991	58.35
Second	294.7	3.991	58.62
Third	296.7	3.991	59.44
Fourth	291.1	3.991	60.81
Fifth	289.2	3.991	59.99
Sixth	287.3	3.991	61.08

foam glass. The six composite methods composed of three different thermal insulation materials are the same as the six composite methods in Figure 1. The thicknesses of the three thermal insulation materials are a, b, and c from inside to outside. Equation (1) is the maximum thermal resistance of the three thermal insulation materials in the first composite method. Formula (2) is the total investment cost (yuan) of the three thermal insulation materials in the first composite method. Equations (3), (4), and (5) are the thicknesses of the three layers of thermal insulation materials, respectively. Taking into account the feasibility of on-site construction, the thickness of each layer of the three-layer insulation material is greater than 0.03m. Formula (6) is that the total thickness of the three-layer insulation material is 0.14m. Formula (7) is the thermal conductivity equation of calcium silicate insulation material. Equation (8) is the thermal conductivity equation of perlite thermal insulation material. Equation (9) is the thermal conductivity equation of the foam glass insulation material. Equation (10) is the heat dissipation loss per unit length of the steam heating pipeline. Equation (11) is the outer surface temperature t_1 of the calcium silicate thermal insulation material. According to *DL/T5072-2007 Design Regulations for Thermal Power Plant Thermal Insulation Paint*, the temperature at the interface between the inner and outer layers of the composite insulation should not exceed 90% of the recommended temperature for the outer insulation material. Because the recommended maximum operating temperature of perlite insulation material is 538°C, the outer surface temperature t1 of the calcium silicate thermal insulation material is less than 484.2°C. Equation (12) is the outer surface temperature t_2 of the perlite thermal insulation material, according to the regulation t2<433.8°C. Equation (13) is the outer surface temperature t3 of the foam glass insulation material. According to *DL/T5072-2007 Design Regulations for Thermal Power Plant Thermal Insulation Paint*, when the ambient temperature is not higher than 27°C, the temperature of the outer surface of the thermal insulation structure of equipment and pipes should not exceed 50°C, so t_3<50°C. The calculation results are shown in Table 11. When the fifth compound method is adopted, maximum thermal resistance, compared with the other five composite methods, the total investment cost of thermal insulation materials is not the highest.

The first composite method objective function:

$$R_{max} = \frac{1}{2\pi\lambda_1} \ln \frac{0.273 + 2a}{0.273} + \frac{1}{2\pi\lambda_2} \ln \frac{0.273 + 2(a+b)}{0.273 + 2a} + \frac{1}{2\pi\lambda_3} \ln \frac{0.273 + 2(a+b+c)}{0.273 + 2(a+b)} \quad (1)$$

$$S_{min} = 750 \times \left\{ 1000\pi \left[(0.273 + 2a)^2 - 0.273^2 \right] \right\}$$

$$+ 550 \times \left\{ 1000\pi \left\{ [0.273 + 2(a+b)]^2 - (0.273 + 2a)^2 \right\} \right\}$$

$$+ 500 \times \left\{ 1000\pi \left\{ [0.273 + 2(a+b+c)]^2 - [0.273 + 2(a+b)]^2 \right\} \right\} \quad (2)$$

The first compound mode constraints:

$$a > 0.03 \tag{3}$$

$$b > 0.03 \tag{4}$$

$$c > 0.03 \tag{5}$$

$$a + b + c = 0.14 \tag{6}$$

$$\lambda_1 = 0.066 + 4.64 \times 10^{-5} \times t_1 + 3.225 \times 10^{-8} t_1^2 \tag{7}$$

$$\lambda_2 = 0.06 + 5.95 \times 10^{-5} \times (t_1 + t_2) + 2.02 \times 10^{-8} \times (t_1 + t_2)^2 \tag{8}$$

$$\lambda_3 = 0.035 + 7.985 \times 10^{-5} \times (t_2 + t_3) + 8.575 \times 10^{-8} \times (t_2 + t_3)^2 \tag{9}$$

$$q = 300 / \left\{ \frac{1}{116.9022} + \frac{1}{2\pi\lambda_1} \ln \frac{0.273 + 2a}{0.273} + \frac{1}{2\pi\lambda_2} \ln \frac{0.273 + 2(a + b)}{0.273 + 2a} \right.$$

$$\left. + \frac{1}{2\pi\lambda_3} \ln \frac{0.273 + 2(a + b + c)}{0.273 + 2(a + b)} + \frac{1}{10.8644 \times [0.273 + 2(a + b + c)]} \right\} \tag{10}$$

$$t_1 = 300 - \frac{q}{2\pi\lambda_1} \ln \frac{0.273 + 2(a + b)}{0.273} < 484.2 \tag{11}$$

$$t_2 = t_1 - \frac{q}{2\pi\lambda_2} \ln \frac{0.273 + 2(a + b)}{0.273 + 2a} < 433.8 \tag{12}$$

$$t_3 = t_2 - \frac{q}{2\pi\lambda_3} \ln \frac{0.273 + 2(a + b + c)}{0.273 + 2(a + b)} < 50 \tag{13}$$

Table 11. Optimization results of six compound methods.

Composite way	a (m)	b (m)	C (m)	Thermal resistance (°C/W)	Total investment (ten thousand yuan)
First	0.038	0.037	0.065	1.67	41.3
Second	0.035	0.034	0.071	1.53	42.2
Third	0.047	0.044	0.049	1.68	42.9
Fourth	0.034	0.034	0.072	1.50	47.7
Fifth	0.036	0.036	0.068	1.72	43.1
Sixth	0.033	0.033	0.074	1.64	48.2

5 CONCLUSION

In this paper, the control variable method is adopted to study the influence of different composite thermal insulation methods on long-distance steam pipelines. The main conclusions can be summarized as follows: (1) For the double-layer composite thermal insulation structure, the composite method of calcium silicate and perlite from the inside to the outside is better. (2) For the three-layer composite thermal insulation structure, the composite method of calcium silicate, perlite and foam glass from the inside to the outside is better. (3) For the three-layer composite thermal insulation structure, 36cm foam glass, 36cm calcium silicate, and 68cm perlite from the inside to the outside are effective.

REFERENCES

Chuanjin Zhou. *Research on long-distance steam pipeline heating technology and its application practice*, 2016(1): 42–44.

Jinping Liu, Fen Hua, Zhi qin Chen. *Optimum Design of Steam Pipeline in Petrochemical Enterprises*, 1997(12): 22–27.

Weidong Cai, Xing hai Zhao, Guo hua Xin. *Analysis of long-distance heat supply by steam in power plant*, 2011, 27(2): 95–98.

Wei Zhang. *Calculation of temperature and pressure drop in industrial long-distance steam pipeline*, 2017.

Xiaolong Bao. *Analysis of Environmental Impact and Energy Consumption Impact of Central Heating Using Cogeneration in Industrial Park*, 2018(10): 32–33.

Yongming Xue, Jiuxi Zhu, Xiaoping Fu. *Study on Calculation Method of Temperature Drop and Pressure Drop in Long-distance Steam Pipeline*, 2018.

Advances in Renewable Energy and Sustainable
Development – Liang & Kasmani (Eds)
© 2023 Copyright the Author(s), ISBN: 978-1-032-39407-7

Preparation and adsorption properties of CaO/modified biochar

Xiaowen Zhang*, Lianghua Chen & Guofeng Yu
College of Technology, Hubei Engineering University, Xiaogan, China

Yu Ding
Hubei Engineering University, Xiaogan, China

ABSTRACT: The pollution of the water environment makes the problem of water resources more and more serious. Developing low-cost and environmentally friendly wastewater treatment agents is the focus of attention. China is rich in biomass resources, but many biomass resources are directly burned or discarded without rational utilization, which not only wastes resources, but also pollutes the environment. Rational use of these biomass resources can effectively reduce their pollution to the agricultural and forestry ecological environment. This paper used rice as the raw material, first synthesized by the high-temperature pyrolysis method, then crushed, dried and calcinated under high-temperature conditions before being ground and sieved to obtain biomass charcoal. Modified biochar was prepared by calcium oxide to achieve the purpose of phosphate adsorption. Modified biochar was characterized by Fourier transform infrared spectroscopy and scanning electron microscopy. The adsorption experiments of calcium-modified biomass charcoal on phosphate were discussed, the effects of adsorption time, temperature, initial concentration, and pH value of the solution on the adsorption effect were studied, and the adsorption mechanism was studied by the adsorption kinetic model. The experimental results show that the optimal adsorption conditions for PO_4^{3-} of modified biomass carbon are: a temperature of 35°C, an initial concentration of PO_4^{3-} at 24 mg·L^{-1}, a pH value of 7, and adsorption time of 1h. Under these conditions, the adsorption amount of modified biochar reaches the maximum value of 33.47 mg·g^{-1}. The adsorption kinetic model fitting results show that the quasi-secondary kinetic equation can describe the adsorption kinetic process well. The adsorption process of modified biomass charcoal to PO_4^{3-} is mainly chemical adsorption.

1 INTRODUCTION

Phosphorus is one of the important indicators affecting water quality and plays a very important role in the growth of plants. Phosphorus exists in many forms in natural water, of which phosphate accounts for the main part. Its main sources include decomposition and geological erosion of rocks and geological minerals, stormwater discharge runoff, atmospheric deposition, industrial discharge and farmland runoff (Zhang & Zhuang 2020). If too much phosphorus is discharged, water bodies become susceptible to eutrophication, and the quality of freshwater resources suffers. Therefore, reasonable treatment and monitoring of phosphate content in water serve as the important issues to solve water pollution and eutrophication.

Biochar is a carbon-rich product obtained by calcining fruit husks, rice husks, etc., at high temperatures under oxygen-limited conditions. However, many biomass resources are directly incinerated or discarded without reasonable utilization, which wastes resources and pollutes the environment. The rational use of these biomass resources can effectively reduce the agricultural

---

*Corresponding Author: jiayou.xiaowen@163.com

DOI 10.1201/9781003349648-3

and forestry environmental pollution. Therefore, a growing number of scientific researchers have conducted various research on biochar, which can not only recycle agricultural waste, but also broaden the practical application scope of various biomass charcoal. Unmodified biochar has a small specific surface area, underdeveloped pore structure, and limited adsorption and removal capacity for various pollutants in water. Modification treatment can improve its biosorption effect to a certain extent (Hu et al 2019).

In this experiment, rice husks were used as raw materials to prepare biochar, and the biochar was modified with calcium oxide to improve its adsorption capacity for phosphate in water, which provided important technical support for promoting phosphorus removal from water.

2 EXPERIMENT

2.1 *Instruments and reagents*

KSL-1100Y Muffle Furnace, Hefei Kejing Material Technology Co., Ltd.; TU-1810 UV-Vis Spectrophotometer, Beijing Puxi General Instrument Co., Ltd.; KYC-100C Constant Temperature Oscillator, Shanghai Fuma Experimental Equipment Co., Ltd.; GYS- M01 Joyoung Mill, Joyoung Co., Ltd.; BSA223S Electronic Balance, Sartorius Scientific Instruments Co., Ltd.; PHS-3C Digital Acidity Meter, Shanghai Honggai Instruments Co., Ltd.; JSM-6510 Scanning Electron Microscope, Japan Electronics Co., Ltd. Company; Nicolet-380 Fourier Transform Infrared Spectrometer, American Thermoelectric Nicoli Company.

Potassium dihydrogen phosphate, antimony potassium tartrate, ammonium molybdate, potassium bromide, calcium oxide, sodium hydroxide, ascorbic acid, disodium EDTA, formic acid, concentrated sulfuric acid, concentrated hydrochloric acid are all analytical reagents, and water is deionized water.

2.2 *Experimental method*

2.2.1 *Preparation of rice husk-based biochar*

After washing the high-quality rice husk, it was pulverized with a Joyoung mill. Putting it in a 105°C oven to dry, a certain amount of high-quality rice husk powder that had been pulverized, dried, and pretreated was put into a crucible, filled, and compacted, and the data was recorded after weighing. Putting the crucible into the muffle furnace, a series of parameters were set to calcine at 250°C, 300°C, 350°C, 400°C, and 450°C, respectively, with a holding time of 2h and a heating rate of 10°C/min. After naturally cooling to room temperature, the powder was taken out and weighed, and the calcined powder was evenly placed in a mortar for multiple grinding, and passed through a 100-mesh sieve to obtain biochar.

2.2.2 *Preparation of calcium-modified biochar*

Biochar (2.000 g), calcium oxide (0.400 g) and distilled water (10.0 mL) were added to the beaker and stirred with a glass rod until they were mixed into a slurry. After filtering with a filtering device, the unreacted calcium oxide was washed away with dilute hydrochloric acid, filtered and washed, put into an oven, and dried at 80°C. It was taken out into a mortar for grinding and passed through a 100-mesh sieve to produce calcium-modified biochar (Liu et al 2020).

2.2.3 *Characterization and testing*

Fourier transform infrared spectrometer was used to test the surface functional groups, a scanning electron microscope was used to test the morphology of the samples, and the concentration of phosphate in the solution was measured by a UV-visible spectrophotometer.

2.2.4 *Plotting the standard curve of phosphate*

The standard working solution of phosphate with a concentration of 0.3 mmol/L was accurately pipetted 0.0, 3.0, 4.0, 5.0, 6.0, 7.0, 8.0, 9.0, and 10.0 mL into 50 mL volumetric flasks. Then, 2.0 mL of ammonium molybdate solution with a concentration of 0.01 mol/L and 3.0 mL of ascorbic acid solution with a concentration of 0.10 mol/L were added to each volumetric flask, diluted with water to the scale, fixed, and shaken well. After standing for 10 minutes, its absorbance was measured at 710 nm, and a standard curve was plotted with absorbance as the ordinate and phosphate mass concentration (mg/L) as the abscissa.

2.2.5 *Phosphate adsorption experiments*

The phosphate solution diluted to a predetermined concentration was placed in a conical flask, and a certain mass of modified biochar was added to the conical flask, placed in a thermostatic oscillator, where the temperature was adjusted, the oscillation frequency was adjusted to a predetermined value, and the oscillation was carried out until a predetermined time. Consequently, taking out the conical flask for extraction and filtering with a 0.4μm filter membrane, the absorbance was measured by adding a color developer, and the remaining concentration of the solution after its adsorption was calculated by substituting the standard curve. The adsorption capacity of PO_4^{3-} was calculated according to Equation (1) (Qin et al 2018).

$$q = (C_0 - C)V/m \qquad (1)$$

Where q is the adsorption capacity, mg/g; C_0 and C are the initial concentration of the solution and the remaining concentration of the solution after adsorption, respectively, mg/L; V is the volume of the adsorption solution, L; m is the dosage of the modified biochar, g.

3 RESULTS AND ANALYSIS

3.1 *Characterization of samples*

3.1.1 *Analysis of fourier transform infrared spectroscopy characterization results (FTIR)*

It can be seen from Figure 1 that the positions of the characteristic absorption peaks of the functional groups of biochar before and after modification are basically the same. 3441 cm^{-1} is the stretching vibration peak of the -OH group, 3129 cm^{-1} is the C-H stretching vibration peak on the carbon-carbon double bond, and near 1604 cm^{-1} is the stretching vibration of C=C and C=O in the aromatic ring vibration, indicating that biochar has a highly aromatic and heterocyclic structure; 1401 cm^{-1} is the deformation vibration of C-H in cellulose and hemicellulose; 1101 cm^{-1} is the stretching vibration absorption peak of C-O bond; 784 cm^{-1} and 707 cm^{-1} are the out-of-plane bending vibration peaks of C-H. The stretching vibration peak of the Ca-O bond at 874 cm^{-1} in Figure 1B indicates that calcium was successfully doped into biochar.

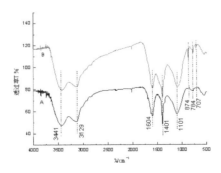

Figure 1. FTIR spectra of unmodified A and modified B biochar.

3.1.2 *Analysis of characterization results of Scanning Electron Microscope (SEM)*

It can be seen from Figure 2A that the unmodified rice husk-based biochar is relatively smooth and has an amorphous scale-like structure.

It can be seen from Figure 2B that the structure of the biochar modified by calcium oxide has changed significantly, and the modified biochar particles are obviously abundant, evenly distributed on the surface and pores of the carbon material, but relatively agglomerated. The aggravation of the surface roughness of biochar can provide a larger specific surface area for its structure and increase its activity. The introduction of Ca^{2+} can bring more positive charges and better promote its binding to PO_4^{3-} through electrostatic adsorption (Liu & Zhao 2019). Particle agglomeration may be caused by their small particle size and high surface energy and activity, forming a stable state (Hua 2021).

Figure 2. SEM images of unmodified A and modified B biochar at 450°C.

3.2 *The effect of adsorption time on adsorption capacity*

After weighing 0.030 g of calcium-modified biochar in a conical flask and adding 50 mL of 24 mg·L^{-1} of phosphate standard working solution with the initial solution pH 7, the solution was shaken for 15, 30, 45, 60, 75 and 120 min at a constant temperature of 35°C and then removed at 250 r/min and let stand for 10min. The absorbance was measured by extraction with a 0.4 μm filter membrane.

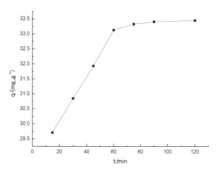

Figure 3. The effect of adsorption time on adsorption capacity.

It can be seen from Figure 3 that the adsorption amount of PO_4^{3-} by calcium-modified biochar increases with the increase of time, the adsorption rate gradually decreases with the increase of time, and the adsorption equilibrium is basically reached after 1 h. It was explained that in the initial stage, the concentration of PO_4^{3-} in the solution is high, the driving force of adsorbate diffusion is large, and there are multiple adsorption sites on the surface of the adsorbent. The PO_4^{3-} diffused to the surface of the biochar could fully contact the active sites on the outer surface of the adsorbent and was rapidly adsorbed. As the adsorption process proceeded, the concentration of PO_4^{3-} in the solution decreased, most of the active sites were occupied, and the adsorption mainly occurred on the inner surface of the biochar. At this time, the adsorption driving force of PO_4^{3-} and the biochar decreased, the adsorption capacity increased slowly, and the adsorption equilibrium was basically reached after 1 h (Zhu 2020). Therefore, the optimal adsorption time in this experiment was 1 h.

3.3 The effect of temperature on adsorption capacity

After weighing 0.030 g of calcium-modified biochar in a conical flask, 50 mL of 24 mg·L^{-1} phosphate standard working solution was added, and the initial solution pH was 7. The solution was shaken at 20, 25, 30, 35, and 40°C for 1 h, then taken out, rotated at 250 r/min, and let stand for 10 min, followed by extraction with 0.4 μm filter membrane to determine the absorbance.

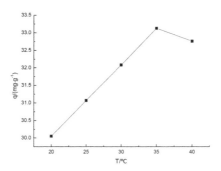

Figure 4. The effect of temperature on adsorption capacity.

It can be seen from Figure 4 that at 20-35°C, the adsorption capacity of calcium-modified biochar to PO_4^{3-} increases with the increase in temperature. At 35-40°C, the adsorption capacity of calcium-modified biochar to PO_4^{3-} decreased with the increase in temperature. When the temperature was 35°C, the adsorption capacity of modified biochar for PO_4^{3-} reached the maximum, which was 33.13 mg·g^{-1}. Therefore, in this stage, the optimum adsorption temperature of calcium-modified biochar for PO43 is 35°C.

3.4 The effect of initial concentration of phosphate on adsorption capacity

After weighing 0.030 g of calcium-modified biochar in a conical flask with the initial solution pH value of 7, 50 mL of standard working solution of phosphate at 12, 16, 20, 24, and 28 mg·L^{-1} were added, shaken at a constant temperature of 35°C for 1 h and then removed, rotated at 250 r/min, left for 10 min, filtered with a filtering device to measure its absorbance.

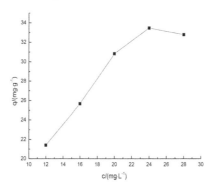

Figure 5. The effect of PO_4^{3-} initial concentration on adsorption capacity.

It can be seen from Figure 5 that with the increase of the initial concentration of PO_4^{3-}, the adsorption capacity of modified biochar gradually increased and then decreased. When the initial concentration of PO_4^{3-} was 24 mg·L^{-1}, the adsorption capacity reached the maximum, which was 33.47 mg·L^{-1}. When the initial concentration of $PO_4^{3-} \leq 24$ mg·L^{-1}, the adsorption rate increased rapidly with the increase of PO_4^{3-} concentration because with the increase of PO_4^{3-} concentration, the driving force of adsorbate diffusion increased, and the number of active sites provided by the modified biochar is much larger than the number of PO_4^{3-}, so the growth rate of adsorption is fast. When the initial concentration of PO_4^{3-}>24 mg·L^{-1}, the adsorption rate decreases because the number of active sites is constant. When the adsorption active site of the biochar was close to saturation, the adsorption rate slowed down and reached the maximum adsorption capacity of the modified biochar. Therefore, the optimal initial adsorption concentration of PO_4^{3-} in this experiment is 24 mg·L^{-1}.

3.5 The effect of ph on adsorption capacity

After weighing 0.030 g of calcium-modified biochar in a conical flask, 50 mL of 24 mg·L^{-1} of phosphate standard working solution was added, and the pH value was adjusted to different levels. The solution was shaken for 1 h at a constant temperature of 35°C and then removed, rotated at 250 r/min, and let stand for 10 min. The absorbance was measured by filtration with an extraction device.

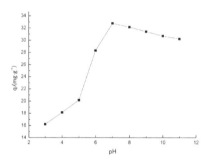

Figure 6. The effect of pH of adsorption solution on adsorption capacity.

As can be seen from Figure 6, when the pH value is 3-7, the adsorption capacity of modified biochar to PO_4^{3-} increases with the increase of pH value and reaches the maximum value at pH value of 7, which is 32.77 mg·L^{-1}, and then gradually decreased with the increase of pH, but was still much larger than the adsorption capacity under acidic conditions. Phosphate in the solution exists in the form of H_3PO_4, $H_2PO_4^-$, HPO_4^{2-}, and PO_4^{3-}, among which $H_2PO_4^-$ is the most likely to replace OH^- and combine with Ca^{2+}, followed by HPO_4^{2-} (Zhang & Wang 2020). When the solution is acidic, it is easy to dissociate H_3PO_4. When the solution is neutral, $H_2PO_4^-$ is the most important form, and it is easy to combine with Ca^{2+} for adsorption. As the pH continued to rise to alkaline, $H_2PO_4^-/HPO_4^{2-}$ decreased and the adsorption capacity of modified biochar for phosphate gradually weakened (Qian & Wu 2020).

3.6 Adsorption kinetics study

Through kinetic studies, the adsorption mechanism of phosphate on modified biochar can be further studied. Therefore, the experimental data are analyzed and processed by pseudo-first-order kinetics, pseudo-second-order kinetics, and intraparticle diffusion equations (Lin et al 2016), and the kinetic equations are as follows:

The pseudo-first-order kinetic equation:

$$\lg(q_e - q_t) = \lg q_e - \frac{k_1 t}{2.303} \qquad (2)$$

The pseudo-second-order kinetic equation:

$$\frac{t}{q_t} = \frac{1}{k_2 q_e^2} + \frac{t}{q_e} \qquad (3)$$

Intraparticle diffusion equation:

$$q_t = k_3 t^{0.5} + C \qquad (4)$$

Where k_1 is the pseudo-first-order kinetic rate constant, min^{-1}; k_2 is the pseudo-second-order kinetic rate constant, g·(mg·min)$^{-1}$; k_3 is the intraparticle diffusion rate constant, mg·(g·min0.5)$^{-1}$. q_t is the adsorption capacity per unit mass of adsorbent at time t, mg·g^{-1}; q_e is the adsorption capacity at equilibrium, mg·g^{-1}.

Figure 7 is the result of fitting the experimental data using pseudo-first-order and pseudo-second-order kinetic equations, and Figure 8 is the result of fitting the experimental data using the intraparticle diffusion equation. It can be seen from the figure that, compared with the other two kinetic fitting models, when the pseudo-second-order kinetic model is used to fit the adsorption of PO$_4^{3-}$ on calcium oxide modified biochar, and the fitting degree of the obtained equation is better, and the correlation coefficient reaches 0.9996, indicating that it is more accurate to use the pseudo-second-order kinetic model to describe the kinetics of the adsorption of PO$_4^{3-}$ on calcium oxide-modified biochar. The pseudo-second-order kinetic equation includes all the reaction processes of adsorption, which can more comprehensively and accurately reflect the adsorption reaction mechanism of PO$_4^{3-}$ on calcium-modified biochar (Jiang et al 2020). Besides, because the main factor affecting the adsorption of the pseudo-second-order kinetic equation is the chemical bond, it shows that the absorption process of calcium-modified biochar to PO$_4^{3-}$ is mainly chemical adsorption (Liu et al 2019).

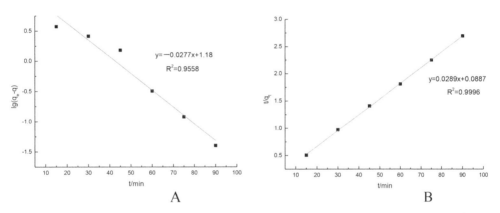

Figure 7. Pseudo-first-order A and pseudo-second-order B kinetic curves of modified biochar to PO$_4^{3-}$.

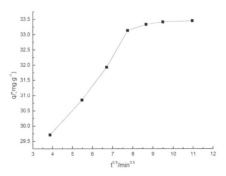

Figure 8. Intraparticle diffusion curve of modified biochar to PO$_4^{3-}$.

3.7 *Adsorption mechanism*

In this paper, the adsorption of PO_4^{3-} on calcium-modified biochar mainly relies on chemical adsorption, and the mechanism includes electrostatic adsorption, Lewis acid-base pairing, and ion exchange coupling. The combination of Ca^{2+}, OH^- and PO_4^{3-} to form calcium phosphate precipitation is the main adsorption mechanism. The main chemical reaction formula is as follows (Liu et al 2018):

$$5Ca^{2+} + 3PO_4^{3-} + OH^- \rightarrow Ca(PO_4)_3(OH) \downarrow \tag{5}$$

$$5Ca^{2+} + 3HPO_4^{2-} + 4OH^- \rightarrow Ca5(PO_4)_3(OH) \downarrow +3H_2O \tag{6}$$

$$5Ca^{2+} + 3H_2PO_4^- + 7OH^- \rightarrow Ca(PO_4)_3(OH) \downarrow +6H_2O \tag{7}$$

The phosphate in the solution exists in the form of H_3PO_4, $H_2PO_4^-$, HPO_4^{2-}, and PO_4^{3-}, among which $H_2PO_4^-$ is the most likely to replace OH- and combine with Ca^{2+}, followed by HPO_4^{2-}. When the solution is acidic, it is easy to dissociate H_3PO_4. When the solution is neutral, $H_2PO_4^-$ is the most important form. With the increase of pH value, $H_2PO_4^-/HPO_4^{2-}$ decreases, and the adsorption capacity of calcium-modified biochar to P gradually weaken. When the pH is stable (9.5~9.7), P in the solution exists in the form of HPO_4^{2-} and PO_4^{3-}, the ion exchange is balanced, and the adsorption is basically completed. There are also Lewis acid-base adsorption and electrostatic adsorption on the surface of biochar, but the adsorption effect is weak.

4 CONCLUSION

This paper calcified the pulverized rice husk-based biomass at 450°C to prepare rice husk-based biochar, which was then modified with calcium oxide to absorb PO43. The results show:

(1) When the temperature is 35°C the initial concentration of PO_4^{3-} is 24 $mg \cdot L^{-1}$, the pH value is 7, and the adsorption time is 1 h, the adsorption capacity of calcium oxide modified biochar reaches the maximum value, which is 33.47 $mg \cdot g^{-1}$.
(2) The fitting results of the adsorption kinetic model showed that the pseudo-second-order kinetic equation could describe the adsorption kinetic process well.
(3) The adsorption process of PO_4^{3-} by modified biochar was mainly chemical adsorption.
(4) In the later period, we will continue to research this topic, such as the reuse of modified biochar and the research on different modification methods.

ACKNOWLEDGMENT

The Scientific Research Program Guidance Project of Hubei Provincial Department of Education (No. B2021459).

REFERENCES

Hu F P, Luo W D, Peng X M, et al. Research progress in modified biochar for the removal of pollutants from water[J]. *Industrial Water Treatment*, 2019, 39(04): 1–4.

Hua L L.Study on phosphate adsorption mechanism of modified biomass carbon[J]. *Anhui Architecture*, 2021, 28(05): 165–167.

Jiang Y H, Li A Y, Deng H, et al. Adsorption performance of magnesium loaded banana straw-based biochar for phosphorus in pig wastewater[J]. *Water Treatment Technology*, 2020, 46(06): 33–38.

Lin L M, Liang X Q, Zhou K J, et al. Preparation of rice straw-derived biochar (RB) activated by CaCl2 and its adsorption on phosphorus[J]. *Acta Scientiae Circumstantiae*, 2016, 36(04): 1176–1182.

Liu F J, Liao Z D, Li X, et al. Adsorption of Cr(VI) on modified bagasse charcoal[J]. *Technology & Development of Chemical Industry*, 2020, 49(09): 1–4.

Liu F J, Zhou Y S, Liang T Y, et al. Research and Development Trend of Plasticizers In China and Abroad[J]. *Technology & Development of Chemical Industry*, 2019, 48(12): 53–57.

Liu X M, Zhao B. Adsorption of Cr(VI) in wastewater by biochar prepared from zinc chloride modified bagasse[J]. *Applied Chemical Industry*, 2019, 48(06): 1354–1358.

Liu X N, Jia B Y, Shen F, et al. Research progress of metal-modified biochar for phosphate adsorption[J]. *Journal of Agro-Environment Science*, 2018, 37(11): 2375–2386.

Qian M, Wu Y. Preparation, Characterization and adsorption properties of biochar derived from phoenix tree leaves[J]. *Anhui Chemical Industry*, 2020, 46(04): 25–32.

Qin F, Wang Y, Hang Y N, et al.[J].Study on adsorption removal of ammonia nitrogen in water by modified straw biochar[J]. *Forest Engineering*, 2018, 34(03): 19–25.

Zhang G, Wang X P. Adsorption properties of Scutellaria residue biochar on tetracycline hydrochloride[J]. *Journal of Leshan Normal University*, 2020, 35(12): 19–26.

Zhang Y H, Zhuang S Y. Characteristics of PO43- adsorption on sulphuric acid-modified bamboo biochar[J]. *Environmental Pollution and Prevention*, 2020, 42(10): 1216–1221.

Zhu X H. *Preparation of biomass carbon from different raw materials and its adsorption properties for methylene blue*[D]. Dalian: Dalian Jiaotong University, 2020.

Advances in Renewable Energy and Sustainable Development – Liang & Kasmani (Eds)
© 2023 Copyright the Author(s), ISBN: 978-1-032-39407-7

Structural design and performance analysis of a two-rotor vertical axis wind turbine applied to a new wind-wave coupled power generator

Guorui Zhu*, Jiefei Gao, Yuxuan Liu, Yuankang Xin & Zixuan Li
International Education College, Wuhan University of Technology, Wuhan, Hubei, China

ABSTRACT: Offshore wind energy is a common source of clean energy. To utilize this energy source more efficiently, we have designed a two-rotor vertical axis wind turbine and a single-point mooring platform. Simulations were performed using XFlow software to compare the performance of the single-rotor vertical axis wind turbine and fixed platform design solution. The design solution was found to have a better distribution of the external flow and vorticity fields, a higher power factor for energy conversion, and a lower shaft load on the turbine. It is a better design solution regarding energy conversion efficiency and service life. The stability of the coupled power generator under wave action was also verified, illustrating the good survivability of the system.

1   INTRODUCTION

For a long time, humans have used fossil fuels as the main source of energy for industrial societies. While promoting the development of civilization, the consumption of such non-renewable fuels has also produced a large amount of carbon dioxide, which has aggravated global warming, acid rain, and other environmental problems. Under the new development pattern, the concept of "carbon peaking and carbon neutral" has been proposed. Replacing traditional energy with clean energy has become the consensus of the whole society.

Offshore wind energy is a typical clean energy source. The mechanism of its generation is the difference in air pressure caused by the temperature difference between sea and land. This characteristic determines the wide distribution of wind energy and the abundance of reserves. Compared to onshore wind energy, offshore wind energy is usually located closer to major economic zones and does not consume land resources. Under the same circumstances, the efficiency of offshore wind turbines can be 20%-40% higher than that of onshore turbines, achieving more efficient use of wind energy (Ye & Zhong 2018).

We usually use wind turbines for wind energy capture. Generally speaking, wind turbines can be divided into floating platforms (FOWT) and bottom-fixed platforms, which, from an economic point of view, are less expensive to build and less damaging to the seabed than bottom-fixed platforms. Generally speaking, horizontal axis wind turbines are widely used, but when they float in the sea, their balance is poor due to their high center of mass position, which can be solved by using vertical axis wind turbines (Lee et al. 2022). This kind of turbine also has the characteristics of low aerodynamic noise and good wind performance, which is suitable for offshore conditions.

The device consists of three major subsystems: a vertical axis wind turbine, a wave energy harvesting system, and a mooring platform. In this paper, we focus on the fan design part, including its structure selection, quantity arrangement, and other aspects, and numerical simulation of the model using XFlow software to compare its performance.

---

*Corresponding Author: 295998@whut.edu.cn

28

DOI 10.1201/9781003349648-4

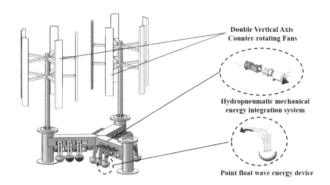

Figure 1. Schematic diagram of power generation platform.

2 DOUBLE ROTOR WIND TURBINE STRUCTURE DESIGN

This part will introduce the wind turbine blade shape design and layout, respectively, and will also introduce the overall layout of the wind turbine.

2.1 *Wind turbine blade shape design*

Wind turbine blade shape design is a relatively complex task involving complex fluid coupling-related content. The common blade forms are Darrieus type, Savonius type, and straight blade type. Considering the blade processing cost and wind energy - electrical energy conversion efficiency, we choose the straight blade as the wind turbine blade form. After consulting relevant information, we found that the National Advisory Committee for Aeronautics (NACA) has designed a series of low-speed airfoil types with lower drag coefficient and highest lift coefficient, which is suitable for offshore wind power generation conditions. In view of this, we chose the NCAA0018 airfoil section data in the design database (National Advisory Committee for Aeronautics 2022) as the design benchmark and built the blade 3D model.

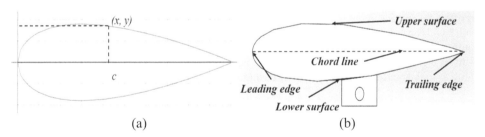

Figure 2. Schematic diagram of our airfoil profile. (a) NCAA0018 airfoil cross-sectional profile; (b) Cross-sectional view of the established wind turbine blade model.

where c is the chord length, x is the chord length coordinate, and y is the distance between the airfoil and the chord at the corresponding x position.

2.2 *Wind turbine blade arrangement*

First of all, we have to choose the number of fan blades. In this regard, we first consider the common three-blade structure. There are several reasons below.

(1) The use of such a uniform arrangement of the blade in the rotation is not easy to deviate from the axis

(2) To better solve the fan rotation of the dynamic balance problem and reduce the operation of the process of trembling and wear on the bearings.
(3) To reduce the manufacturing cost of the fan. If you add a blade, the degree of wind energy utilization is not significant, but a new blade will significantly increase the manufacturing cost of the wind turbine.

2.3 Wind turbine layout design

How to arrange the location of the turbines is also a problem to be considered. In this paper, we use a triangular mooring platform. Accordingly, we should add fans at the three vertices of the triangular mooring platform to achieve the purpose of multi-fan power generation.

In this process, it is natural to think of arranging turbines at all three vertices. But this will cause a problem, that is, the moment imbalance caused by the platform overturning phenomenon.

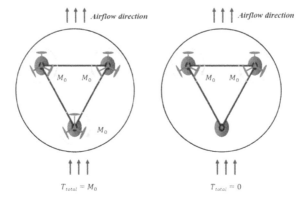

Figure 3. Comparison of force analysis of double and triple fan arrangement form.

We can understand from the analysis results shown in Figure 3 that when two fans are arranged under ideal conditions, the two fans will rotate in opposite directions, generating opposite moments to offset the impact force on the platform. If one fan is added, it will lead to unbalanced torque, which will cause the mooring platform to spin and cause a larger torsional load on the mooring cable during actual operation, thus reducing its service life. Therefore, we shaved off one fan assembly. A mass block is used to balance the third fan at the location where it is placed, thus obtaining a more desirable performance effect in the design.

3 WIND TURBINE PERFORMANCE ANALYSIS

After completing the structural design of the fan, we proceeded to analyze the performance of the fan. Using the hydrodynamic analysis function of XFlow software, we simulated the operation of the fan under the given operating conditions and evaluated the operation of the fan.

3.1 Simulation condition setting

The wind speed at sea is usually higher than the wind speed on land. Here we set the simulated wind speed to 10.8m/s, i.e., a class VI wind standard. To this, 5% turbulence is added to simulate the phenomenon of random disturbance. Specifically, we choose the *RNG* $k-\epsilon$ turbulence model (Wu et al. 2014) for the simulation. Compared with the traditional $k-\epsilon$ model, this method can better describe the phenomena of complex flow while considering the influence of vortices on turbulence and improving the calculation accuracy.

Figure 4. Outflow field simulation results.

Figure 5. Vorticity field simulation results.

Figure 6. Outflow field simulation results.

Figure 7. Vorticity field simulation results.

3.2 *Comparison of single and double rotor fan simulation analysis*

The model is imported into the simulation software, and the corresponding conditions are set and calculated. The simulation results of its external flow field and vorticity field are shown as follows.

We can visually see from the graph that, when all other conditions are equal, the high-speed zone is significantly wider, and the degree of the vortex is relatively less in a twin-rotor fan compared to a single-rotor fan. It can be seen here that the twin-rotor fan has a better wind gain effect and that the wind energy can be used more efficiently using the modified arrangement.

3.3 *Simulation condition setting*

The twin-rotor fan has a higher energy conversion efficiency with respect to the single rotor fan, and has a higher rotational speed and a smaller on-shaft moment for the same input wind power. Similarly, for the simulation input condition of wind speed 10.8m/s with 5% turbulence, the torque variation curves with time for single and dual rotor fans are derived as follows.

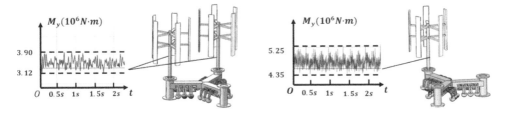

Figure 8. Comparison of torque on the shaft of single and double fans.

From the simulation results, we know that the average on-shaft moment M_s of the single rotor fan is about $4.8 \times 10^6 N \cdot m$, while the peak on-shaft moment M_d of the double rotor fan is about $3.6 \times 10^6 N \cdot m$. On this basis, we calculate the power of both fans. Through the simulation of the velocity flow field size distribution and geometric parameters can be calculated, single and dual rotor fans corresponding to the tip speed ratio of $\lambda_s = 3, \lambda_d = 5$. can be calculated from the expression of the leaf tip speed ratio to obtain the fan rotation of the angular velocity calculated as

$$\omega = \frac{\lambda R}{u_0} \quad (1)$$

The power output of the fan per unit time is calculated as

$$P_t = M \times \omega \quad (2)$$

Substituting the relevant data, we can get that the output power of a single rotor fan is $P_{ts} = 2.7\text{MW}$, while the output power using a double rotor fan is $P_{td} = 6.6$ MW. Substituting into equation (3), we get the energy per unit time flowing through the fan wheel section P_ω

$$P_\omega = \frac{1}{2}\rho A u_0^3 \quad (3)$$

$$C_p = \frac{P_t}{P_\omega} \quad (4)$$

Substituting the values into equation (4), we get the power factor $C_{ps} = 0.45$ for a single rotor wind turbine and $C_{pd} = 0.55$ for a double rotor wind turbine. it can be seen that the peak power factor of the double rotor wind turbine is higher compared with the single rotor wind turbine, and the use of the double rotor arrangement can make use of the wind gain effect to obtain higher power generation while extending its working life, which is a more reasonable wind turbine arrangement scheme (Zhong 2021).

3.4 Impact of single point mooring system on wind turbine

In order to make the system utilize as much wind as possible, we designed a single-point rotating tower-type mooring platform. The working principle is that when the wind direction changes, the triangular platform rotates around a mooring cable fixed at one of the vertices under the action of air flow until the effect of positive wind direction is achieved.

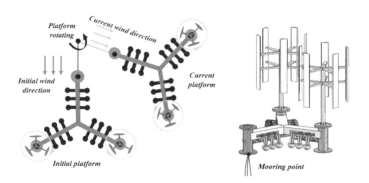

Figure 9. Single point mooring platform rotating in the wind.

Compared to conventional fixed platform structures, moored platforms can reduce the shaft load burden on the turbine and increase power generation. Similarly, the performance of the wind turbine system with and without the single-point mooring system is analyzed. In this case, it can be considered that the wind turbine system cannot be properly aligned with the wind, and the plane where the two turbine axes are located is not perpendicular to the wind speed direction.

Under the given conditions, the simulated moments on the platform are obtained as follows

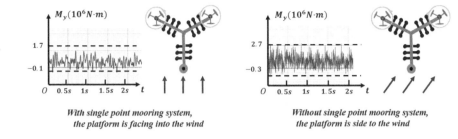

Figure 10. Comparison of moments on the platform in windy and windless conditions.

In comparison with Figure 4, the presence of a single-point mooring system makes the wind turbine smoother against the wind. Compared with the traditional fixed platform structure, the presence of the mooring platform can significantly reduce the shaft load of the turbine and increase the power generation.

4 IMPACT OF SINGLE POINT MOORING SYSTEM ON WIND TURBINE

The application of a wave energy harvesting device can partially convert the impact energy of waves on the platform into wave energy, thus reducing the load on the platform and the turbine and prolonging its service life. In order to analyze the specific impact of the circumscribed buoyancy wave energy generation device on the smoothness of the offshore floating platform, ANSYS-AQWA is used to call out the wavefield in the environment of the offshore floating platform. It is found through simulation that the circumscribed buoyancy can weaken the impact of waves on the platform. The Jonswap spectrum is selected with a meaningful wave height of 5.7m and a spectral peak period of 10.2s, and the platform droop motion response is obtained, as shown in Figure 11.

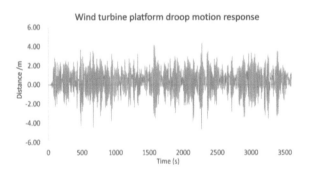

Figure 11. The effect of waves on the stability of co-generation single-point mooring platforms.

The pictures can be analyzed to get that the displacement of the wind turbine platform's vertical swing motion is within 4m, and the platform stability is good. Reducing the wave response of the two degrees of freedom of transverse/longitudinal rocking of the floating platform at sea improves the stability of the platform.

5 CONCLUSION

This paper first discusses a two-rotor vertical axis wind turbine applied to a new wind-wave coupled power generation device and analyzes its structural composition and performance. By analyzing and comparing the external flow field distribution, vorticity field distribution, and power coefficient magnitude of both, it is not only proved that it has higher energy conversion efficiency from the wind gain point of view but also illustrates that the dual-rotor wind turbine has lower on-axis load and longer service life from the force point of view. The effect of the introduction of the single point mooring system on the energy conversion efficiency of the twin-rotor wind turbine system is then analyzed, showing that the automatic wind alignment effect brought by this system can improve the efficiency of the twin-turbine system. The joint design of the twin-rotor wind turbine and the single-point mooring system provides a certain reference for the design work of the new offshore power generation platform at the same time. The joint design of the dual rotor wind turbine and single point mooring system not only improves the overall system stability, but also provides a certain reference for the new offshore power platform design work.

ACKNOWLEDGMENTS

This research is supported by the Academic Affairs Office of Wuhan University of Technology (Theme of the project: National University Student Social Practice and Science Contest on Energy Saving & Emission Reduction)

REFERENCES

Lee H, Poguluri S K, Bae Y H. Development and verification of a dynamic analysis model for floating offshore contra-rotating vertical-axis wind turbine[J]. *Energy*, 2022, 240: 122492.

National Advisory Committee for Aeronautics. *NACA 4-digit airfoil generator*[EB/OL]. [2022-4-18].http://airfoiltools.com/airfoil/naca4digit

Wu Y C, Miao F, Li Y B, Yan Y F. Numerical simulation of vertical plate bypass flow based on RNG k-ε model[J]. *Sichuan Building Science Research*, 2014, 40(04):21–25.

Ye J, Zhong Y J. A review of offshore wind energy utilization and its cost analysis[J]. *Solar Energy*, 2018(06):19–25.

Zhong B W. Experimental study on the power characteristics of straight-bladed vertical axis wind turbine in turbulent environment[D]. *Harbin Institute of Technology*, 2021.

Advances in Renewable Energy and Sustainable
Development – Liang & Kasmani (Eds)
© 2023 Copyright the Author(s), ISBN: 978-1-032-39407-7

Design of small domestic sewage treatment device suitable for homestay

Jialin Cao* & Hao Yang
SinoHydro Foundation Engineering Co., Ltd. (Wuqing, Tianjin), Yongyang West Avenue, Wuqing District, Tianjin, China

Chengming Shi
PowerChina Eco-Environmental Group Co., Ltd. (Bao'an, Shenzhen), Bao'an District, Shenzhen City, China

Feihang Li & Chenghe Sun
SinoHydro Foundation Engineering Co., Ltd. (Wuqing, Tianjin), Yongyang West Avenue, Wuqing District, Tianjin, China

ABSTRACT: In recent years, the homestay industry has developed rapidly. In order to protect the environment, a large amount of domestic sewage produced by it can only be discharged after efficient treatment. According to the discharge characteristics of domestic sewage from home stay, this paper selected a homestay in Zhejiang Province as an example and designed a small domestic sewage treatment device according to local conditions. The device mainly includes three parts: sewage treatment module, thermal insulation module and photovoltaic power generation module, with low overall power consumption. The effluent quality of the unit meets the class I (B) standard in the discharge standard of pollutants for urban sewage treatment plants, and the amount of sludge produced is low. It can be used as fertilizer to realize zero pollution purification of sewage.

1 INTRODUCTION

Compared with traditional hotels, homestays prefer the experience of local conditions and customs, so they are mostly opened in villages with better scenery. For the local environment, the rapid development of home stay operation and subsequent passenger flow will produce a large amount of additional domestic sewage. If the sewage is directly connected to the local sewage treatment pipe network for centralized treatment, the domestic sewage treatment terminal in the village will operate at a high load, and some well covers will even gush sewage (Li 2021). Therefore, efficient and effective treatment of sewage discharged from home stay (Wang 2018) is a problem that must be solved. At present, houses in rural areas in China are basically scattered, and the applied domestic sewage treatment technologies mainly include ecological filters, constructed wetlands, oxidation ponds and membrane bioreactors (Gu et al. 2018; Huang et al. 2018).

In 2016, Zhejiang Province issued the guiding opinions on determining the scope and conditions of homestay (ZHZBF [2016] No. 150), which clearly required that "the location of home stay should comprehensively consider the environmental capacity, strengthen the construction of sewage treatment and other facilities, and ensure up to standard discharge." Take Moganshan Town, where the homestay industry is more advanced, as an example. Since June 2016, the centralized treatment action of home stay sewage has been launched. It is required to complete the installation of sewage treatment facilities within half a year for home stays with more than seven guest rooms

---

*Corresponding Author: 490634311@qq.com

DOI 10.1201/9781003349648-5

and secondary water source protection areas in the town. It can be seen that the homestay needs to formulate reasonable sewage treatment measures according to the local environmental protection standards, the scale of its own Inn and the characteristics of sewage before it can be discharged (Hu et al. 2020). Based on this, this paper selects a homestay in Zhejiang Province as an example and designs a small domestic sewage treatment device according to local conditions.

2 DESIGN CONDITIONS

2.1 *Design object*

Take a homestay in Zhejiang Province as an example. The homestay is designed as three floors with 15 rooms, including 12 guest rooms and three storage rooms. The sewage generation area mainly includes 12 toilets and two kitchens. During daily operation, the daily passenger flow of home stay is about 7-17 people.

According to Zhejiang Statistical Yearbook 2021, the per capita daily domestic water consumption of Zhejiang Province in 2020 is 220.06 liters. Based on the per capita daily domestic sewage volume and 90% of the per capita daily domestic water, it is estimated that the total daily sewage output of the home stay varies from 1.39 tons to 3.37 tons. Considering the characteristics of intermittent discharge of domestic sewage, a sewage treatment device with a design treatment capacity of 3 tons per day is selected.

2.2 *Process analysis*

The sewage discharged from the guest house mainly comes from domestic water, such as toilets, kitchens and sanitation. From the aspect of water quantity, the sewage discharge is small and generally intermittent, and the daily change degree of water quantity is large, showing a trend of more in the day and less at night. From the aspect of water quality, the main pollution indicators are nutrients and bacteria such as cod, nitrogen and phosphorus, which generally do not contain toxic pollutants. Among them, the content of organic matter is high, the concentration of ammonia nitrogen is relatively high, the biodegradability is high, and the composition of heavy metals and harmful pollutants is low. Therefore, it is considered to select sequencing batch activated sludge process as a sewage treatment process.

The sequencing batch reactor (SBR) was proposed in 1970 and was mainly used for wastewater denitrification. Compared with the traditional sewage treatment process, SBR includes homogenization, primary sedimentation, biochemistry, final sedimentation, etc., which has the characteristics of energy conservation, environmental protection and high flexibility.

3 DESIGN OF SEWAGE TREATMENT DEVICE

3.1 *Device composition*

The device mainly includes the sewage treatment module, thermal insulation module and photovoltaic power generation module. The sewage treatment module is mainly used for the biological treatment of domestic sewage discharged intermittently. The thermal insulation module and photovoltaic power generation module can preheat the sewage to maintain the biological activity of activated sludge. A group of reclaimed water reuse tanks shall be set separately to discharge the treated water into the tank for re-sedimentation, and then the clean water shall be pumped out by a submersible pump. Clean water can be used for watering vegetation, greening, etc.

3.1.1 *Sewage treatment module*
The module mainly includes a sewage treatment tank and an intelligent integrated control cabinet. The parts in the treatment tank are connected with the control cabinet through the air hose installed

underground. The treatment tank adopts the sequencing batch activated sludge treatment process, which mainly includes a series of actions such as influent, denitrification, aeration, drainage, sludge reflux and idle, so as to realize the batch treatment of sewage.

The intelligent integrated control cabinet is equipped with a metering pump, which can put chemicals into the treatment tank and react with phosphate in the sewage to form insoluble sediment to remove phosphorus in the sewage. At the same time, an ultraviolet disinfection lamp group can be set in the control cabinet to quickly kill microorganisms and various bacteria in the water by using ultraviolet rays without by-products to further improve the effluent quality.

3.1.2 *Insulation module*

When the sewage treatment plant is in a low-temperature environment, the activity of activated sludge decreases and the treatment effect is not good. Therefore, a thermal insulation module is added, which only operates at low temperatures.

The main body of the device is provided with an insulation interlayer: the thickness is 60 mm, and the material is polyurethane foam. The pipeline is provided with a foam sponge insulation layer with a thickness of 15 mm. A small electric heating water storage tank is set in front of the water inlet pipe of the device. When the tank reaches the preset liquid level, the electric heating pipe starts to work to keep the constant temperature of domestic sewage at 15°C, and the valve is opened to pass the sewage into the sewage treatment device.

If there is no domestic sewage in the electric heating water storage tank, the electric heating device is in the power-saving standby state, and the sewage heated by the electric heating pipe enters the anaerobic tank. If the external ambient temperature does not reach the low-temperature condition, the electric heating function of the small box can be turned off and directly enter the anaerobic tank. The anaerobic tank is equipped with a pressure sensor, which can be used to detect the water level in the anaerobic tank. In the influent stage, if the preset liquid level is not reached, the system will enter a 6-hour economic mode, in which the system maintains sludge activity through occasional aeration. If the preset liquid level is not reached in the anaerobic tank after four consecutive measurements, the water in the reaction tank will be pumped back to the anaerobic tank to maintain sludge activity. If the preset liquid level is reached, enter the denitrification stage. When the activity of activated sludge is in a normal state, the insulation system can be closed. Moreover, a water quality detection device is set at the pre-water outlet to realize the real-time monitoring of the effluent quality.

3.1.3 *Photovoltaic power generation system*

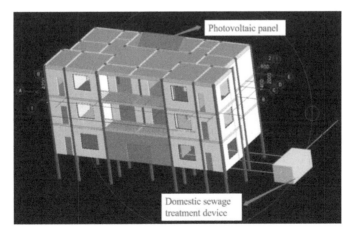

Figure 1. Layout diagram of homestay and sewage treatment plant.

As shown in Figure 1, the sewage treatment device is equipped with a photovoltaic power generation module to realize high purification of domestic sewage with almost zero power consumption. Considering the cost performance and design purpose, the polycrystalline module is used as the photoelectric converter of the photovoltaic power generation system.

Taking the home stay designed in this paper as an example, the energy required is about 1.45 × 108 J. The overall dimension of the selected photovoltaic panel is 1665 mm × 991 mm × 50 mm, single group power of 220 W. The power generation of the photovoltaic panel is about 24kw per hour in autumn and winter. Five photovoltaic panels can be installed, with a total area of 4.95m^2.

3.2 *Design object*

Figure 2 shows the flow diagram of a small domestic sew-age treatment device. The process flow of domestic sewage treatment in the unit is "influent → denitrification → aeration → sedimentation → effluent → sludge dis-charge," which can be repeated 4 to 6 times a day.

The device is equipped with a solar energy device, which can make full use of solar energy resources and reduce power consumption.

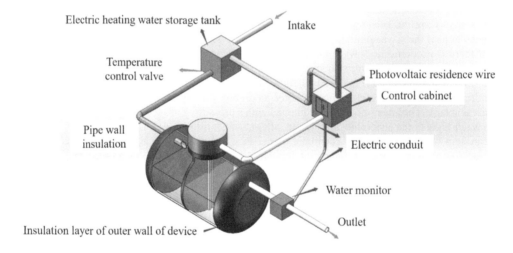

Figure 2. Flow chart of the small domestic sewage treatment device.

3.3 *Sewage discharge grade and purpose after water purification*

The main water quality indexes of the effluent from the unit can meet the class I(B) standard in the pollutant discharge standard for urban sewage treatment plants, which is shown in Table 1. The treated water can be connected to the municipal pipe network for discharge, and the discharged excess sludge can be used for agricultural production and fertilizer.

Considering the possibility of sewage reuse, an additional post-treatment module can also be added to carry out in-depth treatment of the treated water, which can be used for toilet flushing water in the toilet in the guest house, greening irrigation of the courtyard landscape and artificial pond replenishment, as well as car washing and road cleaning. The design of the whole device can make rational use of water resources and is in line with the sustainable development strategy.

3.4 *Device characteristics*

The device can set the cycle and drainage time of the treatment batch according to the living habits of residents in the homestay and the peak water consumption. It has strong impact load resistance and

Table 1. Maximum allowable discharge concentration (daily average value) of basic control items of water quality indicators (unit: mg/L).

Serial number	Basic control items	Class I (B) standard	Serial number	Basic control items	Class I (B) standard
1	Chemical oxygen demand (COD)	60	7	Number of fecal Escherichia coli (PCs./L)	10^4
2	Biochemical oxygen demand (BOD_5)	20	8	Chromaticity (dilution ratio)	30
3	Suspended solids (SS)	20	9	Total phosphorus (P)	1
4	Animal and vegetable oil	3	10	Ammonia nitrogen (in N)	8(15)
5	Petroleum	3	11	pH value	6-9
6	Anionic surfactant	1	12	Total nitrogen (in N)	20

has the advantages of high efficiency, energy-saving and stability. The device has the advantages of low operation noise, convenient installation and a small overall floor area. In addition to being suitable for rural families, homestays and other occasions, it can also expand the treatment scale through flexible combinations for the centralized treatment of domestic sewage in small villages. As the power consumption of the sewage treatment part is small and equipped with a photovoltaic power generation system, the overall power consumption of the device is almost zero, which is in line with the current situation of energy utilization and environmental protection in China. The overall amount of sludge produced by the unit is low, and the discharged sludge can be used for plant fertilization in rural areas, closely following the theme of the times of green energy conservation and low-carbon environmental protection.

4 CONCLUSION

Based on the results and discussions presented above, the conclusions are obtained as below:

(1) A small domestic sewage treatment device for biological treatment is designed according to the characteristics of small and intermittent sewage discharge of homestay.
(2) The effluent quality of the unit meets the class I(B) standard in the discharge standard of pollutants for urban sewage treatment plants, and the amount of sludge produced is very low. It can be used as fertilizer to realize zero pollution sewage purification treatment.
(3) The device that produces little noise has strong impact load resistance and can be buried underground. It optimizes the living environment of the homestay and protects the ecological environment around the homestay.

REFERENCES

Gu Teng, Wu Yong, Wang Xiao-tong. Treatment Of Rural Domestic Sewage Using Biological Aerated Filter and Modular constructed Wetland Combined Process[J]. *Environmental Engineering*, 2018, 36(1): 20–24.

Hu Xiaobo, Luo Hui, Jing Zhao-qian, et al. Research progress of rural domestic sewage treatment technology[J]. *Applied Chemical Industry*, 2020, 49(11):2871–2876.

Huang Yuanyuan, Xu Dongyang, Ji Rongping. Study on Rural Sewage Treatment by Improved Biological Trickling Filter Artificial Wetland[J]. *Technology of Water Treatment*, 2018, 44(5):93–97.

Wang Min. Current Situation and Legal Counter-measures of Rural Domestic Sewage Pollution in China[J]. *Journal of Anhui Agricultural Sciences*, 2018, 46(9): 188–189, 219.

Xiumei Li. Analysis of the current situation and Countermeasures of rural domestic sewage treatment [J]. *Resources Economization & Environmental Protection*, 2021(4):89–90.

Advances in Renewable Energy and Sustainable Development – Liang & Kasmani (Eds)
© 2023 Copyright the Author(s), ISBN: 978-1-032-39407-7

Study on cascade effect of combined sediment retention of gully check dam system in Loess small watershed

Weiying Sun & Pan Zhang*
Key Laboratory of Soil and Water Conservation on the Loess Plateau of Ministry of Water Resources, Yellow River Institute of Hydraulic Research, Zhengzhou, China

ABSTRACT: In order to study the mechanism of joint action between check dam systems in small watersheds and the regulation of flood storage and sediment interception, the Wangmaogou small watershed check dam system was taken as the research object, which is on the basis of systematically analyzing the relationship between the small watershed dam system, the cascading effect of different dam structure combinations is calculated, and the control relationship of shards and layers to sediment between the small watershed dam system and trench unit dam are revealed. The cascading effect of the sediment interception capacity of the check dam system in a small watershed is also elucidated. The research shows that the reasonable configuration of the check dam system in the small watershed can effectively prolong the service life of the backbone dam, and the combined use of the cascade dam system can realize the internal digestion of sediment.

1 INTRODUCTION

A reasonable dam system layout can achieve the relative balance of water and sediment siltation in small watersheds as soon as possible, and ultimately achieve internal digestion of natural precipitation and complete interception of eroded sediment in the watershed (Jin et al. 1995). Therefore, for a dam system in a small watershed, a reasonable dam-reservoir layout can realize the mutual cooperation and division of labor between upstream and downstream, trunk and branch ditches, and give full play to the comprehensive functions and benefits of flood storage, sediment interception, land silting and production of unit dam system (Qin et al. 2004). The spatial layout of the dam system is an important part of the dam system planning, which mainly includes the distribution of dam sites of various scales, the configuration of the proportion of the quantity, and the determination of the dam height (Tian 2005; Wu 1991). Whether the spatial layout of the dam system is reasonable will directly affect the capital investment and benefit performance of the whole dam system. At present, in the research on the spatial layout of dam systems, some scholars have established optimization models for backbone dams and optimized the dam site and dam height, and other scholars have used statistical analysis methods to study the quantitative allocation ratio of dams of different scales in small watersheds (Qin et al. 1994). However, there are few studies on the influence of different layout structures of dam systems in small watersheds on the operation mode and effect of the dam system, especially the joint action mechanism between single dams, between single dams and dam system units, between different dam system units, and between each dam system unit and the whole dam system as well as the regulation effect of flood storage and sediment interception.

Therefore, this paper takes the Wangmaogou small watershed dam system as the research object and analyzes how the upstream and downstream dam reservoirs of the series dam system can achieve layered, fragmented, and segmented interception and siltation of eroded sediment in the region through a reasonable layout, the mutual cooperation, and division of labor, so as to play its cascading regulation effect. And then it provides a scientific basis for the planning, construction,

---

*Corresponding Author: zpyrcc@163.com

DOI 10.1201/9781003349648-6

safe operation, and efficient utilization of soil and water resources of the check dam in the Loess Plateau region.

2 ANALYSIS OF THE RELATIONSHIP BETWEEN DAM SYSTEMS IN SMALL WATERSHEDS

The relationship of small watershed dam systems can be decomposed into sub-dam systems and unit dam systems. A small watershed dam system is often composed of several sub-dam systems. For example, the Jiuyuangou small watershed dam system includes 14 sub-dam systems such as Wangjiagou and Wangmaogou, and each sub-dam system is relatively independent. Generally, there is no water-sediment transfer relationship. The sub-dam system is composed of several unit dam systems. Although each unit dam system controls water and sediment by itself, some unit dam systems also have upstream and downstream relationships. The flood flows downstream after being adjusted and stored by the upstream, and there is a water-sediment transfer relationship between such unit dam systems. For example, Wangmaogou sub-dam system can be divided into four units, namely Guandigou and Matizui.

3 CASCADE EFFECT OF SEDIMENT INTERCEPTION BY DIFFERENT DAM STRUCTURE COMBINATIONS IN WANGMANGOU SAMLL WATERSHED

First, the control section of Guandigou No. 1 dam (dam AC) is restored to the initial state of unbuilt the dam. Then, on the basis of laying the control dam AC, different numbers of small and medium-sized dams are laid at different upstream positions. Under the condition of constant interval soil erosion modulus and different dam system combinations, the amount of sediment intercepted by control dam AC is deduced. Through the influence of the upstream series of dams and reservoirs on the siltation volume and siltation years of the control dams, the cascade effect of joint use and coordinated storage of dam reservoir groups on flood storage and sediment interception of the unit dam system is revealed. The calculation method is as follows:

$$V_{control} = V_{total\ amount\ of\ sediment} - \sum_{i=1}^{n} V_i \qquad i = 1, 2, 3 \dots n \tag{1}$$

$$\bar{V}_{control} = \sum V_i / x_{control} \tag{2}$$

$$y_{siltation} = \frac{V_{total\ control}}{\bar{V}_{control}} \tag{3}$$

$$y_{remaining} = \frac{v_{control\ remaining}}{\bar{v}_{control}} \tag{4}$$

Table 1 shows the cascading characteristics of the sentiment interception for different combinations of dam system structures between the Guandigou No.1 dam area. When there is no dam reservoir upstream of the dam AC, after 44 years of operation, the eroded sediment in the control area is intercepted in the dam AC, with a total volume of 230,700 m^3, accounting for 78% of the total reservoir capacity, and will soon be full of silt and lose flood control capacity. When additional dam A was built upstream to form the combination of dam system AC+A, the dam AC only silted up 165,700 m^3, accounting for 56% of the total reservoir capacity, which extended the silting up period by 19 years. When dam B is added upstream to form a combination of dam system AC+A/B, dam AC intercepts 133,700 m^3 of sediment, accounting for 45% of the total reservoir capacity, and the siltation life is extended by 16 years. And when an additional dam AB is built upstream to form the combination of dam system AC+A/B/AB, the dam AC intercepts 119,700 m^3 of sediment, accounting for 41% of the total reservoir capacity, and the siltation life is extended by 10 years. And

41

when the upstream continues to build additional dam C, forming the combination of dam system AC+A/B/AB/C, which is the current status dam system combination, the dam AC stops 105,700 m^3 of sediment, accounting for 36% of the total reservoir capacity, and the siltation life is extended by 13 years.

Table 2 shows the cascading characteristics of the sentiment interception for different combinations of dam system structures between the Wangmaogou No. 2 dam area. Wangmaogou No. 2 Dam (Dam AD) is located in the middle section of the main ditch of the Small Watersheds, controlling an area of 2.81km². As a controlling backbone dam, it plays a linking role in the transmission of water and sediment in the trench. When the water and sediment in the upstream dam system exceed the reservoir capacity and are discharged, not only should the water and sediment discharged from the upstream dam be stored, but also the water and sediment in the control area of the dam should be stored as much as possible to prevent the loss of water and sediment sand in the dam area. At the same time, the reservoir capacity also affects the operational safety of downstream dams and reservoirs. Therefore, the dam's flood storage and sentiment interception capacity depend on the volume of its own reservoir.

Table 1. The cascading characteristics of the sentiment interception for different combinations of dam system structures between the Guandigou No. 1 dam area.

Single Dam	Reservoir Capacity (m^3)	Silted Volume (m^3)	Dam Structure Combination			
			AC_0	AC_1	AC_2	AC_3
A	13.6	6.5				
B	3.2	3.2	3.2			
AB	1.4	1.4	1.4	1.4		
C	1.4	1.4	1.4	1.4	1.4	
AC	29.4	10.57	10.57	10.57	10.57	10.57
Total Sediment Interception Volume	49.1	23.07	16.57	13.37	11.97	10.57
Proportion (%)		78%	56%	45%	41%	36%
Sedimentation Years (a)		49	68	84	94	107
Remaining Years (a)		5	24	40	50	63

Table 2. The cascading characteristics of the sentiment interception for different combinations of dam system structures between the Wangmaogou No. 2 dam area.

Single Dam	Reservoir Capacity (m^3)	Silted Volume (m^3)	Dam Structure Combination				
			AD_0	AD_1	AD_2	AD_3	AD_4
D1	2	2					
D2	4.51	3.65	3.65				
E1	18.5	9.93	9.93	9.93			
E2	5.07	4.88	4.88	4.88	4.88		
DE	18.5	12.0	12.0	12.0	12.0	12.0	
AD	105.4	28.08	28.08	28.08	28.08	28.08	28.08
Total Sediment Interception Volume	153.98	60.54	58.54	54.89	44.96	40.08	28.08
Proportion (%)		57%	56%	52%	43%	38%	27%
Sedimentation Years (a)		61	63	67	82	92	131
Remaining Years (a)		19	21	25	39	49	86

Table 3 shows the cascading characteristics of the sentiment interception for different combinations of dam system structures between the Wangmaogou No. 1 dam area. Wangmaogou No. 1 dam (Dam AI) is located in the lower section of the main ditch of the small watershed, serving as the outlet of flood cement and sediment in the whole small watershed, control area of 2.62km^2. Built in 1953, it has been in operation for 46 years, for the small watershed dam system construction of the earliest dam control dam. In the initial stage of the dam construction, there was no other dam reservoir upstream as a supplement, and all the sediment in the dam-controlled area entered the dam, resulting in a sharp loss of reservoir capacity. When a check dam was added upstream, the sediment entering the dam decreased and the deposition rate slowed down. Even so, 592,000 m^3 of sediment has been silted up so far, which is close to full and has lost most of its flood control capacity. But as the controlling backbone dam at the outlet of the basin, it is the necessary way for the water and sediment to converge to the next level of the trench in the small watershed. To ensure that the sediment does not cause flood control pressure downstream due to downstream discharge, it is necessary to maintain sufficient reservoir capacity to hold incoming water and sediment from upstream. Therefore, in addition to the expansion of the dam body by raising the capacity and reconstruction of the spillway in a timely manner to discharge excessive flood water. The construction of additional small and medium-sized check dams upstream of the dam can form a structural optimization of the dam system unit with a reasonable layout, relying on mutual cooperation among dams to protect, and the water and sediment are blocked on the spot, and the potential energy after the flood is dispersed and weakened, thereby reducing the flood control pressure of the dam, which is more scientific and reasonable and is more conducive to ensuring the safety of the dam system.

Table 3. The cascading characteristics of the sentiment interception for different combinations of dam system structures between the Wangmaogou No. 1 dam area.

Single Dam	Reservoir Capacity (m^3)	Silted Volume (m^3)	Dam Structure Combination								
			AI	AI$_1$	AI$_2$	AI$_3$	AI$_4$	AI$_5$	AI$_6$	AI$_7$	AI$_8$
G1	2.41	2.41									
G2	5.92	4.72	4.72								
G3	7.33	7.33	7.33	7.33							
G4	15.18	4.91	4.91	4.91	4.91						
H1	2.12	2.12	2.12	2.12	2.12	2.12					
H2	2.64	0.79	0.79	0.79	0.79	0.79	0.79				
H3	2.85	2.66	2.66	2.66	2.66	2.66	2.66	2.66			
I1	2	1.57	1.57	1.57	1.57	1.57	1.57	1.57	1.57		
I2	5.55	2.97	2.97	2.97	2.97	2.97	2.97	2.97	2.97	2.97	
AI	69.83	59.2	59.2	59.2	59.2	59.2	59.2	59.2	59.2	59.2	59.2
Total Sediment Interception Volume		88.68	86.27	81.55	74.22	69.31	67.19	66.4	63.74	62.17	59.2
Proportion (%)		127%	124%	117%	106%	99%	96%	95%	91%	89%	85%
Sedimentation Years (a)		32	32	34	38	40	42	42	44	45	47
Remaining Years (a)		-10	-9	-7	-3	0	2	2	4	6	8

4 RESULTS AND DISCUSSION

Comparing the relationship between the three control dams in the main channel of Wangmaogou small watershed and the dam upstream of the dam shows that, under the premise of a certain area of the control area of the controlled dam-reservoir project, the larger the density of the upstream small and medium-sized dams or the larger the total reservoir capacity, the greater the proportion of the retained flood and sediment to the regional eroded sediment volume. Correspondingly, the less flood and sediment entering the downstream control backbone dam, the slower the siltation, the longer the siltation period, the larger the remaining reservoir capacity, and the stronger the flood control and security capability. The layout of the check dam system and the reasonable arrangement of dam construction sequence can give full play to the cascade regulation of the dam system, greatly increase the effectiveness of the check dam in flood storage and sediment interception, and prolong the service life of the check dam, so as to achieve the internal digestion of flood sediment in small watersheds. Therefore, in the planning of the layout of the small watershed dam system and the timing of dam construction, the density of small and medium-sized dams upstream of the backbone dams should be set scientifically and reasonably, and the dam sites should be selected with reasonable reservoir capacity and flood control standards so that the flood storage and sediment interception capacity of each dam can be fully utilized. Through the mutual coordination, joint operation, and complementary use among the dams, we can realize the " Wheeled storage, wheeled fencing, and wheeled seeding " of flood water, sediment, and dam land, so that the dam system can reach a relatively stable state as soon as possible, realize the safety, efficiency and high productivity of the dam system, and finally give full play to its ecological, economic and social benefits.

ACKNOWLEDGMENTS

This work was financially supported by the National Natural Science Foundation of China (42041006), and the Natural Science Foundation of Henan Province of China (212300410060).

REFERENCES

Jin Minghua, Zhu Mingxu, Bai Fenglin. Optimization Planning Model of Small Watershed Dam System and Its Application. *Yellow River*, 1995(11):29–33.

Qin Hongru, Jia Shunian, Fu Mingsheng. Study on the Construction of Dam Systems in Small Watersheds on the Loess Plateau[J]. *Yellow River*, 2004, 26(1):33–36.

Qin Xiangyang, Zheng Xinmin. Study of Optimum Planning Model of Main Silt Arrester System for Controlling Gully Erosion in Small Watersheds. *Soil and Water Conservation in China*, 1994(1):18–22.

Tian Yonghong. Analysis on Dam System Layout of Jiuyuangou Pilot Small Watersheds. *Soil and Water Conservation in China*, 2005(9):21–22.

Wu Yongchang, Cui Yunpeng. Double Optimization of Dam Height to Retard Mud and Flood for the System of Check Dam. *Journal of Soil and Water Conservation*, 1991, 5(1):19–26.

Advances in Renewable Energy and Sustainable Development – Liang & Kasmani (Eds)
© 2023 Copyright the Author(s), ISBN: 978-1-032-39407-7

The evolution of renewable energy in the United States in a warming climate

Hangzhou Wang*
Western Christian High School, Hull, Iowa

ABSTRACT: In a world where the climate is changing rapidly and the temperature of the earth is increasing due to the carbon emission from human activities, it is exigent to understand the relationships between the various factors that may affect or be impacted by energy consumption. In this article, we analyzed the trends of renewable energy and different types of fossil fuel energy sources in recent years. We also proposed ratio indexes to represent the relationships among those energy sources. Those indexes, combined with our analysis of the seasonal variations of renewable energy consumption, may suggest the efficient exploitation of different energy types at different times. Further, we explored the carbon emissions across different geographical regions in the United States. We found that although the regional total carbon emissions remain similar throughout the years, the total carbon emission is decreasing gradually. Finally, we combined all the factors together and explored their statistical correlation with the United States GDP and global temperature anomaly. We discovered that there is a strong positive correlation between renewable energy consumption and GDP (r=0.964). There is also a strong positive correlation between fossil fuel consumption and CO_2 emission (0.817), and a strong negative correlation between renewable energy consumption and CO_2 emission (–0.888).

1 INTRODUCTION

With the world population growing at an approximately exponential trend and the quality of life constantly improving in general, global energy consumption has been increasing on a large scale. From 2005 to 2017, the world's Total Final Consumption (TFC) of coal increased by 23%, oil and gas increased by 18%, and electricity increased by 41%, causing global fossil CO_2 emissions to grow for three years consecutively (Iea. (n.d.), 2021) As a result, the earth continues to warm rapidly, and we are experiencing unprecedented climate change and natural disasters (Persistent fossil fuel growth ... - iopscience.iop.org. (n.d.), 2021). Earth's temperature has risen by 0.14°F (0.08°C) per decade since 1880, and the rate of warming over the past 40 years is more than twice that: 0.32°F (0.18°C) per decade since 1981 (Dahlman et al. 2018). In North America, there has been an increase in the frequency, intensity, and duration of heat waves in regions that currently experience them and a decrease in snowpack in the western mountains. All of these conditions contribute to the intensifying degradations of ecosystems and unsettling disruptions of human communities. ("What Are the Long-Term Effects of Climate Change?" *What Are the Long-Term Effects of Climate Change? | U.S. Geological Survey,* 2004, https://www.usgs.gov/faqs/what-are-long-term-effects-climate-change) The increase in energy demand in human society and fossil fuel consumption has a solid contribution to the increase in Green House Gas emissions and the subsequent warming of the earth. In fact, approximately 70% of global Green House Gas emissions are due to the burning of fossil fuels for heating, electricity, transport, and industry (Olivier et al. 2017). Fortunately, renewable energy technologies have been developed to reduce society's dependence on traditional energy sources, thus providing an approach to mitigate global warming and ameliorate the catastrophic events brought by it. To promote the transitions to renewable

---

*Corresponding Author: harry828app@gmail.com

DOI 10.1201/9781003349648-7

energy, we need to gain a clear understanding of the latest trends in renewable energy consumption, as well as various types of fossil fuel consumption across the country. We would also need to grasp the relationships between those energy sources, their effects on carbon emissions, and their contributions to global temperature anomalies.

2 FOSSIL FUEL AND RENEWABLE ENERGY CONSUMPTION

2.1 *Consumption trends*

We first collected the monthly data on energy consumption for the three types of fossil fuels (Coal, Natural Gas, and Petroleum) and renewable energy in the United States from the U.S. Energy Information Administration ("Total Energy Annual Data - U.S. Energy Information Administration (EIA)." 2011 *Total Energy Annual Data - U.S. Energy Information Administration (EIA)*, https://www.eia.gov/totalenergy/data/annual/). Calculating the yearly energy consumption and converting them to the same measurement unit Quadrillion BTU, we graphed the total consumption from 2005 to 2020.

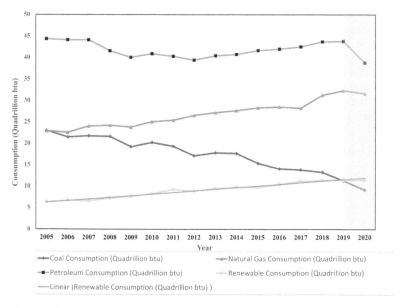

Graph 1. U.S. Fossil Fuels and Renewable Energy Consumption (2005-2020).

As indicated in the figure, the total petroleum consumption remains relatively consistent at approximately

40 to 45 Quadrillion BTU over the years. The first significant dip in total petroleum consumption takes place between 2007 and 2009. One potential major reason would be the financial crisis at that time, which causes an economic recession in the United States. Following the downturn were the skyrocketing unemployment and lower spending incentives, resulting in less demand and consumption of oil for transportation and energy generation. However, after the 4.07 Quadrillion BTU dip, the overall petroleum consumption bounced back in the later years until 2019 to 2020, when an even steeper dip occurs.

With regards to the total coal consumption over the years, it is apparent that it has a gradually decreasing trend. The quickest reduction occurs after 2014, and, similar to petroleum consumption, coal consumption significantly decreased in the year 2020.

Besides petroleum and coal, the other main fossil fuel source, natural gas, has an overall increasing trend. It started to increase at a significantly higher rate until the year 2020 when it dropped by 0.68 Quadrillion BTU. Comparing the three different types of fossil fuel, we noticed that the consumption of the least environmentally friendly fossil fuel, (Leung 2015) coal, drastically decreased over time, whereas that of the cleanest fossil fuel, natural gas took an increasing trend. However, petroleum, whose degree of environmental cleanness falls between the other two fossil fuels, had a total consumption that remained unchanged.

While the consumption of all three fossil fuel sources is declining in 2019-2020, renewable energies maintain their growth trend. As indicated by the line of best fit (green line), renewable energy consumption increased by 0.3808 quadrillion BTU per year on average and the year 2019-2020 followed this general trend. On the other hand, the three fossil fuels, as aforementioned, all experienced a significant decrease in consumption, deviating from each of their former trends. This anomaly may reflect the unusual sharp decline in traditional industries and transportation due to the COVID-19 pandemic. This correlation is further supported by the greatest decrease in the consumption of petroleum, the most popular traditional energy source. Most importantly, we also found that renewable energy was the least impacted energy source during the pandemic. Its robustness and consistency in times of uncertainty add to its environmentally friendly nature, making it the ideal energy source in today's society.

2.2 Ratio index

In addition to visualizing the trends of the energy sources individually in recent years, we also investigated their interconnected relationships. We were specifically interested in how renewable energy consumption changed compared to the three types of fossil fuels. To visualize their relationships, we first proposed four indexes that indicate the proportion of renewable energy consumption in one unit of fossil fuel consumption in a particular year. In percentages, we developed the following measurements: Renewable/Petroleum, Renewable/Coal, Renewable/Natural Gas, Renewable/Fossil Fuel.

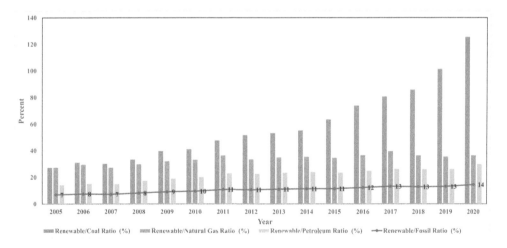

Graph 2. Renewable energy to fossil fuels ratio indexes.

As shown in the graph, renewable energy to coal consumption had approximately exponential growth, renewable energy to natural gas consumption remained relatively the same, and both renewable energy to petroleum and renewable energy to total fossil fuel increased slowly yet consistently. We would later use these ratio indexes to explore their possible correlations with carbon emissions across the country and temperature anomalies around the world.

2.3 *Renewable energy by season*

Besides the total yearly consumption of renewable energy and its correlation with fossil fuels, we also took a deeper look into renewable energy consumption's seasonal changes in the most recent three years (2019, 2020, and 2021). (Note that the average value of the three months is displayed for each season)

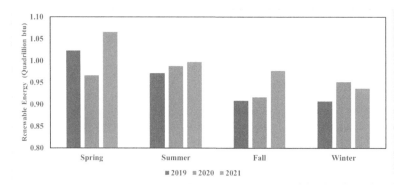

Graph 3. Renewable energy by season.

In general, the renewable energy consumption in spring and summer is the largest, and that in fall and winter is the least. The variation in consumption is most likely due to the amount of renewable energy available to use in those seasons. One of the most predominant renewable energy types, solar energy, is mostly available in summer in the United States. The other popular renewable energy type, wind energy, has the greatest reservation in spring due to the highest wind plant capacity factor in that season. Fall and winter, on the other hand, do not have the seasonal benefits for most renewable energy types.

3 CARBON EMISSIONS

Different types of energy sources would also contribute to CO_2 emissions. Thus, we investigated how much carbon was released into the atmosphere since 2005. For a more compact visualization, we grouped the 51 states into four geographical regions using the divisions (*Census Regions and Divisions of the United States*, 1950 https://www2.census.gov/geo/pdfs/maps-data/maps/reference/us_regdiv.pdf) proposed by the United States Census Bureau: Northeast, Midwest, South, and West. We also graphed the total emission of carbon dioxide. (Note that we only have the total emission data for the years 2019 and 2020).

From the graph, we can see that although the amount of carbon dioxide emitted in each region remained similar throughout the years, the total emitted in the United States fluctuated and eventually decreased slowly. Interestingly, the two most significant decreases in carbon emissions took place during 2007-2009 and 2019-2020, which align with the two major dips in total petroleum consumption we concluded in section 1.1. We will explore their statistical correlations in section 3.2.

4 ENERGY CONSUMPTION, GROSS DOMESTIC PRODUCT, AND GLOBAL MEAN TEMPERATURE ANOMALY

4.1 *Trends with carbon emissions and energy consumption*

To better visualize the growing trends and the relationships, we graphed the time series of these five factors in recent years (from 2005 to 2020): Renewable Energy Consumption, Fossil Fuel

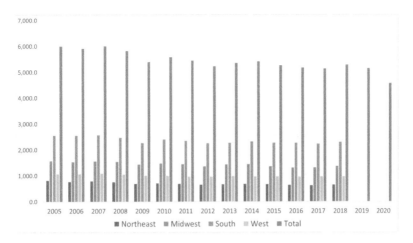

Graph 4. Carbon dioxide emissions by region.

Consumption, Total CO_2 Emission, U.S. GDP, and Global Mean Temperature Anomaly. Note that we first standardized all the data so that the relative values may be demonstrated by the positive and negative signs. In this way, we may also eliminate the influence of the differences in measurement units to obtain a clear comparison and visualization across different factors.

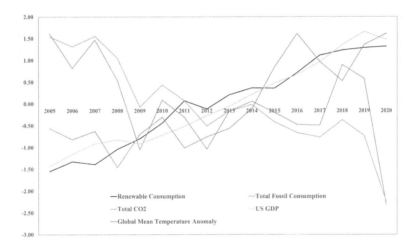

Graph 5. Patterns of the five factors.

From the graph, we may clearly see that both CO_2 emission and Fossil fuel consumption are on a decreasing trend, especially in the year 2019-2020, when they both dropped drastically. On the other hand, the other three factors- renewable energy consumption, global mean temperature anomaly, and U.S. GDP are increasing over time. According to the graph, the two most correlated factors are renewable energy consumption and the U.S. GPD, which have the most similar trends and overlapping areas.

4.2 *Correlation indexes and linear regression models*

We then calculated the correlation coefficient values of the five factors to gain a numerical understanding of the interconnections among them.

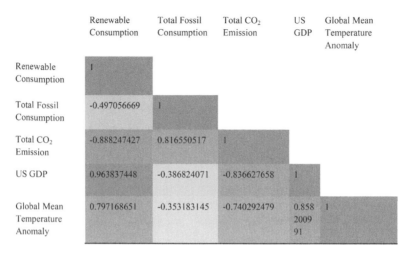

Graph 6. Correlation coefficient matrix.

The most significant correlation coefficient is between renewable energy consumption and the U.S. GDP: with a positive r-value of 0.964, they have a strong positive correlation, as we expected in the former section. The correlation coefficient between total renewable energy consumption and CO_2 emission (-0.888) and that between total fossil fuel consumption and CO2 emission (0.817) form a strong contrast. With similar absolute values, the two correlation coefficients imply that both renewable energy and fossil fuel have strong yet opposite correlations with CO2 emission. In addition, the U.S. GDP and total carbon emission have a strong negative correlation, indicating that CO_2 emission may not only impact the ecological environment but also influence the national economy. We then further established the linear regression model between the U.S. GDP and CO_2 emission.

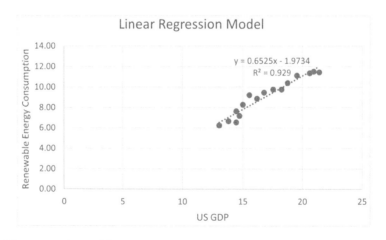

Graph 7. Linear regression model.

Based on our model, our result is greater than 0.25 and thus significant at the 95 percent confidence level.

5 CONCLUSION

In conclusion, we found that the total consumption of renewable energy has been increasing in recent years, while that of fossil fuels varies: coal consumption greatly decreased over time, petroleum consumption remained relatively the same, and natural gas consumption, one of the cleanest forms of fossil fuels, increased.

We also discovered that renewable energy tends to be consumed the most in the first two seasons, possibly due to the seasonal differences in solar and wind energy availability. Knowing the seasonal differences regarding renewable energy, we may be able to adjust to produce and allocate fewer fossil fuels in spring and summer to achieve efficiency.

With regards to the regional carbon emissions across the U.S., it is apparent that the total emissions decreased although the regional levels remained the same.

Finally, combining all the factors with temperature anomaly and the U.S. GDP, we found that there is a strong, positive correlation between renewable energy consumption and the U.S. GDP. There is also a contrasting and strong correlation between CO_2 emission and the two types of energy sources. Finally, GDP is negatively correlated with CO_2 emission, suggesting the dangerous economic impact of releasing too much carbon from anthropogenic activities.

REFERENCES

"Total Energy Annual Data - U.S. Energy Information Administration (EIA)." Total Energy Annual Data - U.S. Energy Information Administration (EIA), https://www.eia.gov/totalenergy/data/annual/.

"What Are the Long-Term Effects of Climate Change?" What Are the Long-Term Effects of Climate Change? | U.S. Geological Survey, https://www.usgs.gov/faqs/what-are-long-term-effects-climate-change.

Census Regions and Divisions of the United States. https://www2.census.gov/geo/pdfs/maps-data/maps/reference/us_regdiv.pdf.

Dahlman, Rebecca Lindsey and LuAnn. *"Climate Change: Global Temperature."* Climate Change: Global Temperature | NOAA Climate.gov, https://www.climate.gov/news-features/understanding-climate/climate-change-global-temperature#:~:text=August%2012%2C%202021-,Highlights,land%20areas%20were%20record%20warm.

Iea. (n.d.). Data tables – data & statistics. IEA. Retrieved October 11, 2021, fromhttps://www.iea.org/data-and-statistics/data-tables?country=WORLD&energy=Balances&year=2018.

Leung, Guy CK. *"Natural gas as a clean fuel."* Handbook of clean energy systems (2015): 1-15.

Olivier, J. G. J., Peters, J. A. H. W., & Schure, K. M. (2017). *Trends in global emissions of CO2 and other greenhouse gases: 2017 Report* (PBL report no. 2674). PBL Netherlands Environmental Assessment Agency, Bilthoven, the Netherlands.

Persistent fossil fuel growth ... - iopscience.iop.org. (n.d.). Retrieved October 11, 2021, from https://iopscience.iop.org/article/10.1088/1748-9326/ab57b3/meta.

Advances in Renewable Energy and Sustainable Development – Liang & Kasmani (Eds)
© 2023 Copyright the Author(s), ISBN: 978-1-032-39407-7

Optimization control strategy of waste heat utilization in data center with renewable energy

Liangliang Zhu*, Pengpai Feng & Kui Wang
Nari Group Corporation/State Grid Electric Power Research Institute, Nanjing, China
State Grid Electric Power Research Institute Wuhan Efficiency Evaluation Company Limited, Wuhan, China

Tianheng Chen
State Grid Tianjin Electric Power Company, Tianjin, China

Yi Ding
State Grid Tianjin Electric Power Research Institute, Tianjin, China

Ye Li
State Grid Tianjin Electric Power Company, Tianjin, China

ABSTRACT: With the rapid development of artificial intelligence, cloud computing, and big data, large data centers are constantly emerging. A large amount of waste heat generated by data centers during operation is directly discharged, which will result in serious energy waste. Taking a data center in Tianjin as an example, this paper established the heating system framework with renewable energy utilization based on waste heat recovery. To further reduce the system operation cost, a power grid demand response mechanism based on flexible load regulation was introduced. The trend of peak shaving benefit and heating cost of system under different user subsidy prices was studied. The results show that the peak shaving benefit increases firstly and then decreases with the increase of user subsidy price, and when the user subsidy price is 0.15 CNY/kWh, the maximum peak shaving benefit is obtained. The system heating cost decreases firstly and then increases with the increase of the user subsidy price, which will reach the lowest level when the user subsidy price is 0.17 CNY/kWh.

1 INTRODUCTION

Data centers gather a large amount of computing equipment, storage equipment, network equipment, and other IT equipment, which make it an infrastructure and service platform for centralized processing, storage, transmission, exchange, and management of information. According to relevant statistics, data centers currently account for nearly 3% of the world's electricity supply. At the same time, with further growth of data process load, the energy consumption of data centers will continue to increase, which means there is great potential for energy conservation and emission reduction (Huang et al 2020).

Data centers typically operate around the clock, and most of the consumed power is eventually converted into heat, which is released into the atmosphere through refrigeration equipment. If abundant waste heat generated during operation can be recovered, it will be able to meet part of the building heating demand, which will significantly improve the comprehensive efficiency of energy utilization, and achieve the purpose of energy conservation and carbon reduction. Many researchers had carried out related research. Mahdi (Deymi et al. 2019) used an air source heat pump to recover waste heat from a data center in Mashhad, Iran, and saved 35,000 m^3 of natural gas, 20.8 MWh of power, and $25,000 per year, respectively. Lv (2019) recovered heat from chilled water or cooling

---

*Corresponding Author: zll_seu@163.com

52   DOI 10.1201/9781003349648-8

water of the data center in Hangzhou, China for heating, and formulated corresponding control strategies. The total operation cost, standard coal quantity, and annual emission of CO_2 of the system were reduced by 290,000 CNY, 89,133 kg, and 222 t per year, respectively. Luo (2019) recovered and utilized waste heat of the data center through heat pump units to provide heat in winter and annual domestic hot water for offices. Compared with boilers, 910,000 CNY was saved during the heating season, and the annual operation saving ratio could reach 66%.

The existing waste heat utilization method of data centers is mainly to recover waste heat of cooling water for residential heating or domestic hot water, and less consideration is given to responding to the peak shaving demand of the power grid. This paper introduced the power grid demand response mechanism based on flexible load regulation, studied the effect of user subsidy price on system peak shaving benefit and heating cost, and finally proposed an optimal pricing strategy.

2 SYSTEM DESCRIPTION

Taking a data center in Tianjin as the research object, the economy of the energy supply system based on cooling system waste heat recovery is discussed. The overall system framework is shown in Figure 1. The power of the system is supplied by a municipal power grid and renewable energy, and low-level waste heat is raised to heating temperature by a heat pump to meet the heating demand of surrounding residents. The electric boiler is configured as the standby heat source to maintain the stability of the energy supply. In order to further reduce energy cost, the energy supply system participates in the peak shaving task delivered by the power grid to obtain the benefit, which attracts users to participate in demand response through subsidy price, and finally realize peak shaving of power grid under flexible load control.

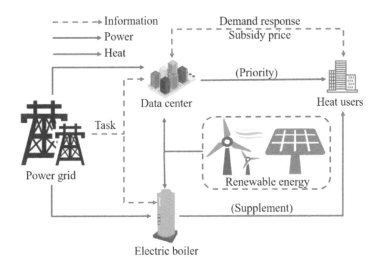

Figure 1. The framework of the waste heat utilization system in the data center.

3 MODEL DESCRIPTION

The energy supply system recovers waste heat of cooling water from refrigeration units in the data center to meet the heating demand of surrounding residents. The supply and return temperatures of cooling water and heating water are set to 37/32°C and 50/40°C, respectively. The followings are the mathematical models for relevant equipment in this waste heat recovery system.

3.1 Data center energy consumption model

Hourly energy consumption data of a data center in Tianjin in December was collected, as shown in Figure 2. The hourly energy consumption of the data center shows a trend of obvious periodic change. The highest hourly energy consumption is 12,211 kWh and the lowest value is 10,538 kWh. The energy consumption is relatively stable and at a high level, which can meet the requirement of a waste heat recovery system.

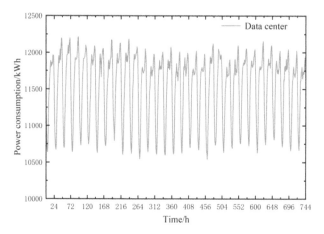

Figure 2. Hourly energy consumption data of a data center in Tianjin in december.

3.2 User response model

The corresponding subsidy can be obtained while the user can actively reduce the demand for heating in the peak period of power consumption, i.e., participating in the demand response. The maximum demand response capacity that can be obtained at each moment is the load reduction by adjusting the normal indoor temperature to 18°C. However, due to the different heating and consumption habits of each user, not all users would like to reduce their load by 100% for demand response, and they will adjust their response ratio according to the user subsidy price provided by the load supplier.

The Weber Fechner law can be applied to quantify the psychological activities of users when they participate in market-oriented transactions. It can be considered that the user's response ratio Z is the logarithmic function of subsidy price $P_{B,i}$:

$$Z = K \ln P_{B,i} + C \qquad (1)$$

where K and C are the response characteristic coefficients of users.

According to the response ratio, the user's real-time response amount can be calculated, which is the load reduction amount.

$$L_X = Z \cdot (L_{now} - L_{18}) \qquad (2)$$

where L_{now} is the load for maintaining the current indoor temperature. L_{18} is the load for maintaining the indoor temperature at 18°C. L_{now} and L_{18} can be calculated by setting the heating temperature in the heat load calculation model.

3.3 Heat pump unit model

When the backwater temperature is lower than the heating set point of the heat pump unit, the compressor runs. The supply water temperature of the heat pump unit can be calculated according to the following formula (Shen et al. 2022):

$$\begin{cases} T_G = T_b + \frac{\Delta t P_{HP}}{c_w m_w}, T_b < T_q \\ T_G = T_b, T_b \geq T_q \end{cases} \tag{3}$$

where T_q is the heating set point of the heat pump, °C. T_b is the backwater temperature, °C. P_{HP} is the heating power of the heat pump, kW.

The heat pump power consumption can be calculated according to the coefficient of performance (*COP*). Under the premise of fixed supply and return water temperature, this paper mainly discusses the influence of unit load rate on *COP*.

The correction coefficient of the heat pump load rate on *COP* can be calculated as follows (Huang 2018):

$$\varepsilon_{loadrate} = 0.2061 R_{load}^3 + 0.0043 R_{load}^2 + 0.7818 R_{load} + 0.0102 \tag{4}$$

where $\varepsilon_{loadrate}$ is the correction coefficient of heat pump load rate on *COP*. R_{load} is the unit load rate.

Therefore, the comprehensive *COP* expression of the heat pump is as follows:

$$COP_{HP} = COP_{HP,d} \varepsilon_{loadrate} \tag{5}$$

where COP_{HP} is the actual *COP* of the heat pump. $COP_{HP,d}$ is the *COP* of the heat pump under design conditions.

The power consumption of the heat pump can be calculated as follows:

$$e_{HP} = \Delta t P_{HP} / COP_{HP} \tag{6}$$

where e_{HP} is the power consumption of the heat pump.

3.4 *Electric boiler model*

Due to the energy loss in the process of electrothermal conversion, the calculation method of the heating capacity of an electric boiler is as follows:

$$Q^k = \eta \cdot \Delta t \cdot P_{HP}^k \tag{7}$$

where Q^k is the heating capacity of the electric boiler at k moment, kJ. *Pk HP* is the power of the electric boiler at k moment, kW. η is the electrothermal conversion efficiency, %.

3.5 *Renewable energy unit model*

The output power of photovoltaics depends on solar radiation intensity, and its energy conversion model is as follows:

$$EP_{pv}^t = IC_{pv} df_{pv} (SI_{pv}^t / SI_{pv}^{ra})(1 + t_c(T_{pv}^t - T_{pv}^{ra})) \tag{8}$$

where IC_{pv} represents the installed capacity of photovoltaics, kW. SI_{pv}^t represents the solar radiation intensity, W/m². SI_{pv}^{ra} represents the rated solar radiation intensity, W/m². df_{pv} represents the power reduction factor of photovoltaics, and the value is 0.9. t_c represents the temperature coefficient, which ranges from 0.004°C to 0.006°C. T_{pv}^t represents the surface temperature of the photovoltaic panel at t moment, °C. T_{pv}^{ra} represents the surface temperature of the photovoltaic panel under the rated test condition, °C.

The power of the wind turbine is closely related to the wind speed, and its energy conversion model is as follows:

$$EP_{wt}^t = \begin{cases} 0, & v^t \leq v^{min} \text{ or} v^t \geq v^{max} \\ \frac{(v^t)^2 - (v^{min})^2}{(v^{ra})^2 - (v^{min})^2} IC_{wt}, & v^{min} \leq v^t \leq v^{ra} \\ IC_{wt}, & v^{ra} \leq v^t \leq v^{max} \end{cases} \tag{9}$$

where v^{min}, v^{max} and v^{ra} represent cut-in, cut-out, and rated wind speed, respectively, m/s. IC_{wt} represents the installed capacity of the wind turbine, kW.

4 RESULTS AND DISCUSSION

The installed configuration of the waste heat recovery heating system of the data center is shown in Table 1. Assuming that the heating area of each house is 100 m², the system can provide domestic heating for 1300 houses, and the calculation of relevant residents' heat load is completed by DeST. In this study, the system optimization control for one month as the heating cycle is carried out. According to the control strategy, the subsidy price varies between 0.1 CNY/kWh and 0.2 CNY/kWh. The trend of energy consumption, peak shaving benefit, and heating cost of the system under different subsidy prices is shown in Figure 3.

Table 1. Installed configuration of waste heat recovery heating system in the data center.

Heat pump heating capacity (kW)	Electric boiler power (kW)	Wind power capacity (kW)	Photovoltaic capacity (kW)
8614	10000	500	800

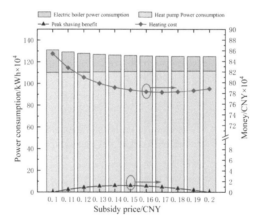

Figure 3. The change in energy consumption and heating cost of waste heat recovery heating system in the data center.

As shown in Figure 3, with the increase in subsidy price, the total energy consumption of the waste heat recovery system decreases gradually. When the subsidy price is 0.13 CNY/kWh, with the further increase in subsidy price, the reduction of energy consumption tends to be flat. The reason is that at the beginning, with the increase in subsidy price, the proportion of users participating in load reduction gradually increases, which reduces the total heat load of the system, and the corresponding system energy consumption will decrease. With the further increase in subsidy price, the total heat load continues to decrease, but the lower load rate will have a negative impact on the heat pump performance, which will increase the power consumption of unit heating capacity. Therefore, even if the system heat load is reduced, the system energy consumption will not decrease significantly. From the peak shaving benefit curve in Figure 3, it can be seen that with the increase in subsidy price, the peak shaving benefit initially shows an upward trend, and the maximum benefit is obtained when the subsidy price is 0.15 CNY/kWh. As the subsidy price continues to increase, the increase of users participating in load reduction slows down, while a higher subsidy price for users leads to the decrease of peak shaving benefit actually obtained by load suppliers. Under the comprehensive influence of system energy consumption and peak shaving benefit, with the increase of subsidy price, the system heating cost firstly decreases and then increases. When the subsidy price is 0.17 CNY/kWh, the system heating cost reaches the lowest level.

When the subsidy price is 0.17 CNY/kWh, the load reduction and renewable energy utilization are shown in Table 2. At the above subsidy price, the system cuts peak load by 6.23% of the total heating load, and the utilization rate of renewable energy reaches 14% of the total system energy consumption.

Table 2. Load reduction and renewable energy utilization of waste heat recovery heating system in data center.

Category	Value
Subsidy price (CNY/kWh)	0.17
Peak shaving load (kWh)	326,667
Peak shaving ratio (%)	6.23
Renewable energy consumption (kWh)	175,938
Proportion of renewable energy (%)	14.04

5 CONCLUSION

A large amount of waste heat generated by data centers during operation has not been utilized effectively, which will result in serious energy waste. Therefore, it is important to effectively recycle the waste heat in data centers. Aiming at solving this problem, this paper established the heating system framework with renewable energy utilization based on waste heat recovery and proposed an optimal subsidy price strategy to reduce heating costs by encouraging users to participate in demand response. The main conclusion can be summarized as follows:

(1) With the increase in subsidy price, the peak shaving benefit of the system shows the trend of increasing firstly and then decreasing, and the maximum peak shaving benefit is obtained when the subsidy price is 0.15 CNY/kWh.
(2) The heating cost of the system shows a downward trend firstly and then an upward trend with the increase of subsidy price. When the subsidy price is set at 0.17 CNY/kWh, the cost of energy supply reaches the lowest level, which is 781,847 CNY.
(3) The excessively high subsidy price reduces the actual peak shaving benefit obtained by suppliers, and on the other hand, it will make the heat pump unit work at a lower load rate, which reduces the performance of the unit and increases the heating cost. Therefore, setting an appropriate subsidy price will make the overall economy of the system optimal.

In terms of future work, the research on introducing heat storage into the system should be carried out to further reduce heating costs.

ACKNOWLEDGMENTS

This research is supported by the State Grid Corporation of China Science and Technology Project "Research on Precise Cooling and Waste Heat Utilization Technology of Data Center Based on Data Load Interaction (5400-202112159A-0-0-00)".

REFERENCES

Deymi Dashtebayaz M, Valipour Namanlo S. Thermoeconomic and environmental feasibility of waste heat recovery of a data center using air source heat pump[J]. *Journal of Cleaner Production*, 2019, 219: 117–126.

Huang P, Copertaro B, Zhang X, et al. A review of data centers as prosumers in district energy systems: Renewable energy integration and waste heat reuse for district heating[J]. *Applied energy*, 2020, 258: 114109.

Huang R. The Research and Optimization about cooling system solutions of large data center[D]. *Chongqing University*, 2018.

Luo Y. Energy-saving research on waste heat recovery and utilization in the large data center[J]. *Energy Conservation*, 2019, 38(08):46–48.

Lv M, Chen J, Wang F. Design and Application of Waste Heat Utilization System in Data Center[J]. *Construction Science and Technology*, 2019(12): 55–57.

Shen R, Zhong S, Wen X, et al. Multiagent deep reinforcement learning optimization framework for building energy system with renewable energy[J]. *Applied Energy*, 2022, 312: 118724.

Advances in Renewable Energy and Sustainable
Development – Liang & Kasmani (Eds)
© 2023 Copyright the Author(s), ISBN: 978-1-032-39407-7

Allocation optimization of combined cooling and heating system based on high-voltage and high-temperature composite phase change heat storage device

Meixiu Ma*, Wei Kang, Jibiao Hou, Lixiao Liang & Zhanfeng Deng
State Key Laboratory of Advanced Transmission Technology (State Grid Smart Grid Research Institute Co., Ltd.), Beijing, China

Na Zhang
State Grid Shanxi Electric Power Research Institute, Taiyuan, China

Mengdong Chen
State Key Laboratory of Advanced Transmission Technology (State Grid Smart Grid Research Institute. Co., Ltd.), Beijing, China

ABSTRACT: With the rapid development of the social economy, the demand for power is increasing sharply, and the contradiction between power supply and load is becoming more and more serious. The paper takes the combined cooling and heating system as the research object, load aggregator is introduced into the system, so that power grid companies and wind power plants can establish contact with heat users through the load aggregator. The investment payback period is introduced as an economic indicator to analyze the capacity allocation of high-voltage and high-temperature composite phase change heat storage devices. The example analysis shows that in winter and summer, the cost saved by users through load reduction can reach 19.6% and 25.1%, respectively. The model can achieve a multi-party win-win with the aggregator as the core. The system can achieve the result of the shortest investment payback period when the capacity is 90MWh.

1   INTRODUCTION

The development of renewable energy is the only way to solve the energy and environmental crisis. Wind energy, as a clean and environmentally renewable energy, has been widely concerned and utilized by researchers. However, renewable energy has its own shortcomings such as intermittent and energy flow density. If the multi-energy complementary form is adopted, the complementarity between the power grid and new energy will be brought into play and the reliability and sustainability of power supply on the load side will be greatly improved. Integrating demand response into the dispatching of the cold-heat co-generation system and fully mobilizing load-side resources to promote wind energy dissipation are helpful to realize the economic and safe operation of the power grid.

A lot of research has been done on abandoned wind and load side resources of the power systems. Helseth et al. (2013) proposed an adjustment strategy to reduce the wind abandoning rate based on the integrated energy network of wind hydrogen storage. Hydrogen and heat were used as energy storage media, and electricity, heat, and gas networks were combined. Hananeh et al. (2014) established two load response models and proposed a two-stage stochastic programming model. Then analyzed the role of PDR and IDR in suppressing the uncertainty of predicted power

---

*Corresponding Author: mameixiu@163.com

DOI 10.1201/9781003349648-9

of wind power. The results showed that load response plays a positive role in participating in wind power consumption. Li Hongzhong et al. (2019) proposed the concept of generalized energy storage resources, mainly including cooling, heat storage, and power storage. Based on the energy complementary characteristics, an optimal dispatching model including energy storage was built. In this case, not only the operating cost of the system can be reduced, but also the wind power grid can be effectively improved. Two operation schemes are proposed by using the electric quantity of the valley section. Li Junhui et al. (2018) put forward the optimized control strategy of the heat storage electric boiler to track the power of wind discarded, which effectively reduces the heating load of the thermoelectric unit and improves the level of wind power absorption. Li Xiaojun et al. (2016) elaborated on the feasibility and economy of wind power heating and established the economic value model and environmental value model of power grid enterprises. The example analysis shows that heating with wind power discarded can effectively alleviate wind discards, improve the utilization efficiency of wind power, and have good social benefits. Bie Chaohong et al. (2014) introduced price demand response into the power system with wind power, and considered customer satisfaction constraints to reduce the impact of the intermittent output of wind power on system dispatch.

Based on the above research, this paper establishes a combined cooling and heating system based on high-voltage and high-temperature composite phase change heat storage. In order to further play the role of peak shaving and valley filling, the load aggregator is introduced to form a quarternary interaction model with the power grid, wind power plant, and heat users. Finally, the investment payback period is introduced to analyze the capacity allocation of phase change heat storage devices as an economic index.

2 MODEL DESCRIPTION

2.1 *System introduction*

This system model and energy and information flow process are shown in Figure 1. The wind power plant generates electricity and heats the heat storage device at midnight. In the heating condition, the electric boiler provides heat to users during the valley period, while the heat storage device supplies the stored heat to users during the peak period, and the electric boiler performs auxiliary heating. In the refrigeration condition, the Libr absorption refrigerator is driven by the heat storage device as the heat source, which provides cooling for the user.

In order to further play the role of demand response peak shaving and valley filling, a load aggregator is introduced into the combined cooling and heating system, so that the power grid company and wind power plant can establish contact with heat users through the load aggregator. Load aggregators can integrate users' demand response resources and enable them to participate effectively in the demand response market. At the same time, it can further encourage users to reduce their own load by offering subsidized prices and maximizing the potential of load regulation. On the other hand, load aggregators assist users in demand response by controlling heat release in peak hours and heat storage in low hours, so as to realize time transfer of grid load and increase wind power consumption.

2.2 *Constraint condition*

This paper puts forward the following constraints on the operation of the system.

2.2.1 *Constraints on phase change regenerative electric boiler systems*
(1) Capacity constraints of high-voltage and high-temperature composite phase change heat storage device:

$$0 \leq S_t \leq S \tag{1}$$

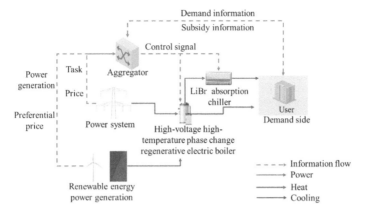

Figure 1. Schematic diagram of energy flow and information flow of the system.

Where S_t is the capacity state of the phase change heat storage system at time i, S is the maximum capacity of the phase change heat storage device.

(2) Heating constraints of high-voltage and high-temperature composite phase change heat storage device:

$$0 \leq Q_e \leq Q_{e\,max} \quad (2)$$

$$0 \leq Q_h \leq Q_{h\,max} \quad (3)$$

Where Q_e is the heat storage power, Q_{emax} is the maximum heat storage power of the phase change heat storage device, Q_h is the heating power, and Q_{hmax} is the maximum heating power of the phase change heat storage device.

(3) Capacity state constraints of high-voltage and high-temperature composite phase change heat storage device. For the heat storage device, it is generally required to restore the heat storage to the initial state after operating a cycle:

$$S_{start}^t = S_{end}^t \quad (4)$$

Where S_{start}^t is the heat storage capacity at the beginning of the cycle; S_{end}^t is heat storage capacity at the end of the cycle.

(4) Operating power constraints of an electric boiler:

$$0 \leq P_{eb}^t \leq P_{eb,max} \quad (5)$$

Where P_{eb}^t is operating power of electric boiler at time t; $P_{eb,max}$ is maximum operating power of the electric boiler.

2.2.2 *Wind power plant revenue risk constraints*

In this study, wind power will be used for heat storage only when the price of wind power is lower than that of valley power. If the bundling price is higher than the valley price, the wind power plant will not benefit. According to this, the risk constraint model of the wind power plant is as follows:

$$\begin{cases} E_{W,i} = P_{W,i} \cdot (L_{C,i} + \sum_{m=1}^{M} N_{m,i}(L_{m,i} - L_{X,m,i})), P_{W,i} \leq P_i \\ E_{W,i} = 0, P_{W,i} > P_i \end{cases} \quad (6)$$

Where $E_{w,i}$ is the income of wind power plant at time i; $P_{w,i}$ is the preferential wind power price at time i; P_i is peak valley electricity price; $L_{c,I}$ is the heat storage capacity of the aggregator at time

i; $N_{m,i}$ is the number of users participating in the response; $L_{X,m}$ is the user load; $L_{X,m,i}$ is the user's active load reduction.

2.2.3 *Power grid security constraints*

The main purpose of the power grid to participate in the cooling and heating system is to reduce the operational risk caused by the high peak-valley difference. Therefore, the risk constraint of the power grid is to ensure that the reduction ratio of user load is not zero in the operation process. The constraint is:

$$\theta \geq \varepsilon \tag{7}$$

Where ε is a minimum acceptable reduction.

According to the above formula, during the operation of the combined cooling and heat supply system, four parties need to operate under corresponding constraints to ensure that the probability of risks is minimized while obtaining benefits for themselves.

3 CASE ANALYSIS

3.1 *Constraint condition*

Taking 100 hotel buildings in a city in northern China as the research object, each building has a total construction area of 12292 m^2, a total of 10 floors and a floor height of 3.3 m, of which the first floor is public areas such as halls, shops, and restaurants, the second floor is functional areas such as gym, conference room and multi-function hall, and more than three floors are guest rooms. Time of use electricity price is adopted in the calculation, and the specific electricity price information is shown in Table 1.

Table 1. TOU power price information.

	Peak power period	Valley power period	Flat power period
Time	9:00-12:00 17:00-23:00	0:00-8:00	8:00-9:00 12:00-17:00 23:00-24:00
Electricity price /(Yuan/kWh)	1.17	0.414	0.78

3.2 *System evaluation indicator*

In order to investigate the economy of the system, the dynamic investment payback period is selected as the system evaluation indicator, which is calculated according to the following formula:

$$\sum_{t=0}^{n_p} \frac{f_{in} - f_{out}}{(1 + i)^t} = 0 \tag{8}$$

Where f_{in} is annual revenue; f_{out} is annual expenditure; i is a discount rate.

(1) Investment cost

Investment cost includes equipment investment cost and maintenance cost. The investment cost of equipment mainly includes an electric boiler, heat storage device, and Libr absorption chiller. The annual operation and maintenance cost is the product of the total equipment investment cost and the annual operation and maintenance rate. The maintenance rate is 1.5%. The investment cost is as follows:

$$C_{inv} = p_{ceb} \cdot V_{eb} + p_{cht} \cdot V_{ht} + p_{lb} \cdot V_{lb} \tag{9}$$

Where, C_{inv} is the total cost of equipment investment; p_{ceb} is unit electric boiler capacity cost, V_{eb} is electric boiler construction capacity; p_{cht} is the price per unit capacity of heat storage device, V_{ht} is the construction capacity of heat storage device; p_{lb} is the capacity cost of lithium bromide absorption chiller, V_{lb} is the construction capacity of bromine chiller.

(2) Electricity cost

The total electricity cost is shown in the formula below:

$$C_{DL} = C_d \cdot W_Z + + \sum_{i=1}^{24} C_{SC,i} \cdot W_{SC,i} \tag{10}$$

Where C_{DL} is the electricity cost in a heating period, C_d is the unit price of wind abandoned power, W_Z is the total wind abandoned absorption of electric boiler system, $C_{SC,i}$ is the unit price of wind abandoned electric quantity absorbed by heat storage device in period i, $W_{sc,i}$ is the total amount of wind abandoning power absorbed by the heat storage device in period i.

(3) Heating income

Heating income is generally charged according to the residential area, and the calculation formula is as follows:

$$S_{hot} = S \cdot C_{hot} \tag{11}$$

Where, S_{hot} is total heating income in winter, ten thousand yuan; S is a total heating area, 10,000 /m^2; C_{hot} is unit area heating income, ten thousand Yuan/ten thousand m^2.

3.3 *Result analysis*

Figure 2 shows the original load power, reduced load, and charge/discharge power curve of high-voltage and high-temperature composite heat storage device of hotel building within 24 hours on winter and summer design days. It can be seen that under the heating condition, the abandoned wind power generation is used to heat the heat storage device in the low period at night. At this time, the electric boiler is heating the user. During the peak period, the user participates in the demand response to reduce the load. The heat storage device releases heat to supply the stored heat to the user, and the electric boiler is used for auxiliary heating; during normal electricity hours, the heat storage device does not participate in the heating supply. Through comparative calculation, it can be concluded that in winter and summer, the cost saved by users through load reduction can reach 19.6% and 25.1% respectively.

Figure 3 shows the variation curve of heat consumption and subsidy of heat users with subsidy price. As can be seen from Figure 3, user 1 is sensitive to the change in subsidy price, it fully participates in the demand response and the response ratio reaches the maximum when the subsidy price is larger than 0.1 Yuan/kWh. Then its heat consumption and subsidy are monotonically affected by the subsidy price. When the subsidy reaches 0.2 Yuan/kWh, the user's expenditure is only reduced by 4% and the subsidy is doubled. User 2 is not sensitive to price changes. it is willing to participate in the response when the subsidy price exceeds 0.1 Yuan/kWh. Its heat consumption and subsidy are affected not only by the response proportion but also by the subsidy price. And user 2 will fully participate in the response, then its expenditure will be reduced by 55% when the subsidy reaches 0.2 Yuan/kWh. At the same time, the comparison shows that the expenditure of user 1 is 53.6% lower than that of user 2 when the subsidy price is 0.1 Yuan/kWh. The expenditure of user 1 is only 4.7% lower than that of user 2 when the subsidy price is 0.2 Yuan/kWh, and the subsidy of user 2 is 9.1% lower than that of user 1.

After obtaining the cooling and heating load of the hotel, this paper proposes four kinds of heat storage device capacities, where Option 1, Option 2, and Option 3 represent 80MWh, 90MWh, 100MWh, and 110MWh of the heat storage device capacity respectively. The initial investment in combined cooling and heating system includes the cost of an electric boiler, a Libr absorption refrigerator, and a heat storage device. The details of each item in the calculation are shown in Table 2. And uses the dynamic investment payback period as the system evaluation indicator. Table

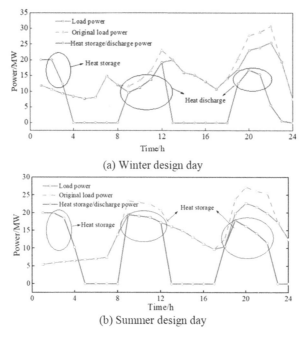

Figure 2. Original load power, reduced load power, and heat storage/discharge power.

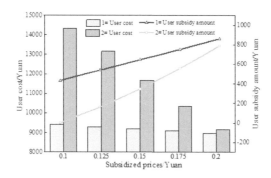

Figure 3. Changes in user expenses and subsidies with subsidy price variation.

3 shows the aggregator income of the system, the income of the wind power plant, and the income of the power grid under different capacities.

Table 2. Specific parameter information.

Name	Value
Electric boiler price	50 (Ten thousand Yuan/MWh)
heat storage device price	5 (Ten thousand Yuan/MWh)
Libr absorption refrigerator	20 (Ten thousand Yuan)
Maintenance rate	1.5%
Wind power price	0.28 (Yuan/MWh)
Heating price	30 (Yuan/m^2)

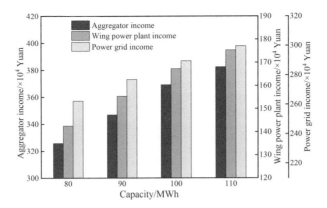

Figure 4. Load aggregator, wind power plant, and power grid income.

According to the formula, the payback period of Option 1 Option 2, Option 3, and Option 4 are 6.9 years, 6.6 years, 7.5 years, and 7.8 years, respectively. Based on the above analysis, the capacity configuration of Option 2 is the best option for the combined cooling and heating system.

4 CONCLUSION

In order to give full play to the role of peak filling of the heat storage device, a combined cooling, and heating system based on high-voltage and high-temperature composite phase change heat storage is designed in this paper. The third-party load aggregator is introduced into the system to further play the role of demand response and reduce the peak valley difference. According to the case analysis, the capacity configuration of high-temperature phase change heat storage devices is further studied on the basis of ensuring a win-win for all parties and taking the dynamic payback period as the economic indicator. The conclusions are as follows:

(1) The phase change heat storage device stores heating in the valley power period and releases heating in the peak power period. At the same time, users participate in demand response for load reduction. The cost saved by users through load reduction was 19.6% and 25.1% in winter and summer, respectively.
(2) The impact of subsidy price changes on user spending and subsidies of different types of demand response users was explored. Users who are not sensitive to the change in subsidy price will spend more on heat consumption and receive fewer subsidies than sensitive users. When the subsidy price is 0.2 yuan/kWh, the cost of user 2 is only 4.7% higher than that of user 1, and the subsidy is 9.1% lower than that of user 1.
(3) Four-parties interaction model with load aggregator as the intermediary was constructed, and the investment payback period was introduced as the economic indicator to analyze the heat storage device. The results show that the four-party interaction model can achieve a multi-party win-win with aggregators as the core. The investment payback period is 6.6years when the capacity is 90MWh.

ACKNOWLEDGMENTS

This work was supported by the technology projects of the state grid corporation of China. (The key technology research and prototype development of high voltage and high-temperature composite phase change heat storage heating device, NO. 5500-201958507A-0-0-00).

REFERENCES

Bie, C, H. Optimal dispatching of wind power system considering demand response. *Automation of Electric Power Systems* 38: 115–120+159. 2014.

Hananeh, F. (2014) The role of demand response in single and multi-objective wind-thermal generation scheduling: A stochastic programming. *Energy* 64(1): 853–867.

Helseth, A. (2013) A model for optimal scheduling of hydro thermal systems including pumped-storage and wind power. IET Generation, *Transmission & Distribution* 7(12).

Li, H, Z. (2019) Research on optimal operation of regionally integrated energy system considering broad energy storage. *Power System Technology* 43: 3130–3138.

Li, J, H. (2018) Design of optimum planning platform for electric boiler with heat storage for lifting wind power dissipation. *Acta Energiae Solaris Sinica* 39: 3270–3276.

Li, X, J. (2016) Promoting electric energy replacement and evaluating economic value and environmental benefit of wind power waste air heating to power grid enterprises. *North China Electric Power* 04: 1–5.

Advances in Renewable Energy and Sustainable
Development – Liang & Kasmani (Eds)
© 2023 Copyright the Author(s), ISBN: 978-1-032-39407-7

Adsorption removal of Cu in wastewater with waste motherwort by citric acid modification

Chinghua Liao*
School of Resource & Chemical Engineering, Sanming University, Fujian, China
Cleaner Production Technology Engineering Research Center of Fujian Universities, Sanming, China

Haiqi Lin
School of Resource & Chemical Engineering, Sanming University, Fujian, China

Chien Hung*
Huaian Huaibo Technology Co., Ltd., Jiangsu, China
School of Chemistry and Chemical Engineering, Huaiyin Normal University, Jiangsu, China

Qiyong Li, Shengchung Chen & Chihung Wu
School of Resource & Chemical Engineering, Sanming University, Fujian, China
Cleaner Production Technology Engineering Research Center of Fujian Universities, Sanming, China

ABSTRACT: Due to frequent industrial production activities, the heavy metal wastewater in water pollution is one of the most serious concerns of the world. General treatment methods are not economical and further generate a huge quantity of toxic chemical sludge. It is also called hazardous waste. Therefore, it is very important to find highly efficient and low-cost methods to treat heavy metals wastewater. And using agricultural waste as bioadsorbents is one good material. In this study, the motherwort waste was used as the bio-adsorbent to perform the copper removal treatment in wastewater. Results showed the waste motherwort modified with citric acid has benefits adsorption efficiency of copper ion in wastewater.

1 INTRODUCTION

Nowadays, industrial development is very fast that produces environmental pollution to affects human health and society (Lakherwal 2014). Particularly heavy metals and dye-contaminated wastewater are the most serious problems in China (Liao et al. 2020). However, heavy metal wastewater was mainly through the discharge of industrial effluents, metallurgy, fuel consumption, and discharge of municipal wastewater entering the environment (Crini 2005; Farhadi et al. 2021; Kanagaraj et al. 2014). Releasing heavy metal ions containing industrial wastewater is a major concern in the water environment day by day (Singh et al. 2011). Because the heavy metal ions in wastewater are strong toxicants, they are non-biodegradable. Moreover, heavy metal ions are also carcinogenic (Srivastava et al. 2011).

The methods of removal treatment of heavy metals in wastewater include chemical precipitation, ion exchange, reverse osmosis, electro-dialysis, filtration, flocculation, floatation, and adsorption (Banerjee 2020). However, these removal treatment methods exist some limitations such as high operational cost, hazardous sludge production, and energy requirements. Especially, when the heavy metals concentration was below 100 mg/L, the cost of removal treatment would be higher (Marín-Rangel et al. 2012; Mishra et al. 2012; Nguyen et al. 2013; Tamjidi et al. 2019; Volesky

---

*Corresponding Authors: chliaodr@outlook.com and chiywn@hotmail.com

DOI 10.1201/9781003349648-10

& Holan 1995). Comparatively, the adsorption method does not have the reverse sides of high operational cost, hazardous sludge production, and energy requirements instead of low energy consumption, ease of operation, and high removal treatment efficiency (Chakraborty et al. 2022; Tamjidi et al. 2019).

For this reason, in all low-cost adsorbents, both natural and biological adsorbents could be used to treat wastewater. Zeolites, clay, chitosan, and red mud are used as natural adsorbents. Bioadsorbent resources are made from agricultural and animal waste (Anastopoulos et al. 2017; Bhatnagar & Anastopoulos 2017; Foroutan et al. 2017; Mo et al. 2018). These techniques have been evaluated for the removal treatment of heavy metals from wastewater (Osman et al. 2010), especially when adsorbents are derived from cellulosic materials (Coelho et al. 2007). Adsorption by agricultural waste is a reversible reaction of heavy metals with biomass. Because it has the ion-exchange capacity and adsorption characteristics (Osman et al. 2010). These are derived from their constituent polymers and structures such as functional groups that facilitate metal complexation which helps for the sequestering of heavy metal ions in wastewater (Bailey et al. 1999; Hashem et al. 2005; Hashem et al. 2007; Laszlo & Dintzis 1994; Sud et al. 2008). Agricultural waste could be said to be a green adsorption material.

In the culture of Chinese traditional medicine, some Chinese herbal medicines have a widely extensive diet like motherwort. Motherwort is mostly used as a decoction in Chinese medicine for promoting blood circulation and removing blood stasis (Zhou et al. 2021). It would produce a large amount of motherwort waste after being extracted. If it does not treat or recycled, it is going to be a biomass resource waste.

2 MATERIAL AND METHOD

2.1 *Material*

In this study, the motherwort (scientific name: Leonurus artemisia) was used as the bioadsorbent. As motherwort has good health for women once a month to drink in China. Therefore, the materials of motherwort were collected from the Chinese medicine store after motherwort was extracted. Therefore, it has a very large quantity every month. It confirms the basic requirements of bioadsorbents for the low-cost and large quantities of agricultural waste.

Then, this research prepared some reagents like citric acid (Analytical Reagent, AR), sodium hydroxide (NaOH) (Analytical Reagent, AR), copper sulfate solution ($CuSO_4$) (Analytical Reagent, AR), sulfuric acid (H_2SO_4) (Analytical Reagent, AR), and copper standard solution (1,000 mg/L). These were purchased from Sinopharm Chemical Reagent Co., Ltd., Xilong Scientific Co., Ltd., and Tanmo Quality Inspection Technology Co., Ltd., respectively.

2.2 *Method*

The collected waste Motherwort was washed and cut into small sections and put into the oven of 60°C till all dry. Because the finer the particle size of this waste Motherwort, the larger the specific surface area is. That may adsorb more copper ions (Cu^{2+}) in wastewater. Therefore, all the dried waste Motherwort was ground, screened, and passed through the 50 mesh (<0.297 mm).

According to Ma's study, the agricultural waste (orange peel) modified with citric acid could get good adsorption efficiency for hexavalent chromium (Cr^{6+}) (Ma et al. 2018). Therefore, this study tried to modify the waste Motherwort powder with citric acid to test the adsorption efficiency of the copper ion (Cu^{2+}) in wastewater. The 5% of citric acid was used to modify the waste Motherwort powder and the incubation setting was stirred at 160 rpm for 4 hours in a constant speed oscillator (Changzhou Langyue Instrument Manufacturing Co., Ltd.). Then, the deionized water was used to wash modified materials until the neutral pH value.

The evaluated factors for the tests of the adsorption efficiency include pH values, the concentrations of copper ion wastewater, bioadsorbent doses, and stirring time as well as the comparisons between the modification and non-modification. The solution of each adsorption test was filtrated with a 0.45 μm filter membrane. The filtrates were analyzed by using atomic absorption spectrophotometer (AAs) (Beijing PERSEE General Instrument Co., Ltd.) to determine the copper ion concentration in wastewater. The equation (1) was used to calculate the adsorption efficiency of copper ions in wastewater.

$$\text{Adsorption efficiency (\%)} = \left(\frac{\text{start of Cu weight} - \text{end of Cu weight}}{\text{start of Cu weight}}\right) \times 100\% \quad (1)$$

3 RESULTS AND DISCUSSIONS

3.1 *Effect of pH adjustment*

To find the highest adsorption of copper without copper precipitation, the pH values from 4 to 6 were selected to evaluate their effect on copper ion precipitation in wastewater. Figure 1 shows the ratio of Cu precipitation at different pH values. After 10 min incubation, it found 0% of Cu precipitation at pH 4, and with the increasing pH values, the ratio of Cu precipitation also increased (Figure 1).

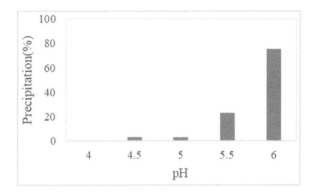

Figure 1. The ratio of Cu precipitation at different pH values.

Figure 2 shows the effects of different incubation times on the ratio of Cu precipitation at pH 4. The result showed that as the incubation time increases, the ratio of the Cu precipitation increases gradually. The Cu precipitation ratio was 0%, 1.01%, 3.03% and 6.63% at the 0, 10-, 60-, 90-, and 120-min incubation time, respectively (Figure 2). The optimum condition of pH 4 and incubation time of 10 min was selected, which has 0% Cu precipitation in wastewater.

3.2 *Effect of adsorption time*

In Figure 3, the efficiency of copper adsorption using the citric acid modified or unmodified bioadsorbents was tested at different stirring times. It shows that the bioadsorbent (waste Motherwort) modified with citric acid has more adsorption efficiency for copper than that of the unmodified bioadsorbent. The modified bioadsorbent has 71.53% of adsorption efficiency for copper at 1 hour stirring time. When the stirring time increased, the efficiency of copper adsorption was increased. But after 1 hour, both modified and unmodified bioadsorbents of copper adsorption efficiency were increased very slowly, which may indicate that both achieved the balance of copper adsorption. To have the maximum copper adsorption, the condition of four hours of stirring time was selected as an optimum condition of stirring time for further experiments.

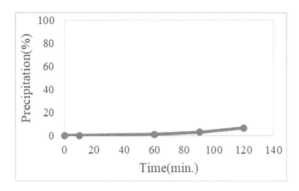

Figure 2. Different incubation times for Cu precipitation at pH 4.

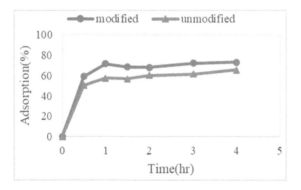

Figure 3. Effects of different stirring times on the Cu adsorption efficiency tests.

3.3 *Effect of adsorbent dose*

The bioadsorbent doses of 0.5, 1, and 2 g in the 50 ml copper wastewater were tested for copper adsorption efficiency at pH 4 for 4 hours of stirring time (Figure 4). The result presents that the copper adsorption efficiencies were 48.17 % (modified) and 50.01% (unmodified) under the condition of bioadsorbent dose of 0.5 g. Using 1 g of modified and unmodified bioadsorbents doses, the copper adsorption efficiencies were 73.16% and 65.70%, respectively. And using 2 g of bioadsorbent dose, the copper adsorption efficiency for the modified bioadsorbent could achieve 91.22%, which is higher than that of the unmodified bioadsorbent of 73.28%. The result indicated that adding more bioadsorbent (waste motherwort) to wastewater has better efficiency for copper adsorption. Therefore, bioadsorbent of 2 g was selected in further experiments.

3.4 *Effect of wastewater concentration*

The copper starting wastewater concentrations of 90 mg/L, 150 mg/L, 200 mg/L, 300 mg/L, and 400 mg/L were selected to evaluate the effect of wastewater concentration (Figure 5). The results show that the copper adsorption efficiency with a 2 g dose of bioadsorbent is highest under the copper starting wastewater concentration of 90 mg/L, which are 91.22% and 73.28% for modified and unmodified bioadsorbents, respectively. It also shows as the copper starting wastewater concentration increased, the copper adsorption efficiency decreased. According to this experiment inference, the used waste motherwort has a benefit adsorption efficiency of copper on 90 mg/L of starting wastewater concentration.

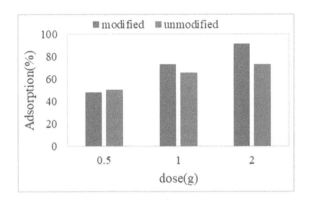

Figure 4. Different adsorbent doses of Cu adsorption.

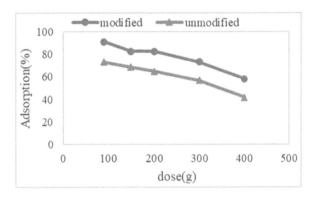

Figure 5. Effects of different wastewater concentrations on Cu adsorption efficiency.

4 CONCLUSION

In accordance with this study's experiments such as pH adjustment, stirring time, added bioadsorbent dose, and starting copper concentration in wastewater, it is concluded that the optimal condition of adsorption of copper ion in wastewater was using the modified waste motherwort (bioadsorbent) by citric acid, 2 g of bioadsorbent, pH 4 with 90 mg/L of copper wastewater for 4 hours of stirring, which achieves to 91.22% adsorption efficiency.

ACKNOWLEDGMENTS

This paper was supported by the Introduced High-Level Talents Research Start-Up Funding Project of Fujian Sanming University (Grant No. 20YG08; 20YG09); Fujian Provincial Department of Scicnce and Technology pilot project (Grant No. 2020Y0090; 2021Y0075); and Sanming City Guiding Science and Technology Project (Grant No. 2020-G-60; 2021-G-3; 2021-G-4).

REFERENCES

A. Bhatnagar, and I. Anastopoulos, "Adsorptive removal of bisphenol A (BPA) from aqueous solution: a review," *Chemosphere*, vol. 168, pp. 885–902, 2017.

A. Farhadi, A. Ameri, and S. J. P. C. R. Tamjidi, "*Application of Agricultural Wastes as a Low-cost Adsorbent for Removal of Heavy Metals and Dyes from Wastewater: A Review Study,*" vol. 9, no. 2, pp. 211–226, 2021.

A. Hashem, E. Abdel-Halim, K. F. El-Tahlawy, et al., "Enhancement of the adsorption of Co (II) and Ni (II) ions onto peanut hulls through esterification using citric acid," *Adsorption Science Technology*, vol. 23, no. 5, pp. 367–380, 2005.

A. Hashem, R. Akasha, A. Ghith, et al., "Adsorbent based on agricultural wastes for heavy metal and dye removal: A review," *Energy Educ. Sci. Technol*, vol. 19, no. 69, pp. e86, 2007.

B. Volesky, and Z. Holan, "Biosorption of heavy metals," *Biotechnology Progress*, vol. 11, no. 3, pp. 235–250, 1995.

C.-H. Liao, J.-Y. Shiu, C.-E. Hung et al., "*Treatment of Chromium in Wastewater by using Loquat Leaves as Biosorbent*," vol. 2, no. 3, pp. 6–10, 2020.

D. Lakherwal, "Adsorption of heavy metals: a review," *International Journal of Environmental Research and Development*, vol. 4, no. 1, pp. 41–48, 2014.

D. Sud, G. Mahajan, and M. Kaur, "Agricultural waste material as potential adsorbent for sequestering heavy metal ions from aqueous solutions–A review," *Bioresource Technology*, vol. 99, no. 14, pp. 6017–6027, 2008.

G. J. P. i. p. s. Crini, "*Recent developments in polysaccharide-based materials used as adsorbents in wastewater treatment*," vol. 30, no. 1, pp. 38–70, 2005.

H. E. Osman, R. K. Badwy, and H. F. Ahmad, "Usage of some agricultural by-products in the removal of some heavy metals from industrial wastewater," *Journal of Phytology*, vol. 2, no. 3, 2010.

I. Anastopoulos, A. Bhatnagar, B. H. Hameed et al., "A review on waste-derived adsorbents from sugar industry for pollutant removal in water and wastewater," *Journal of Molecular Liquids*, vol. 240, pp. 179–188, 2017.

J. A. Laszlo, and F. R. Dintzis, "Crop resides as Ion-exchange materials. Treatment of soybean hull and sugar beet fiber (pulp) with epichlorohydrin to improve cation-exchange capacity and physical stability," *Journal of applied polymer science*, vol. 52, no. 4, pp. 531–538, 1994.

J. Kanagaraj, T. Senthilvelan, R. C. Panda, et al., "*Biosorption of trivalent chromium from wastewater: an approach towards Green chemistry*," vol. 37, no. 10, pp. 1741–1750, 2014.

J. Mo, Q. Yang, N. Zhang, et al., "A review on agro-industrial waste (AIW) derived adsorbents for water and wastewater treatment," *Journal of environmental management*, vol. 227, pp. 395–405, 2018.

K. J. A. S. E. Banerjee, "*Bioadsorbents as Green Solution to Remove Heavy Metals from Waste Water-Review J*," vol. 6, pp. 43–45, 2020.

M. MA, P. Wang, and L. M. Qiao, "Removal of Chromium Cr (VI) from Water by Citric Acid Modified Orange Peel," *Journal of Anhui Agricultural Sciences*, vol. 46, no. 36, pp. 64–66, 2018.

R. Chakraborty, A. Asthana, A. K. Singh, et al., "Adsorption of heavy metal ions by various low-cost adsorbents: a review," *International Journal of Environmental Analytical Chemistry*, vol. 102, no. 2, pp. 342–379, 2022.

R. Foroutan, A. Oujifard, F. Papari, et al., "Calcined Umbonium vestibulum snail shell as an efficient adsorbent for treatment of wastewater containing Co (II)," 3 *Biotech*, vol. 9, no. 3, pp. 1–11, 2019.

R. Singh, N. Gautam, A. Mishra, et al., "Heavy metals and living systems: An overview," *Indian Journal of pharmacology*, vol. 43, no. 3, pp. 246, 2011.

S. E. Bailey, T. J. Olin, R. M. Bricka, et al., "A review of potentially low-cost sorbents for heavy metals," *Water Research*, vol. 33, no. 11, pp. 2469–2479, 1999.

S. Tamjidi, H. Esmaeili, and B. K. Moghadas, "Application of magnetic adsorbents for removal of heavy metals from wastewater: a review study," *Materials Research Express*, vol. 6, no. 10, pp. 102004, 2019.

T. C. Coelho, R. Laus, A. S. Mangrich, et al., "Effect of heparin coating on epichlorohydrin cross-linked chitosan microspheres on the adsorption of copper (II) ions," *Reactive and Functional Polymers*, vol. 67, no. 5, pp. 468–475, 2007.

T. Nguyen, H. Ngo, W. Guo, et al., "Applicability of agricultural waste and by-products for adsorptive removal of heavy metals from wastewater," *Bioresource Technology*, vol. 148, pp. 574–585, 2013.

V. M. Marín-Rangel, R. Cortés-Martínez, R. A. Cuevas Villanueva et al., "As (V) biosorption in an aqueous solution using chemically treated lemon (Citrus aurantifolia swingle) residues," *Journal of food science*, vol. 77, no. 1, pp. T10–T14, 2012.

V. Mishra, C. Balomajumder, and V. K. Agarwal, "Kinetics, mechanistic and thermodynamics of Zn (II) ion sorption: a modeling approach," *Clean–Soil, Air, Water*, vol. 40, no. 7, pp. 718–727, 2012.

V. Srivastava, C. Weng, V. Singh, et al., "Adsorption of nickel ions from aqueous solutions by nano alumina: kinetic, mass transfer, and equilibrium studies," *Journal of Chemical & Engineering Data*, vol. 56, no. 4, pp. 1414–1422, 2011.

Zhou Ying, Tang Deng-Feng, Zhang Wen-ting, et al., "Quality Analysis of Yimucao (Leonuri Herba) Materials Based on the National Survey of Chinese Medicine Resources," *Guiding Journal of Traditional Chinese Medicine and Pharmacology*, vol. 27, no. 2, pp. 36–40, 2021.

Advances in Renewable Energy and Sustainable
Development – Liang & Kasmani (Eds)
© 2023 Copyright the Author(s), ISBN: 978-1-032-39407-7

Preparation of hydrophobically modified PVA/sponge and its properties for oil/water separation

Mouyuan Yang, Qin Liu* & Junkai Gao
School of Naval Architecture and Maritime, Zhejiang Ocean University, Zhoushan, China

ABSTRACT: At present, oily wastewater and oil spills pose a great threat to the ecosystem and human life. To solve these serious problems, a lot of efforts have been made to develop new oil-water separation materials. Porous oil-absorbing materials, especially three-dimensional (3D) porous materials with easy preparation and inherent hydrophobicity, have attracted wide attention worldwide. Due to the inherent hydrophobicity of polyvinyl alcohol, it is difficult to achieve oil-water separation. In this paper, we propose a method to modify the surface hydrophilicity to hydrophobicity by constructing a rough surface structure and by applying a low surface energy coating such that the sponge has high oil absorption capacity and good water resistance during the oil-water separation procedure. Sponge pyrolysis was used to prepare polyvinyl alcohol. The sponge was characterized by scanning electron microscope, transmission electron microscope, X-ray diffraction analysis, Brunauer-Emmett-Teller measurement, X-ray photoelectron spectroscopy, and other techniques. The results showed that sponges had good hydrophobicity and excellent lipophilicity.

1 INTRODUCTION

Over the past few years, leaks of petrochemical products during production and transportation and the emission of oil-bearing outlet water from chemical units have caused serious environmental degradation of the ocean. Oil spills pose a serious threat to the ecological environment system and human life, and even endanger human life (Journal of Materials Chemistry A 2009). The oil absorption method based on new materials has been widely concerned by researchers in the field of oil-water separation technology because of its advantages of low cost, simple operation, and environmental protection. Three-dimensional (3D) porous material with high porosity is one of the potential candidates for oil-water separation (Lei et al. 2016). Among these 3D porous materials, carbon-based materials have attracted a lot of interest from researchers because of their low-cost, fast adsorption speed, strong hydrophobicity, etc. However, compared with traditional oil absorbing materials, such as foam, sponge (Liang et al. 2019), and aerogels (Zhuang et al. 2020), low adsorption capacity is the shortcomings of carbon-based materials and has hindered its applications in oil-water separation, so the defect of low adsorption capacity is a problem that should be urgently solved. The solution to this problem is to prepare oil-absorbing materials with high adsorption capacities, which has become a hot topic in the field of oil-water separation.

Sponges are one of the attractive three-dimensional functional wettability materials because they have many advantages, such as low cost, mass production, low density, good absorption capacity, high specific surface area, etc. However, the hydrophobicity of the sponge itself hinders its applications in oil-water separation. Accordingly, scientists can transform the surface hydrophilicity into a hydrophobic property by creating a rough surface structure and coating it with a low surface

---

*Corresponding Author: zjou_emglq@163.com
*These authors contributed equally: Mouyuan Yang, Qin Liu.

DOI 10.1201/9781003349648-11

energy material. This will enable the sponge to absorb oil well and repel water during the separation of oil and water. Using polyvinyl alcohol (PVA) as the adhesive is also one of the commonly used methods for modification by a hydrophobic agent.

In the paper, a melamine sponge (MS) was soaked in PVA solution and immersed in a solution of octadecyl trichlorosilane (OTS)/ethanol to obtain hydrophobic and lipophilic microporous adsorption materials. Hydrophobic modified PVA/sponge (OTS/PVA/MS) was studied using SEM, FTIR, BET, contact angle measurement, and experiments to determine surface morphology, chemical composition, specific surface, and pore size distribution, wettability, oil absorption, and recycling performance. The results show that the adsorbent has a better contact angle, high adsorption capacity, and good cycle characteristics, which can quickly and efficiently treat oil spill pollution.

2 EXPERIMENT INSTRUMENTS, MATERIALS AND METHODS

2.1 *Instruments*

The DF-101s collector constant temperature heating magnetic stirring and the YN-ZD-5 stainless steel electric distilled water heater were purchased in Shanghai Lichen Bangxi Instrument Technology Co., Ltd. The constant temperature blower dryer was obtained from Shanghai Bangxi Instrument Technology Co., Ltd. Model CP214 Electronic Balance was purchased in Shanghai Orhouse Instrument Co., Ltd. The FEG-250 scanning electron microscope was obtained from FEI, America. Nova2000e BET specific surface area and pore size analyzer were obtained from Conta, USA. The Nicolet6700 Fourier transform infrared spectrometer was obtained from Thermo Scientific. The OCA-20 Water Contact Angle Test was obtained from German Date Physics Instruments.

2.2 *Materials*

Melamine sponges were purchased from Melamine Sponge. PVA and the Trichorooctadecylsilane were supplied by Guoyao Group Chemical Reagent Co., Ltd. All the commercially available reagents are used as received without further purification.

2.3 *Preparation of PVA/sponge*

The process is as follows: Firstly, take 0.5 g melamine sponge placed in 250 ml beaker, which was submerged in enough distilled water, and take out with tweezers after 2 hours of immersion, then put it in a 70°C oven to dry 24 hours. Repeat these steps twice to remove insoluble impurities. And then, add 0.1 g pure polyvinyl alcohol into a 100 ml beaker and stir for 3 hours in a water bath at 80°C to fully dissolve polyvinyl alcohol and prepare a 1% polyvinyl alcohol solution. Lastly, add the sponge to the polyvinyl alcohol solution which was mentioned before, stir for 2 hours, removed it with tweezers, dried in a 70°C oven for 12 hours to obtain a hydrophobic polyvinyl alcohol sponge.

2.4 *Preparation of OTS/PVA/MS*

The above PVA/sponge was placed in the mixture of trichlorooctadecylsilane and ethanol (the volume ratio of trichlorooctadecylsilane to ethanol was 1:99). After standing for 3 hours, the PVA/sponge was taken out with tweezers drying oven at 60°C for 12 hrs to obtain OTS/PVA/MS.

2.5 *Characterization*

The SEM images of OTS/PVA/MS at different cross sections and different resolutions were taken by a scanning electron microscope (FEG-250; FEI, USA). The pore size distribution of PVA/sponge was tested by the pressure pump method.

2.6 Oil absorption performance test of OTS/PVA/MS

After PVA/sponge was dried, the wettability of oil and water was tested without tableting (normal). The average values were measured multiple times, and the angle of measurement method was used to measure the contact angles between oil and water.

We added 50 ml each of diesel oil, lubricating oil, vacuum pump oil, and vegetable oil to the four same beakers, along with the same quantity of OTS/PVA/MS. After 20 minutes, the adsorption was completed, and the remaining oil on the surface was wiped with filter paper. The adsorption capacity of the adsorbent was measured according to the following equation:

$$q = \frac{[m_a - (m_0 + m_b)]}{m_b}$$

Where q is the adsorption capacity (g), m_a is the total weight (g), m_b is the original weight of the adsorbent (g), and m_0 is the number of the press cloth.

3 RESULT ANALYSIS AND DISCUSSION

3.1 Surface morphology of OTS/PVA/MS

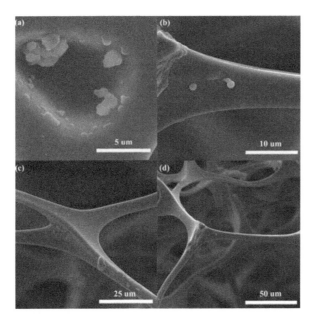

Figure 1. SEM images of OTS/PVA/MS.

The SEM images of OTS/PVA/MS at different magnifications are shown in Figure 1. In Figure 1(a), there are many fine pores distributed on the sponge skeleton, which provide the necessary adsorption sites for oil adsorption and improve the adsorption rate of OTS/PVA/MS for different oil products. The pores and fine particles on the skeleton may be the residual product of the hydrophobic agent-modified sponge. It can not only improve the hydrophobicity of the sponge but also increase the roughness of the skeleton, thus improving the adsorption capacity. It can be seen from Figure 1(b) that the diameter of a single skeleton is 6–10 microns. It can be seen from Figures 1(c) and 1(d) that the three-dimensional porous network structure of the sponge has a large number of voids between different skeletons, which is conducive to storing large amounts of oil in the sponge.

3.2 Pore structure of OTS/PVA/MS

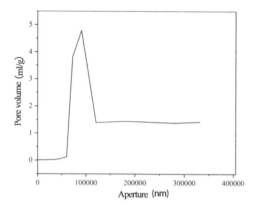

Figure 2. The pore diameter distribution of OTS/PVA/MS.

The pore size distribution of OTS/PVA/MS is shown in Figure 2. It can be seen that when the pore size is less than 60 um, the pore volume is less than 0.14 ml/g. When the pore size is 60–89.58 um, the pore volume increases rapidly. When the pore size is 89.58–121.4 um, the pore volume decreases rapidly, but the overall pore volume maintains higher water quality. After 121.4 um, the pore volume has no significant change. The above results showed that the pore size distribution of OTS/PVA/MS was wide, mainly concentrated in 60–121.4 um, and the average pore size was 89.58 um, much larger than 50 nm, which belonged to the large pore range. Usually, the pore size distribution directly affects the pore volume, thereby affecting the adsorption capacity. Macroporous materials have higher adsorption capacity than mesoporous and microporous materials.

3.3 Effect of polyvinyl alcohol concentration on the contact angle of OTS/PVA/MS

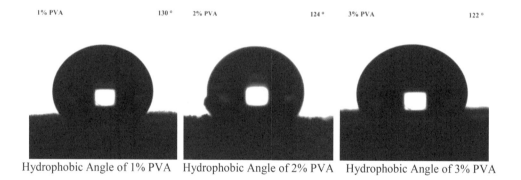

Hydrophobic Angle of 1% PVA   Hydrophobic Angle of 2% PVA   Hydrophobic Angle of 3% PVA

Figure 3. Water contact angle of different PVA concentration.

According to the method described in Section 2. 3, polyvinyl alcohol (PVA) liquor with a potency of 1%, 2%, and 3% were prepared, and the OTS/PVA/MS was prepared after hydrophobic modification. The sponge was placed in a contact angle measuring instrument. In the non-compression state, the relatively flat surface was selected as the water contact surface to measure the contact angle of OTS/PVA/MS in the air. The contact angle results are shown in the following figures. When the concentration of PVA solution was 1%, 2%, and 3%, the corresponding water contact angles were 130°C, 124°C, and 122°C, respectively, and the hydrophobicity gradually decreased,

Figure 4. Oil contact angle (a) and Contrast images of wettability (b) of OTS/PVA/MS.

indicating that the concentration of PVA had a great influence on the water contact angle. The higher the concentration of PVA was, the smaller the water contact angle was and the worse the hydrophobicity was. Therefore, the most suitable PVA concentration is 1%.

Figure 4(a) shows the contact angle between OTS/PVA/MS and oil, from which it can be made out that the oil contact angle is 0°, indicating that OTS/PVA/MS has obvious superoleophilicity. To facilitate the observation of oil adsorption, Sudan Red III was used to dye the lubricating oil red. The comparison results of oil-water contact angles are shown in Figure 4(b). It can be seen that the colorless water droplets on the sponge still maintain the spherical state, while the red oil droplets have already fallen into the sponge, indicating that OTS/PVA/MS has good lipophilic and hydrophobic properties.

3.4 *Oil absorption performance test*

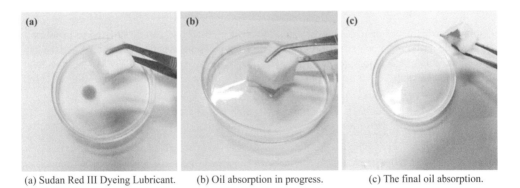

(a) Sudan Red III Dyeing Lubricant.   (b) Oil absorption in progress.   (c) The final oil absorption.

Figure 5. Oil absorption process of OTS/PVA/MS.

To better observe the oil absorption process of the sponge, Sudan III was used to dye the lubricating oil. The specific adsorption process is shown in Figures 5(a), (b), and (c) above. The OTS/PVA/MS was gently clamped with tweezers, and slowly close to the red oil droplets on the water surface. The oil droplets spread rapidly on the OTS/PVA/MS. Ten seconds later, the OTS/PVA/MS absorbed oil and the water surface was almost clear. It can be seen that OTS/PVA/MS has a good adsorption rate for floating oil.

To further research the maximum adsorption capacity of OTS/PVA/MS, the oil absorption capacity of four different oils at room temperature and anhydrous conditions were tested. As shown in

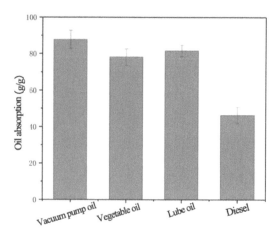

Figure 6. The adsorption capacity of OTS/PVA/MS with different oils.

Figure 6, the adsorption capacity of OTS/PVA/MS for vacuum pump oil, peanut oil, lubricating oil, and diesel oil was 87.7 g/g, 78.2 g/g, 81.6 g/g, and 46.3 g/g, respectively. According to the adsorption capacity, the order was diesel oil < vegetable oil < lubricating oil < vacuum pump oil. The change in adsorption capacity was consistent with the change in oil viscosity. The greater the viscosity of oil, the higher the adsorption capacity (Zhao et al. 2016). In addition, the adsorption capacity was also related to the pore volume, macroporous OTS/PVA/MS showed better adsorption capacity. Moreover, the adsorbent's ability to adsorb quality oil is an important evaluation criterion of its final adsorption capacity. Adsorbents with a low density usually have a higher capacity to absorb oil than those with high densities (Pham & Dickerson 2014).

4 CONCLUSION

In summary, this paper provides an easy and low-cost method that can be utilized to produce a polyvinyl alcohol sponge, which features a three-dimensional porous network structure, and the gap between the network skeleton provides space for oil spill storage. The OTS/PVA/MS has excellent oil absorption capacity, and the adsorption capacity can still reach 88% of the initial value after ten times of repeated use.

REFERENCES

Journal of Materials Chemistry A. (2009) A water solvent-assisted condensation polymerization strategy of superhydrophobic lignocellulosic fibers for efficient oil/water separation[J]. *Journal of Materials Chemistry A.*, 7.

Lei, Z, Deng, et al. (2016) Ambient-temperature fabrication of melamine-based sponges coated with hydrophobic lignin shells by surface dip adsorbing for oil/water separation[J]. *Rsc Advances*.

Liang L, Dong Y, Liu Y. (2019) Modification of Polyurethane Sponge Based on the Thiol-Ene Click Reaction and Its Application for Oil/Water Separation[J]. *Polymers.*, 11(12): 2072.

Pham V H, Dickerson J H. (2014) Superhydrophobic Silanized Melamine Sponges as High Efficiency Oil Absorbent Materials. *ACS Applied Materials & Interfaces.*, 6(16): 14181.

Zhao Y, Miao X, Lin J. (2016) A hierarchical and gradient structured superabsorbent comprising three-dimensional interconnected porous fibers for efficient oil spillage cleanup[J]. *Journal of Materials Chemistry A.*, (4): 9635–9643.

Zhuang J, Dai J, Ghaffar S H. (2020) Development of highly efficient, renewable and durable alginate composite aerogels for oil/water separation. *Surface and Coatings Technology.*, (388): 125551.

Advances in Renewable Energy and Sustainable
Development – Liang & Kasmani (Eds)
© 2023 Copyright the Author(s), ISBN: 978-1-032-39407-7

Effects of wood vinegar on the growth of oil sunflower (*Helianthus annuus* L.) seedlings in a salt-affected soil of Yellow River Delta, China

Shuai Wu, Yanfei Yuan & Qiang Liu *

Institute of Coastal Environmental Pollution Control, Key Laboratory of Marine Environment and Ecology, Ministry of Education, College of Environmental Science and Engineering, Ocean University of China, Qingdao, China

ABSTRACT: Wood vinegar has attracted extensive attention because of its multiple benefits in soil improvement. However, the effects of different wood vinegar on the growth of oil sunflower in salt-affected soils are not clear. In this study, a woody waste-derived wood vinegar was applied to a coastal salt-affected soil collected from the Yellow River Delta to investigate its effects on the growth of oil sunflower seedlings through a pot experiment. The results showed that the addition of wood vinegar respectively increased plant height, plant diameter, fresh biomass, root length, and root surface area of oil sunflower seedlings by 6.1–9.0%, 5.3–7.5%, 12.8–141.0%, 16.1-89.0% and 38.8-80.7%, which may be due to the active compounds in wood vinegar promoting seedling growth. The three kinds of wood vinegar could effectively promote the growth of oil sunflower seedlings, and FWV600 had the greatest promoted effect. These research results will provide a theoretical basis and technical support for the development of green technology to remediate the degraded salt-affected soil, thus ensuring the sustainable development of agriculture.

1 INTRODUCTION

The coastal "blue carbon" ecosystems are of great significance for ensuring global food security and mitigating global climate change (Tang et al. 2018). The Yellow River Delta is a typical coastal wetland in China and an important part of the "blue carbon" ecosystem (Wang et al. 2011). However, the soil salinization in the Yellow River Delta is becoming increasingly serious. The main problems in salt-affected soil are high soil pH and salt content, low soil organic matter, and low nutrient content (Tedeschi et al. 2011). A large amount of soluble salt or exchangeable sodium in salt-affected soils will lead to plant physiological drought and a decline in the ability to absorb nutrients, resulting in low soil primary productivity. Therefore, there is an urgent need for eco-friendly technologies to restore these degraded salt-affected soils.

Traditional improvement methods for salt-affected soils include hydraulic engineering measures (e.g., leaching and draining off the salt) (Sharma & Minhas 2005), chemical improvement measures (e.g., application of gypsum or organic fertilizer) (Oo et al. 2013), and agronomic and biological measures (e.g., planting salt tolerant plants or applying microbial agents) (Ravindran et al. 2007). To some degree, these conventional methods can alleviate soil salinization, but they have several drawbacks, including low efficiency, long cycle time, high costs, and the potential for secondary pollution (Liu et al. 2019). Wood vinegar is an acidic by-product produced in the production of biochar (Pan et al. 2017). It has the potential to reduce the pH of salt-affected soils, improve the nutrient availability of salt-affected soils and promote plant growth (Pan et al. 2017). However, the effects of wood vinegar on plants were inconsistent. It is found that wood vinegar may promote or inhibit plant growth through direct or indirect effects, which depend on concentration, preparation

_____
*Corresponding Author: liuqiang906@163.com

DOI 10.1201/9781003349648-12

of raw materials, preparation temperature, and plant type of wood vinegar. For example, Pan et al. (Lu et al. 2019) showed that poplar wood vinegar can promote plant height, root length, root volume, and root tip number may be due to organic acids increasing cell activity, thereby promoted seed germination and root growth (Mungkunkamchao et al. 2013). However, Mungkunkamchao et al. (Jeong et al. 2015) believed that the application of eucalyptus wood vinegar had little effect on the total dry weight of tomato plants, the number of fruits, the fresh weight, and the dry weight of fruits. These effects of wood vinegar are closely related to its raw materials and preparation conditions. Whereas, it is not clear how different types of wood vinegar affect plant growth.

Therefore, bamboo wood vinegar (BWV110), reed wood vinegar (RWV450) and fruitwood vinegar (FWV600) were prepared by slow pyrolysis at 110°C, 450°C and 600°C, which were used as salt-affected soils amendments in this study. A pot experiment was conducted in the salt-affected soils collected from the Yellow River Delta to explore the effects of BWV110, RWV450 and FWV600 on oil sunflower (*Helianthus annuus* L.) growth. This study will provide important data support for the development of salt-affected soil remediation technology and provide a theoretical basis for the further utilization of salt-affected soils.

2 MATERIALS AND METHODS

2.1 *Soil and amendment*

Soil samples were taken from the Dongying halophyte garden (118.49°E, 37.46°N), which is in the Yellow River Delta, China. The surface soil of 0–20 cm was collected by a five-point sampling method. We dried soil samples indoors before cultivation, removed visible contaminants such as stones, mixed them evenly, and passed the mixture through a 2 mm mesh sieve.

In this study, BWV110 was purchased from Jinan Shunke Chemical Technology Co. Ltd., and RWV450 and FWV600 were purchased from Jinan XinLongSheng Biotechnology Co. Ltd. Specifically speaking, BWV110 was prepared by slow pyrolysis at 110°C, RWV450 was prepared by slow pyrolysis at 450°C and FWV600 was prepared by slow pyrolysis at 600°C. After storing in dark for 6 months, the wood vinegar was filtered with a 0.45 μm cellulose acetate membrane. The wood vinegar was diluted with water at the ratio of 1:300 (v/v) for further experiments.

2.2 *Pot experiment*

Three kinds of diluted wood vinegar were watered at 40% of soil water holding capacity (WHC), Hereinafter, referred to as BWV110, RWV450 and FWV600, respectively. The salt-affected soil without wood vinegar was used as a blank control and was recorded as CK. In this experiment, oil sunflower (*Helianthus annuus* L.) was selected as the test plant. Five seeds were sown in each flowerpot, and the oil sunflower seeds were buried 2-3 cm in the soil. During planting, all potted plants were maintained at 40% of the maximum WHC of each treated soil with distilled water. After 10 days of culture, the branches and roots of oil sunflower were harvested respectively. The plant height and stem diameter of oil sunflower were measured with a vernier caliper (500-173, Mitutoyo, Japan) and ruler respectively. After that, the twigs and roots were washed with Milli-Q water to remove soil and other debris, and the fresh weight of the twigs and roots was weighed by electronic balance (TP-214, Denver Instrument, USA). Root morphology was analyzed by using the root morphology scanner (Epson Scanning, Japan).

2.3 *Data analysis*

The data of this study were processed and plotted using Excel 2016 software. SPSS 24.0 (IBM-SPSS, USA) was used for a one-way analysis of variance (ANOVA). Duncan test ($P < 0.05$) was used for significant difference analysis. The significant differences between different treatment groups ($P < 0.05$, $n = 4$) were expressed in lowercase letters.

3 RESULTS AND DISCUSSION

3.1 *Effects of wood vinegar on plant height and diameter of oil sunflower seedlings*

The effects of different types of wood vinegar on the plant height and diameter of oil sunflower seedlings are shown in Figure 1. Compared with CK, the addition of BWV110, RWV450, and FWV600 increased the plant height of oil sunflower by 9.0%, 6.1%, and 8.1%, respectively, and there was no significant difference. The promoted effect on plant height of oil sunflower was BWV110 > FWV600 > RWV450. Compared with CK, the addition of BWV110, RWV450, and FWV600 significantly increased the plant diameter of oil sunflower by 6.2%, 7.5%, and 5.3%, respectively, but no significant difference was observed among the three different wood vinegar. The promoted effect on the plant diameter of oil sunflower was RWV450 > BWV110 > FWV600. This showed that the addition of wood vinegar can effectively promote the plant height and diameter of oil sunflower seedlings. This is consistent with previous studies. For example, Jeong et al. showed that the application of wood vinegar can improve the dry weight of roots, plant height, panicle number, and yield of rice, and produce a certain yield increase effect in the seedling stage (Litalien & Zeeb 2020). The application of wood vinegar led to the increase of plant height and diameter of oil sunflower for the following reasons. Firstly, wood vinegar contains Ca^{2+} and Mg^{2+} and other nutrient ions, which can directly promote plant growth (Lashari et al. 2015). Secondly, wood vinegar can keep the soil weakly acidic and wood vinegar is rich in active compounds (such as carboxylic acids, phenols, and lactones) (Pan et al. 2017), which may also be the reason the application of wood vinegar promotes the growth of oil sunflower seedlings.

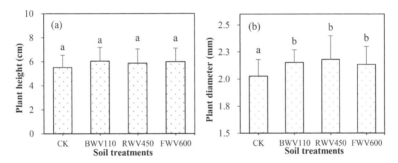

Figure 1. Effects of different wood vinegar on plant height and diameter of oil sunflower per plant. (a) Plant height; (b) plant diameter. Error bars represent standard deviations of the means (n = 4). Different small letters indicate significant differences among different treatments (Duncan's test, $P < 0.05$).

3.2 *Effects of wood vinegar on the biomass of oil sunflower seedlings*

The effects of different types of wood vinegar on the biomass of oil sunflower seedlings are shown in Figure 2. Compared with CK, the addition of BWV110, RWV450, and FWV600 increased the fresh weight of aboveground parts of oil sunflower by 12.9%, 3.2%, and 12.8%, respectively, and the differences in the fresh weight of aboveground parts of oil sunflower were non-significant for the three kinds of wood vinegar treatment. But no significant difference was observed among the three different wood vinegar. The promoted effect on the fresh weight of aboveground parts of oil sunflower was BWV110 > FWV600 > RWV450. Compared with CK, the addition of BWV110 significantly reduced the fresh weight of oil sunflower underground by 34.2%. The addition of RWV450 and FWV600 increased the fresh weight of oil sunflower underground by 2.6% and 58.6%, respectively, and the differences in the fresh weight of underground parts of oil sunflower were non-significant for the RWV450 treatment. Compared with BWV110, the addition of RWV450 and FWV600 significantly increased the fresh weight of underground parts of oil sunflower by 55.8% and 141.1%, respectively. The promoted effect on the fresh weight of underground parts of oil sunflower was FWV600 > RWV450. These results showed that the addition of wood vinegar could promote the aboveground and underground biomass of oil sunflower plants. Lashari et al. (Pan et al. 2017) have shown that the addition of wood vinegar can enhance the availability of N and P in the soil, leading to an increase in plant biomass.

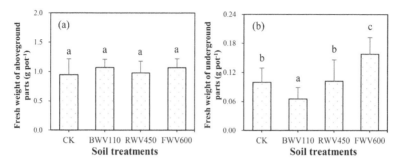

Figure 2. Effects of different wood vinegar on the biomass of oil sunflower per pot. (a) Fresh weight of aboveground parts; (b) fresh weight of underground parts. Error bars represent standard deviations of the means (n = 4). Different small letters indicate significant differences among different treatments (Duncan's test, $P < 0.05$).

3.3 *Effects of wood vinegar on root parameters of oil sunflower seedlings*

The effects of different types of wood vinegar on root length and root surface area of oil sunflower seedlings are shown in Figure 3. Compared with CK, the addition of BWV110, RWV450, and FWV600 increased the root length of oil sunflower by 16.1%, 50.4%, and 89.0%, respectively. and the addition of BWV110 had no significant difference in the root length of oil sunflower. Compared with BWV110, the addition of RWV450 and FWV600 significantly increased the root length by 29.3% and 47.9%, respectively. The promoted effect of different types of wood vinegar on the root length of oil sunflower was FWV600 > RWV450 > BWV110. Compared with CK, the addition of RWV450 and FWV600 significantly increased the root surface area of oil sunflower by 38.8% and 80.7%, respectively. Compared with BWV110, the addition of RWV450 and FWV600 significantly increased the root surface area of oil sunflower by 39.3% and 81.4%, respectively. The promoted effect of different types of wood vinegar on the root surface area of oil sunflower was FWV600 > RWV450 > BWV110. This showed that the addition of wood vinegar can effectively promote the root growth of oil sunflower plants. This is consistent with previous studies. For example, Pan et al. (2017) showed that the application of poplar wood vinegar could increase root length, root surface area, root volume and root tip number due to active compounds such as organic acids in poplar wood vinegar promote the growth of plant roots.

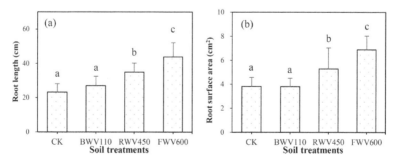

Figure 3. Effects of different wood vinegar on root parameters of oil sunflower per plant. (a) Root length; (b) root surface area. Error bars represent standard deviations of the means (n = 4). Different small letters indicate significant differences among different treatments (Duncan's test, $P < 0.05$).

4 CONCLUSIONS

In this paper, a pot experiment was conducted in a salt-affected soil collected from the Yellow River Delta to explore the effects of BWV110, RWV450 and FWV600 on oil sunflower (*Helianthus annuus* L.) growth. The main conclusions can be summarized as follows: (1) BWV110, RWV450 and FWV600 can promote the plant height and diameter of oil sunflower seedlings. (2) BWV110, RWV450 and FWV600 promoted the biomass of oil sunflower seedlings. (3) BWV110, RWV450 and FWV600 promoted root growth of oil sunflower seedlings. (4) The promoted effect on oil sunflower may be because of the active compounds such as organic acids in wood vinegar. In terms of future work, the practical application effect of wood vinegar needs the support of field large-scale experimental results. At present, it only stays in the laboratory simulation stage. Therefore, long-term field experiments should be carried out to explore the improvement effect of wood vinegar on plant growth in salt-affected soils.

ACKNOWLEDGMENTS

This work was supported by the National Science Fund for Distinguished Young Scholars of Shandong Province (Grant No. ZR2021JQ13) and Shandong Key Research and Development Program-Science and Technology Innovation Project (Grant No. 2018CXGC0304).

REFERENCES

Jeong, K.W., Kim, B.S., Ultra, V.U., Lee, S.C. Effects of Rhizosphere Microorganisms and Wood Vinegar Mixtures on Rice Growth and Soil Properties. *The Korean Journal of Crop Science* 2015, 60: 355–365.

Lashari, M.S., Ye, Y., Ji, H., Li, L., Kibue, G.W., Lu, H., Zheng, J., Pan, G. Biochar-manure compost in conjunction with pyroligneous solution alleviated salt stress and improved leaf bioactivity of maize in saline soil from central China: a 2-year field experiment. *J. Sci. Food Agr.* 2015, 95: 1321–1327.

Litalien, A., Zeeb, B. Curing the earth: A review of anthropogenic soil salinization and plant-based strategies for sustainable mitigation. *Sci. Total Environ.* 2020, 698: 134235.

Liu, B., Cai, Z., Zhang, Y., Liu, G., Luo, X., Zheng, H. Comparison of efficacies of peanut shell biochar and biochar-based compost on two leafy vegetable productivity in infertile land. *Chemosphere* 2019, 224: 151–161.

Lu, X., Jiang, J., He, J., Sun, K., Sun, Y. Effect of Pyrolysis Temperature on the Characteristics of Wood Vinegar Derived from Chinese Fir Waste: A Comprehensive Study on Its Growth Regulation Performance and Mechanism. *ACS Omega* 2019, 4: 19054–19062.

Mungkunkamchao, T., Kesmala, T., Pimratch, S., Toomsan, B., Jothityangkoon, D. Wood vinegar and fermented bioextracts: Natural products to enhance growth and yield of tomato (Solanum lycopersicum L.). *Sci. Hortic-Amsterdam* 2013, 154: 66–72.

Oo, A.N., Iwai, C.B., Saenjan, P. Soil Properties and Maize Growth in Saline and Nonsaline Soils using Cassava-Industrial Waste Compost and Vermicompost with or Without Earthworms. *Land Degrad. Dev.* 2013, 26: 300–310.

Pan, X., Zhang, Y., Wang, X., Liu, G. Effect of adding biochar with wood vinegar on the growth of cucumber. IOP Conference Series: *Earth and Environmental Science* 2017, 61: 012149.

Ravindran, K.C., Venkatesan, K., Balakrishnan, V., Chellappan, K.P., Balasubramanian, T. Restoration of saline land by halophytes for Indian soils. *Soil Biol. Biochem.* 2007, 39: 2661–2664.

Sharma, B.R., Minhas, P.S. Strategies for managing saline/alkali waters for sustainable agricultural production in South Asia. *Agr. Water Manage.* 2005, 78: 136–151.

Tang, J., Ye, S., Chen, X., Yang, H., Sun, X., Wang, F., Wen, Q., Chen, S. Coastal blue carbon: Concept, study method, and the application to ecological restoration. *Sci. China Earth Sci.* 2018, 61: 637–646.

Tedeschi, A., Lavini, A., Riccardi, M., Pulvento, C., D'Andria, R. Melon crops (Cucumis melo L., cv. Tendral) grown in a mediterranean environment under saline-sodic conditions: Part I. Yield and quality. *Agr. Water Manage.* 2011, 98: 1329–1338.

Wang, M., Qi, S., Zhang, X. *Wetland loss and degradation in the Yellow River Delta*, Shandong Province of China. Environ. Earth Sci. 2011, 67: 185–188.

Advances in Renewable Energy and Sustainable Development – Liang & Kasmani (Eds)
© 2023 Copyright the Author(s), ISBN: 978-1-032-39407-7

Quantitation and extraction of flavonoids from Okra flowers

Ziping Zhu*, Linhua Zhao, Zhenda Xie & Na Li
School of Life Science, Taizhou University, Taizhou, China

ABSTRACT: The content of flavonoids in okra (Abelmoschus esculentus) flowers was determined and the extraction process was studied. Results showed that the absorbance was high and stable after adding $NaNO_2$, $Al(NO_3)_3$, and NaOH successively in the extracting solution and standing for 4, 4, and 10 min, respectively. Rutin was taken as the standard substance, and the flavonoid content in okra flowers was determined to be (3.42 ± 0.04) %. A high yield of flavonoids of (3.28 ± 0.03) % can be obtained after extraction at 40°C, ethanol concentration of 60%, and the material-liquid ratio of 1:90 (g/mL) for 1 h. Factors are listed in descending order as ethanol concentration, material-liquid ratio, extraction temperature, and extraction time according to their influences on the yield of flavonoids.

1 INTRODUCTION

Okra (Abelmoschus esculentus) is a kind of annual herb belonging to the Malvaceae. It likes warm climates, and is mainly distributed in tropical and subtropical regions (Nwangburuka et al. 2013; Adelakun et al. 2009), and grown widely in Africa, Asia, southern Europe, and America (Khomsug et al. 2010). People generally eat fruit pods of okra, while the leaves, buds, and flowers of okra are also edible. Okra is a kind of highly nutritive plant that can be used as a vegetable, medicine, and flower (Liu et al. 2006). Okra is monoclinous and has solitary and gynodioecious flowers that generally bloom under each axil above the third true leaf. It generally blooms from bottom to top and two flowers bloom every morning, wither in the afternoon, and drop in the next day. Okra is characterized by a long flowering phase, large yield, and rich resources. Among existing flowers discovered, okra flowers contain the highest content of flavonoids (12.50–37.60 mg/g), higher than the total flavonoid content in its fruits (8.49–23.84 mg/g) and far higher than the isoflavone content in soybean (0.455–6.680 mg/g) (Xue et al. 2013). Therefore, okra flowers can be used as a new source of natural flavonoid compounds, and studying the quantitation method and extraction process of flavonoid compounds in okra flowers can lay a basis for the utilization of okra flowers.

2 EXPERIMENTAL METHODS AND RESULTS

2.1 Determination of the flavonoid content

2.1.1 Determination of the maximum absorption wavelength

Okra flowers of 1.0 g were placed in a conical flask, in which 15 mL ethanol solution (60%) was added to extract flavonoids for 1 h by water bath at 80°C. After suction filtration, the flavonoids were extracted again from the filter residues. The two filtrates were merged and adjusted to 50 mL using 60% ethanol, thus obtaining the sample solution. Rutin of 25.0 mg was dissolved with 60%

---

*Corresponding Author: zhuzp123@sohu.com

84   DOI 10.1201/9781003349648-13

ethanol and adjusted to 250 mL, thus obtaining the 100 ug/mL standard solution. Then, 0.5 mL sample solution and 0.5 mL standard solution were separately added in 10 mL volumetric flasks, in which 0.3 mL NaNO₂ solution (5%) was added, and the flask was shaken up and allowed to stand for 6 min. This was followed by the addition of 0.3 mL Al(NO₃)₃ solution (10%), shaking up and standing for 6 min, and then the addition of 4 mL NaOH solution (1 mol/L) and shaking up. After that, 60% ethanol was used to adjust the volume. After standing for 15 min, the solution was subjected to full-band scanning and the maximum absorption wavelengths were measured to be 500 and 510 nm. Because flavonoid compounds in the sample solution are not a single substance, the sample solution may differ from the standard solution in the maximum absorption wavelength. To truly reflect the flavonoid content in the sample solution, 500 nm was adopted as the absorption wavelength during determination.

2.1.2 *Drawing of standard curves*
Standard solution of 0, 0.5, 1, 1.5, 2, 2.5, and 3 mL was separately added in 10 mL volumetric flasks for solution preparation and determination following the method in Section 2.1.1. The standard curve y=0.0117x-7×10-5 (R^2 = 0.9963) was attained by regression taking the rutin concentration as the abscissa and the absorbance as the ordinate. The standard curve was of high linearity when the mass concentration of rutin was in the range of 0-30 ug/mL.

2.1.3 *Influences of coloration time on absorbance*
Some 0.5 mL sample solution was poured into a 10 mL volumetric flask, in which 0.3 mL NaNO₂ solution (5%) was added and then the flask was shaken up and allowed to stand for 0, 2, 4, 6, 8, 10, and 15 min. Then, a 0.3 mL Al(NO₃)₃ solution (10%) was added to the flask, which was shaken up and stood for 0, 2, 4, 6, 8, 10, and 15 min. Finally, 4.0 mL NaOH solution (1 mol/L) was added. After standing for 0, 2, 4, 6, 8, 10, 15, and 20 min, the absorbance was measured, as shown in Figure 1.

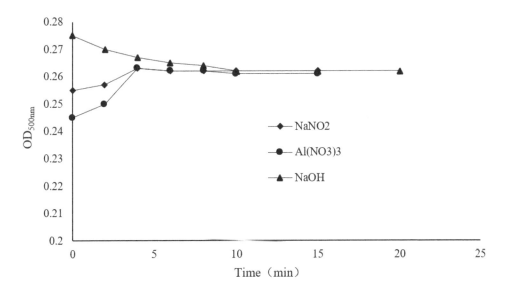

Figure 1. Influences of coloration time on absorbance.

It can be seen from the figure that the absorbance is the maximum when the solution stands for four min after adding NaNO₂ and Al(NO₃)₃ solution, and the absorbance tends to stabilize 10 min after adding NaOH solution. The result indicates that the coloration reaction stabilizes then.

2.1.4 *Detection method and determination of the flavonoid content*

Five test tubes were taken, in which 0.3 g of okra flowers and 4.5 mL ethanol solution (60%) were added and subjected to a water bath at 80°C to extract flavonoids for 1 h. The filtrate was taken after filtration. The flavonoid contents were measured to be 3.35, 3.31, 3.34, 3.37, and 3.38, with a mean value of 3.35, standard deviation (SD) of 0.027, and relative standard deviation (RSD) of 0.81% after coloration under the coloration conditions optimized in Section 2.1.3. The results indicate that the method is of high reproducibility. Okra flowers of 0.3 g were placed in the test tubes, in which 4.5 mL ethanol solution (60%) was added, followed by a water bath and extraction at 80°C for 1 h. After that, it was detected that the filtrate contains 0.289 mg of flavonoids. Then, the flavonoid content was detected again after adding 0.121 mg of rutin to the filtrate. The detection results of the three repeated experiments are 0.4076, 0.4082, and 0.4069 mg, based on which the recovery rates were calculated to be 98.05%, 98.49%, and 97.49%, with an average of 98.01%, SD of 0.501%, and RSD of 0.51%. This indicates a high recovery rate and the detection method is reliable. After being extracted with 4.5 mL ethanol solution (60%) three times, flavonoids in 0.3 g of okra flowers have been completely extracted out. The flavonoid content was detected by combining the extracting solution obtained three times and the detection results in three repeated experiments were 3.41%, 3.46%, and 3.39%. The average flavonoid content in okra flowers is (3.42 ± 0.04)%.

2.2 *Flavonoid extraction*

2.2.1 *Optimization of single factors for extraction*

Okra flowers of 1.0 g were added in nine conical flasks, in which 20, 30, 40, 50, 60, 70, 80, 90, and 100 mL ethanol solution with the volumetric concentration of 40% was added, respectively. After immersing the flowers for 12 h at room temperature, the solution was filtered. The flavonoid content in the filtrate was then measured to calculate the yield. According to the results in Figure 2(A) that the yield of flavonoids increases with the growth of the volume of ethanol solution, while the yield increases slightly when the volume reaches 90 mL. Therefore, three volume levels (80, 90, and 100 mL) were selected to perform orthogonal experiments. Okra flowers of 1.0 g were added in five conical flasks, in which 40 mL ethanol solution with the volumetric concentrations of 0, 20%, 40%, 60%, and 80% was added to immerse the flowers for 12 h at room temperature. The filtrate was taken to detect the flavonoid content and calculate the yield. As shown in the results in Figure 2(B), the yield of flavonoids grows with the increasing ethanol concentration. The yield of flavonoids is high when the volumetric concentration of ethanol is 60%, while it decreases if the concentration is increased further. This indicates that the volumetric concentration of 60% of ethanol is favorable for the dissolution of flavonoids. Therefore, three concentration levels of 40%, 60%, and 80% were selected to carry out orthogonal experiments. Okra flowers of 1.0 g and 40 mL ethanol solution with the volumetric concentration of 40% were added in five conical flasks to immerse the flowers for 0.5, 1.0, 1.5, 2.0, and 2.5 h at room temperature, followed by filtration. Then, the filtrate was taken to detect the flavonoid content and calculate the yield. The results in Figure 2(C) show that the yield of flavonoids increases with the prolonging extraction time, while it grows slowly when the extraction time is longer than 2 h. Therefore, the extraction time should not be too long. Three levels of 1, 1.5, and 2 h were selected to conduct orthogonal experiments. Okra flowers of 1.0 g and 40 mL ethanol solution with the volumetric concentration of 40% were added in five conical flasks to immerse the flowers at 25, 30, 35, 40, and 45°C for 1.5 h. After filtration, the flavonoid content in the filtrate was detected and the yield was calculated. As shown in Figure 2(D), the yield of flavonoids grows with the rising extraction temperature while it does not increase any more after rising to 40°C. Therefore, three temperature levels of 35°C, 40°C, and 45°C were selected to carry out orthogonal experiments.

2.2.2 *Orthogonal experiments*

The L9(43) orthogonal experiments were designed according to the experimental results of single factors. The experimental results are listed in Table 1.

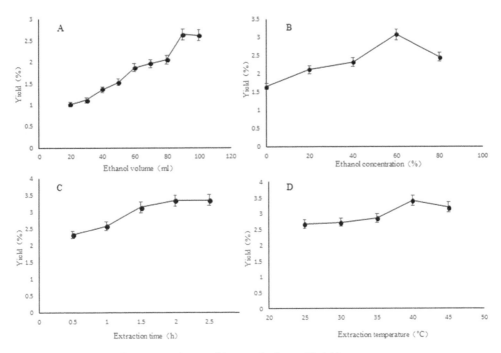

Figure 2. Effects of different extracting conditions on the flavonoid yield.

Table 1. Design and results of orthogonal experiments.

Test serial number	A	B	C	D	Flavonoid yield (%)
1	1	1	1	1	1.85
2	1	2	2	2	3.12
3	1	3	3	3	1.98
4	2	1	2	3	2.08
5	2	2	3	1	3.31
6	2	3	1	2	3.14
7	3	1	3	2	2.01
8	3	2	1	3	3.16
9	3	3	2	1	2.74
K1	2.317	1.980	2.717	2.633	
K2	2.843	3.197	2.647	2.757	
K3	2.637	2.620	2.433	2.407	
R	0.526	1.217	0.284	0.350	

It can be seen from Table 2 that the optimal extraction process of flavonoids from okra flowers is A2B2C1D2. The four factors are listed in descending order as B, A, D, and C regarding their influences on the extraction rate of flavonoids, and all factors do not reach the significance level.

2.2.3 *Verification experiments*

Okra flowers of 1.0 g were weighed and placed in a conical flask, in which 90 mL ethanol solution (60%) was added to extract flavonoids at 40°C for 1.0 h. After suction filtration, the absorbance of the filtrate was detected and the yield of flavonoids was calculated to be (3.28±0.03)% (Mean±SD, n = 3).

3 CONCLUSIONS

Based on the results and discussions presented above, the conclusions are obtained as below:

(1) When the flavonoid content in okra flowers was detected, the absorbance was high and stable when adding $NaNO_2$, $Al(NO_3)_3$, and NaOH successively in the extracting solution and standing for 4, 4, and 10 min, respectively. Thereafter, the flavonoid content was detected to be $(3.42\pm0.04)\%$ when rutin was taken as the standard substance.
(2) A high yield of flavonoids of $(3.28\pm0.03)\%$ was obtained when extracting flavonoids at 40°C, ethanol concentration of 60%, and the material-liquid ratio of 1:90 (g/mL) for 1 h. Factors were listed in descending order as ethanol concentration, material-liquid ratio, extraction temperature, and extraction time according to their influences on the yield of flavonoids.

ACKNOWLEDGMENT

This work was financially supported by the Taizhou Science and Technology Project (Grant No. 21nya18).

REFERENCES

Adelakun, O.E., Oyelade, O.J., Ade-Omowaye, B.I.O., Adeyemi, I.A., Van de Venter, M. (2009) Chemical composition and the antioxidative properties of Nigerian Okra Seed (Abelmoschus esculentus Moench) Flour. *Food and Chemical Toxicology*, 47: 1123–1126.

Khomsug, P., Thongjaroenbuangam, W., Pakdeenarong, N., Suttajit, M., Chantiratikul, P. (2010) Antioxidative activities and Phenolic content of extracts from Okra (Abelmoschus esculentus L.). *Research Journal of Biological Sciences*, 5: 310–313.

Liu, D.X., Ye, H.L., Liu, G.D. (2006) Review of the Application Value and Cultivation Technique of Okra. *Journal of Anhui Agricultural Science*, 34: 3718–3720, 3725.

Nwangburuka, C.C., Chibundu, E.N., Oyekale, K., Anokwuru, C.P., Ivie, E.K. (2013) Cytomorphological and antifungal analysis of Acalypha wilkesiana, Moringa oleifera extracts, and sodium hypochlorite on Abelmoschus esculentus L. Moench. treated seeds. *Nature and Science*, 11: 32–39.

Xue, Z.Z., Liu, S.Y., Yang, Y.H. (2013) Research Progress on Application Value and Development of Okra. *Storage and Process*, 13: 58–60.

Advances in Renewable Energy and Sustainable
Development – Liang & Kasmani (Eds)
© 2023 Copyright the Author(s), ISBN: 978-1-032-39407-7

Detection and analysis of six toxic elements in *Codonopsis lanceolata* in different areas of Jilin Province in China

Fenglin Li*
College of Food Engineering, Jilin Agricultural Science and Technology University, Jilin, China
Jilin Brewing Technology Innovation Center, Jilin, Jilin, China

Zhimin Liu
College of Food Engineering, Jilin Agricultural Science and Technology University, Jilin, China

Li Li, Zhongkui Lu & Wuyang Hua
College of Food Engineering, Jilin Agricultural Science and Technology University, Jilin, China
Jilin Brewing Technology Innovation Center, Jilin, Jilin, China

ABSTRACT: With the rapid development of urbanization and industrialization in the past few decades, the problem of soil contamination with toxic elements has become increasingly serious. The accumulation of toxic metals in plants can endanger human health through the food chain. The aim of the study was to evaluate the edible safety of *Codonopsis lanceolata Bench.* et Hook (*C. lanceolata*) from different areas of Jilin Province in China. The wet digestion method was used to pre-treat *C. lanceolata*, and then the samples were analyzed for lead (Pb), cadmium (Cd), chromium (Cr) and copper (Cu) by graphite furnace atomic absorption spectrometry (GFAAS) method, for mercury (Hg) by atomic fluorescence spectrometry (AFS) method, and for arsenic (As) by inductively coupled plasma mass spectrometry (ICP-MS) method. The results showed that the contents of Pb, Cd, Cr, Cu, Hg and As in the measured samples were all less than the stipulated limits of the national food safety standards of China, which indicated that the edible safety of *C. lanceolata* in different areas of Jilin Province in China was guaranteed to a certain extent.

1   INTRODUCTION

Codonopsis lanceolata Bench. et Hook (*C. lanceolata*) is a perennial herb of the *Codonopsis* genus *Platycodonaceae*, mainly distributed in North Korea, Russia, Japan and Northeast China. The roots of *C. lanceolata* contains abundant pharmaceutical substance and has long been used as a traditional herbal medicine in China for health maintenance, such as reducing swelling, moistening the lungs, and resolving phlegm. Modern chemical and pharmacological studies have confirmed that *C. lanceolata* is rich in alkaloids, flavonoids, triterpenoids, polysaccharides, and other bioactive ingredients. Hot-water extracts of *C. lanceolata* have the functions of antioxidants, anti-tumor, anti-inflammatory, anti-hypoxia, liver injury protection, and enhanced immune activity (Bae et al. 2020). In recent years, the wild resources of *C. lanceolata* have become increasingly scarce. For this reason, after long-term artificial breeding and cultivation, the researchers have summed up the artificial high-yield cultivation technology of C. lanceolate, which makes it adapted to natural conditions. *C. lanceolata* is planted artificially in Jilin Province, and the cultivated area increases yearly.

It is well known that the food chain has a bioaccumulation effect; each time a chemical passes through a biological organism, its concentration increases significantly. Bioaccumulation means an

---

*Corresponding Author: 568169115@qq.com

DOI 10.1201/9781003349648-14

increase in chemical concentration in a biological organism over time compared to the concentration of the chemical in the environment (Maria Rita & Francesca 2014). At the top of the food chain, through their regular diet, the human may accumulate a much greater concentration of chemicals than ones present in lower positions of the food chain. With the rapid development of urbanization and industrialization in the past few decades, soil contamination with toxic elements has become one of the most dreaded global challenges of environmental degradation. Lead (Pb), cadmium (Cd), chromium (Cr), copper (Cu), mercury (Hg), and arsenic (As) are non-degradable pollutants with a strong accumulation of harm to the human body, mainly chronic poisoning and long-term effects (Young & Myong-Keun 2015). In this study, graphite furnace atomic absorption spectrometry (GFAAS), atomic fluorescence spectrometry (AFS), and inductively coupled plasma mass spectrometry (ICP-MS) were used to determine the Pb, Cd, Cr, Cu, Hg and As in *C. lanceolata* from different areas of Jilin Province in China, to determine whether it meets the national food safety standards, which will provide technical reference to ensure the edible safety of and the supervision of *C. lanceolata* product processing.

2 MATERIAL AND METHODS

2.1 *Plant material and reagents*

Four kinds of fresh *C. lanceolata* from different areas of Jilin Province in China were purchased from Ji'an City, Tonghua City, Huinan County and Liuhe County and identified by Dr. Li Y. X. (College of Chinese Medicine, Jilin Agricultural Science and Technology University, Jilin, China). The identity was confirmed by comparing it with a voucher specimen available in the herbarium of Jilin Agricultural Science and Technology University. Standard solutions of cadmium and copper (1000 mg/L) were obtained from Tanmo Quality Inspection Technology Co., Ltd. (Changzhou, China). Standard solutions of lead, chromium, mercury and arsenic (1000 mg/L) were purchased from National Iron and Steel Materials Testing Center (Beijing, China). All other chemicals and reagents are of excellent grade and were purchased from local suppliers. Double deionized water was used for all dilutions.

2.2 *Instrumentation*

The analysis of lead, cadmium, chromium and copper was performed in a TAS-990 atomic absorption spectrophotometer (Beijing Purkinje General Instrument Co., Ltd). Mercury determinations were performed with a PF52 atomic fluorescence spectrophotometer (Beijing Purkinje General Instrument Co., Ltd). Arsenic determinations were performed with a NexION 1000 inductively coupled plasma mass spectrometry (PerkinElmer Life and Analytical Sciences, Shelton, Conn). The optimum operating parameters are given in Table 1 and Table 2.

2.3 *Sample preparation for wet digestion*

Fresh *C. lanceolata* was washed with water, pulverized into powder by a grinder, and dried in a constant temperature drying oven at 95°C for 1.75 h. The dried powder is stored in an airtight bag for later use. 1.0 g of *C. lanceolata* powder in a 150 mL conical flask was added by 20.0 mL of nitric acid and 1.0 mL of perchloric acid and was soaked overnight. Then the sample was digested, and the initial temperature was set from about 80°C to 100°C. After the reddish-brown smoke gradually dissipated, the temperature was raised to about 150°C. When the mixture is transparently colorless or slightly yellow, the conical flask is filled with white smoke, and 3 mL of pure water is added to remove excess acid. The mixture was heated again until the bottle was filled with white smoke again, then cooled, and 1.0% nitric acid was added to make up the volume to a 10 mL colorimetric tube, and the impurities in the digestion solution were filtered off with a needle filter. In the determination of arsenic, perchloric acid was not added to the sample during digestion.

The digested solutions were used for GFAAS, AFS and ICP-MS analysis, and the same digestion procedure was applied for the preparation of calibration solutions. The results were reported as the average of three repeated measurements, and all digestions were conducted in triplicate.

Table 1. Instrumental parameters for the toxic elements using GFAAS.

Instrumental parameters	Pb	Cd	Cr	Cu
Wavelength (nm)	283.3	228.8	357.9	324.8
Slit width (nm)	0.2	0.4	0.4	0.4
Lamp current (mA)	6.0	2.0	4.0	3.0
Drying temperature (°C)	105	105	105	105
Ashing temperature (°C)	750	700	990	800
Atomization temperature (°C)	2300	2300	2700	2350

Table 2. Instrumental parameters for the toxic elements using AFS and ICP-MS.

AFS parameters	Hg	ICP-MS parameters	As
Negative high pressure (v)	250	Radio frequency power (W)	1600
Gas flow (mL/min)	300	Nebulizer gas flow (L/min)	1.02
Lamp current (mA)	40	Auxilary gas flow (L/min)	1.2
Atomization temperature (°C)	200	Plasma gas flow (L/min)	15
		Sampling depth (nm)	7
		Analysis chamber vacuum (Torr)	4.3×10^{-7}
		Sampling cone	Nickel

2.4 *Analytical working solutions*

Concentrations of 0.0 ng/mL, 0.5 ng/mL, 1.0 ng/mL, 1.5 ng/mL, 2.0 ng/mL, and 3.0 ng/mL for Cd were prepared with 1% nitric acid solution. Concentrations of 0 ng/mL, 5.0 ng/mL, 10.0 ng/mL, 20.0 ng/mL, 30.0 ng/mL, and 40.0 ng/mL for Pb, Cr, and Cu were prepared with 5% nitric acid solution. Concentrations of 0.0 ng/mL, 0.1 ng/mL, 0.2 ng/mL, 0.4 ng/mL, 0.6 ng/mL, 0.8 ng/mL, and 1.0 ng/mL for Hg were prepared with 1% nitric acid solution. Concentrations of 0.0 ng/mL, 5.0 ng/mL, 10.0 ng/mL, 20.0 ng/mL, 60.0 ng/mL, and 100.0 ng/mL for As were prepared with 1% nitric acid solution. These solutions were prepared using appropriately diluted dilutions of the stock standard solutions.

3 RESULTS AND DISCUSSION

3.1 *Linear relationship and limit of detection*

Different concentrations of lead, cadmium, chromium, copper, mercury, and arsenic standard solutions were determined. And the longitudinal signal value was obtained by taking the concentration (ng/mL) as the abscissa. At the same time, the reagent blank solution was measured eleven times in parallel. The limit of detection (LOD) is used to express the lowest concentration of an analysis that can be detectable by a specific instrument, method or sample, and the limit of quantification (LOQ) is the lowest level that an analysis can be quantified with any degree of certainty. LOD and LOQ were calculated as ten and three times the standard deviation (SD) of the blank solution, respectively. The linear relationship and LOD values are shown in Table 3. The linear relationship of the six elements was good within the injection concentration range, and a good linear relationship was obtained within the injected sample concentration range. The LOD values were 0.0004 mg/L to 0.003 mg/L, and the LOQ values were 0.008 mg/L to 0.01 mg/L.

Table 3. Linear relationship and limit of detection of six toxic elements.

Element	Linear regression equation	Correlation coefficient (R^2)	Linear range (ng/mL)	LOD (mg/L)	LOQ (mg/L)
Pb	A= 0.0046C + 0.0182	0.9989	0.0-40.0	0.002	0.005
Cd	A=0.0379C + 0.0047	0.9991	0.0-3.0	0.0004	0.001
Cr	A=0.0219C + 0.0451	0.9980	0.0-40.0	0.002	0.007
Cu	A=0.0034C + 0.0305	0.9986	0-40.0	0.003	0.01
Hg	A=1205C − 25.047	0.9969	0.0-1.0	0.002	0.008
As	A=4945.2C + 4719.6	0.9996	0.0-100.0	0.001	0.004

3.2 *Precision test*

A sample of the digestion solution was repeatedly measured six times, and the arithmetic mean and SD of Pb, Cd, Cr, Cu, Hg and As were calculated from these results. The relative standard deviations (RSD) were calculated to assess whether they were within limits established by the AOAC (2002). It can be seen from Table 4 that the RSDs of the six elements range from 2.35% to 12.80%, which indicates that the precision of the instrument is good.

Table 4. The precision of six toxic elements.

Element	Measured value (mg/kg)						Arithmetic mean (mg/kg)	RSD (%)
Pb	0.209	0.242	0.216	0.177	0.188	0.218	0.208	11.06
Cd	0.071	0.067	0.067	0.067	0.069	0.068	0.0682	2.35
Cr	0.102	0.126	0.117	0.096	0.109	0.114	0.111	9.73
Cu	2.018	2.350	2.147	2.111	2.094	2.347	2.1778	6.37
Hg	0.0027	0.0027	0.0028	0.003	0.0028	0.003	0.0028	3.57
As	0.26	0.32	0.25	0.32	0.32	0.25	0.2867	12.80

3.3 *Spike and recovery test*

Spike and recovery assay is used to determine whether the detection of analysis is affected by sample matrix and differences in the standard curve diluents. The accuracy of the experimental method was assessed by spike and recovery tests in the samples. To assess the accuracy of the method, measurements were made with the standard solution of Pb (20.0 ng/mL), Cd (1.5 ng/mL), Cr (20.0 ng/mL), Cu (20.0 ng/mL), Hg (0.6 ng/mL), and As (40.0 ng/mL), added to six reagent blanks respectively. It can be seen from Table 5 that the recoveries of the spiked samples were in the range of 88.6 - 91.6%. The results obtained in the accuracy study were suitable because recoveries were within the range proposed by AOAC (2002) (70 - 125%) over the concentration range studied (International 2002). which indicated that good recoveries could be obtained using this method.

3.4 *Result of the sample measurement*

GFAAS is simpler, less expensive, quicker and more accurate than neutron activation or emission spectrometric technique. The absorbance signals obtained for lead in the optimizing conditions presented ensures a well-defined profile and a low background. AFS has better sensitivity than many atomic absorption techniques and offers a substantially longer linear range. An atomic fluorescence spectrometer is capable of measuring mercury at parts per trillion (ppt) level using the unique vapor hydride generator. ICP-MS is a new elemental analysis technology capable of detecting most of the periodic table of elements at trace and ultra-trace levels, which combines the high-temperature ionization characteristics of the ICP ion source with the sensitive and fast scanning advantages of the four-pole mass spectrometer (Li et al. 2020). As shown in Table 6, six toxic elements in *C. lanceolata* obtained from Huinan County were 0.19 mg/kg, 0.076 mg/kg, 0.11 mg/kg, 2.3 mg/kg, 0.0030 mg/kg and 0.25 mg/kg, respectively. The contents of the above-mentioned six toxic elements in *C. lanceolata* obtained from Liuhe County were 0.20 mg/kg, 0.078 mg/kg, 0.12 mg/kg, 2.3 mg/kg, 0.0029 mg/kg and 0.23 mg/kg, respectively. The quantitative results show that the contents of six toxic elements in *C. lanceolata* from different areas of Jilin Province were all less than the limits stipulated by the National Food Safety Standards of China (GB 2762-2017).

Table 5. Recoveries of six toxic elements.

Element	Background value (ng/mL)	Value of standard solution added (ng/mL)	Measured value (ng/mL)	Recovery (%)
Pb	5.109	20.0	23.429	91.6
Cd	0.835	1.5	2.238	90.5
Cr	4.973	20.0	22.913	89.7
Cu	6.824	20.0	24.984	90.8
Hg	0.0379	0.6	0.5695	88.6
As	4.38	40.0	40.916	91.3

Table 6. Contents of toxic elements in *C. lanceolata*.

Elements	Ji'an City	Tonghua City	Huinan County	Liuhe County
Pb (mg/kg)	0.28	0.24	0.19	0.20
Cd (mg/kg)	0.085	0.068	0.076	0.078
Cr (mg/kg)	0.16	0.18	0.11	0.12
Cu (mg/kg)	0.0029	0.0028	0.0030	0.0029
Hg (mg/kg)	0.26	0.28	0.25	0.23
As (mg/kg)	2.1	2.1	2.3	2.3

4 CONCLUSIONS

The main conclusions can be summarized as follows: (1) Methods of GFAA, AFS and CP-MS provide good precision, ease of operation, and good recovery compared to other methods. (2) The contents of Pb, Cd, Cr, Cu, Hg and As in *C. lanceolata* from different areas of Jilin Province were all less than the limits stipulated by the National Food Safety Standards of China. (3) The edible safety of *C. lanceolata* to consumers could be guaranteed. In terms of future work, the toxic element contents of *C. Lanceolata* from more areas of Jilin Province need to be determined, and the measures and methods to avoid the pollution of toxic elements in the soil need to be studied.

ACKNOWLEDGMENTS

This work was supported by the Key R&D Program of the Ministry of Science and Technology of China (2021YFD1600903), the Science and Technology Project of Jilin Province (20200402065NC), and the Research Team of Nutrition and Health Food of Jilin Agricultural Science and Technology University.

REFERENCES

Bae WJ, Bo-Ra Y and Jiwoo L 2020 Biomol. *Ther*. 22 246.
International A 2002 AOAC 38.
Li F. L, Li L and Jiang X *2020 E3S Web Conferences.* 185 4019.
Maria Rita C and Francesca P 2014 J. *Environ. Manage*. 133 378.
Young S and Myong-Keun Y 2015 *Soil.Sediment. Contam*. 24 423.

Advances in Renewable Energy and Sustainable Development – Liang & Kasmani (Eds)
© 2023 Copyright the Author(s), ISBN: 978-1-032-39407-7

Different approaches to animal testing from scientific and ethical perspectives

Ziqi Liu*
Hiroshima University, Higashihiroshima, Japan

ABSTRACT: Animal testing has been a heated topic in recent years and there is a growing concern for animal protection and animal rights. The purpose of this article is to summarize the history and contemporary debate on the utilization of living animals in scientific experiments, thus helping readers make their own informed stances. The importance of these debates is analyzed from scientific, practical, ethical, and philosophical perspectives, and the future of animal testing is discussed. This article demonstrates the pros and cons of animal testing and the current need for it and concludes with the proposal and analysis of two promising alternatives.

1 INTRODUCTION

Animal testing can be defined as the use of non-human animals in scientific studies, such as rats and insects, to evaluate the physiological or behavioral effects of particular elements on the animals being tested and to compare them to the status of the same kind of organisms in their natural habitat. Its purposes cover a wide range, such as studying basic biology and disease, evaluating the efficacy of novel therapeutics, and determining the reliability of consumer health and safety, as well as industrial products such as cosmetics, household cleaners, and food additives, agricultural chemicals, etc. Conducting research on live animals can be traced back to at least 500 BC. Today, as society grows and research needs increase, the use of animals in experiments increases. The research directed by People for the Ethical Treatment of Animals (PETA) in 2015 showed that over the last 15 years, the use of animals in research at major government-sponsored institutions has surged by about 73% (Animal experimentation up 73 percent, study says 2015). In addition, each year, it is estimated that approximately 115 million animals are utilized in scientific research across the world (Humane Society International 2012). A significant amount of animal testing is also used to test whether medical products are safe enough for human trials. Vaccines, for example, cannot be developed successfully without animal research, and the COVID-19 vaccine alone has saved hundreds of thousands of lives in less than a year. Thus, it is an undeniable fact that the contribution and sacrifice of animals in experiments have helped science to make great advancements, but as science and society have progressed, more people have become concerned about the cruelty of some experiments and have become aware of the rights of animals, thus popularizing the question whether animal experiments should still exist and whether they can be replaced.

Although animal experiments have been around for a long time, there is still room for improvement in this area. An article published in Science Daily suggests various approaches to enhance animal studies so that relevant therapeutic results can be better anticipated (Animal testing essential to medical progress but protocols could be improved 2017). Firstly, more research is needed and techniques need to be developed in order to enhance and minimize the usage of animals in accordance with the 3R rule. In addition, to minimize bias, criteria created for better clinical testing

---

*Corresponding Author: b201865@hiroshima-u.ac.jp

DOI 10.1201/9781003349648-15

should be included in animal testing. Furthermore, more research could be conducted to better translate animal findings into human cases.

The aim of this paper is to summarize and outline the different views currently held on animal testing, discuss its alternatives and future directions, and finally raise and discuss two issues. Next, this paper will describe the history and laws of animal testing, and looks at animal experimentation in terms of science and practicality, ethics and philosophy. Finally, two of the more promising alternatives are presented, and the question of whether animal experimentation can really be abolished and the ethics of researchers are discussed.

2 HISTORY OF THE DEBATE

The study of the ethics of animal experimentation did not emerge anywhere and has been the subject of centuries of philosophical as well as biological inquiry, however, there is no precise date for its commencement. With that said, 'Animal Liberation', a book by Peter Singer, an Australian philosopher, in 1975 can be considered a milestone. This book summarizes some of the arguments that had been put forward before and have since led to many new discussions. The book refers in particular to the question of how, if the possession of a greater level of intelligence does not enable one human being to exploit another human being for his own purposes, can it entitle a human being to make use of a non-human being for the same purpose (Singer 1990), which provokes discussion to this day. In addition, there are data showing that there has been a growing discussion of the ethics of animal experimentation since the 1980s. According to MEDLINE®, which is widely utilized by health experts, the number of articles has increased in recent years, as seen in Figures 1, 2, and 3. In Figure 1, it can be seen that in 1984 there was hardly any article with 'animal welfare' as a keyword, but within just 10 years the number of articles had increased by over 200. Figure 2 shows that there was also an almost continuous increase in the articles with 'animal' and 'ethics' as their keywords between the years 1984 to 1993. Figure 3 shows an overall upward trend in the number of articles with 'animal rights' as their keywords from 1982 to 1996, with two significant increases in 1990 and 1993, most likely due to the new regulations enacted in those years by the Animal Welfare Act.

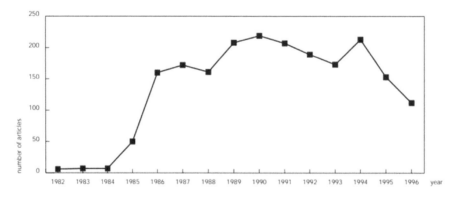

Figure 1. Number of papers on MEDLINE with the keyword 'animal welfare' (Paixão & Schramm 1999).

3 CURRENT LAWS AND REGULATIONS ON ANIMAL TESTING

Although animals have long been used for research and teaching, clear ethical guidelines and related legislation have only been developed in the last few decades. It is important to recognize these regulations in order to better understand the debate. Each country has its own laws that are

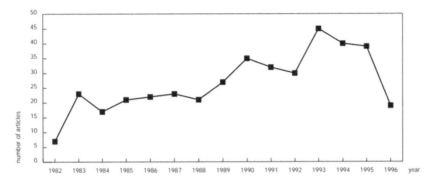

Figure 2. Number of papers on MEDLINE with the keywords 'animal' and 'ethics' (Paixão & Schramm 1999).

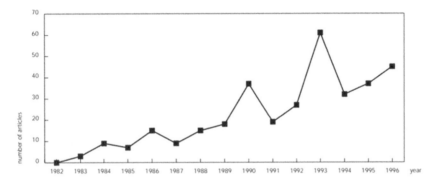

Figure 3. Number of papers on MEDLINE with the keyword 'animal rights' (Paixão & Schramm 1999).

strictly adhered to by universities or research institutions in that region. For example, both the Animal Welfare Act in the UK and the Twenty-Eight Hour Law in the US protect the welfare of animals. In addition, there are many regulatory organizations, such as the animal care committee, that monitor the behavior of researchers through sudden unannounced checks of the condition in the institutions.

Although the regulations might be different in various countries or regions, arguably the best known and the most utilized set of principles are The Three Rs (3Rs). The 3Rs, which stand for reduction, refinement, and replacement, are ethical standards in respect of the utilization of animals in product testing as well as scientific studies. They were initially articulated in 1959 by WMS Russell and RL Burch (Hubrecht & Carter 2019). The principle recommends that living sentient vertebrates be replaced by phylogenetically more primitive forms of life, such as the more degenerate metazoan microorganisms and endoparasites, or with computer simulations (Fernandes & Pedroso 2017). The principle of reduction states that experiments or research studies should be conducted on the fewest animals possible, whereas the principle of refinement emphasizes that the strategies and technologies utilized should minimize the animals' suffering and pain at all phases of the study. In addition, although the use of invertebrates in animal experiments is mainly unprotected by legislation, as a large number of these animals are considered to be non-sentient beings, some countries, such as the UK, New Zealand, and Norway, have laws that protect certain invertebrate species when they are involved in animal experiments.

4 SCIENTIFIC AND PRACTICAL APPROACH TO ANIMAL TESTING

By approaching the issue of animal experimentation scientifically, the aim is to analyze whether, ethical orientations aside, animal experimentation is necessary or not. As early as the nineteenth century, two heated opinions already emerged in the scientific field. While animal testing has grown more widespread and institutionalized, there was also an increase in startling concern for animal welfare (Paixão & Schramm 1999).

Opponents of animal testing claim that since alternatives, such as modelling, animal testing are available, and there is no longer a need to experiment with live animals. In addition, since no animal's DNA is 100% human-like, animal testing cannot always reliably predict the outcomes. In one research published in the British Medical Journal in 2006, six potential therapies for five different disorders, including stroke, hemorrhage, and head injury, were evaluated. Researchers analyzed all animal data for these six different treatments with all human data and discovered that only half of the time did the animal data match the human data (Perel et al. 2007). Besides, some drugs may be harmful to humans but not to animals and vice versa. For example, data show that 95 percent of medications tested were proven to be effective and safe on animals, but failed in human clinical trials (Stafl 2019). In addition, scientists have discovered several therapies for Alzheimer's disease in animals (Franco et al. 2014), but these advances do not necessarily transfer to people. To this end, rats have been cured several times, but humans are still unable to do so. Furthermore, because of the reasons stated above, the use of animals is a waste of their suffering and lives, and a waste of money for researchers considering the low cost of experimentation via stem cells or synthetic organs.

Although there is some strong evidence on the opposing side, many researchers still believe that animal experimentation is vital for scientific advancement. The use of animals is essential to biomedical research as animal models aid in ensuring the efficacy and safety of novel medicines. Studying them can provide scientists with insights into human biology and health because animals and humans are biologically highly similar, with rats matching over 98% of their DNA as an example. Besides, animals, as well as humans, are susceptible to a lot of the same health conditions, such as diabetes, cancer, and heart disease. In addition, because the life cycle is shorter than that of humans, animal models thus are able to be examined over generations and all around their life cycle, which is crucial for understanding the way diseases develop and interact with the entire living biological system.

It is true that alternative methods can reduce the use of animal experimentation and thus improve animal ethics. However, this denomination has caused much debate, with some researchers preferring the term complementary methods (Adolphe 1995), implying that they are not a complete replacement for animal experimentation. Although there have been some phenomenal advancements with computational modeling and tissue culturing techniques, the technology hadn't advanced to the point where the animal can be completely removed from the research process without causing significant hindrance if therapeutics for human use were to be created. More importantly, in order to make the computer models completer and more accurate, scientists need more data from animal studies, as humans do not now have a complete understanding of how animal cells and tissues function. In fact, before a new drug treatment can be put on the market, there is a legal requirement to have data from animal trials. As an example, the Food and Drug Administration (FDA) in the United States mandates animal testing to assure the safety of numerous medications and equipment. So, from an economical perspective, using animal testing could avoid lawsuits and save money for remaking the drug if it failed the human trial. Besides, since there are institutions purely based on animal testing, it would be economically challenging to completely stop animal testing.

5 ETHICAL AND PHILOSOPHICAL APPROACH TO ANIMAL TESTING

For centuries, Western philosophy has been exploring the question of animal ethics. Aristotle argued for a hierarchy of animals headed by humans because they could reason and had 'rational souls'. Descartes believed that animals were mere 'mechanisms' and could not perceive emotions. According to Kant, animals can suffer, but because they lacked moral autonomy, they also possess no moral status. These famous theories led to a traditional perception that animals were inferior to humans and lacked sentience. However, with the lively discussion of the issue and the development of science, it has been proven that animals are not incapable of perception. As a result, more and more ethical discussions have taken place and a number of animal protection laws have been established. For example, the Society for the Prevention of Cruelty to Animals (SPCA) was originally organized in 1822 in the U.K. In 1831, the English physiologist Marshall Hall planted the first seeds for today's code of animal ethics (Cheluvappa et al. 2017). In general, the arguments involved in the discussion are from two perspectives. On one side, there are people who defend anthropocentric perspectives, while those who defend sentient perspectives are on the other (Paixão & Schramm 1999).

6 ALTERNATIVES, FUTURE WORK AND QUESTION

Aside from the debate, finding an alternative is the top priority for research. The essence of animal testing is to test drugs or technologies on animals to examine the safety of items meant for human consumption. Often, the number of organ responses that need to be tested for a product is not large, or at least often the whole animal body is not required, which indicates that only animal organs are really needed in the test. Furthermore, the results of animal organ testing do not always correlate with human organs, so the establishment of separate human organs for research is the future of animal testing. In fact, back in 2011, the US-National Research Council already set out a new vision for 'Toxicology in the Twenty-first Century', the purpose of which is to utilize human cells and tissues instead of living animals for in vitro toxicity testing (Liebsch et al 2011).

A relatively new research result, human organ-on-a-chip, was introduced in 2019, a revolutionary technology that has the potential to transform the pharmaceutical research and development industry. Organ-on-a-chip mimics the physiological structure as well as the function of human organs and tissues, creating a controlled microenvironment with a vascular perfusion-like scenario that constitutes a microfluidic platform (Franzen et al. 2019). This research has at least two revolutionary benefits. Firstly, it eliminates experimental errors due to animal genes not exactly matching those of humans because human cells are used in this technology instead of animal cells. Secondly, it allows the activity of the drug to be tracked in real time. In spite of the possible success of R&D efficiencies, most of the studies to date have been directed at the biophysical possibilities of organ-on-a-chip, leaving the economic implications unexplored (Franzen et al. 2019). Therefore, it is estimated that it will take some time and a lot of research before this technology is fully marketed and completely replaces animals in experiments.

Another similar and promising technology is iPS cells (Induced pluripotent stem cells). One of the many exciting clinical applications of stem cells lies in regenerative medicine. Scientists have been examining the utilization of pluripotent stem cells as a substitute for animal testing to evaluate drug toxicity, such as iPS cells (Center for iPS Cell Research and Application, Kyoto University 2022). It is anticipated that iPS cells could help investigate illness origins, create novel medications, and be used for cell transplantation treatment and other kinds of regenerative medicine. For example, certain diabetes treatments that entail regenerative medicine require transplanting cells with the capability of regulating blood sugar levels, or in instances of trauma when a nerve has been damaged, where regenerative medicine involves transferring nerve cells that can be utilized to reestablish damaged connections, iPS cells might be able to help to make these transplants (Center for iPS Cell Research and Application, Kyoto University 2021). However, in 2021, the Fujibuchi lab reported that there is still a long way to go before stem cells can be utilized as a complete alternative for animal testing for medication toxicology studies. This is because drug toxicity databases based on other cell types show that cells are more sensitive to drugs (Center for iPS Cell Research and Application, Kyoto University 2021). In addition, when evaluating pluripotent stem cells and living animals, the toxicity of the eight medications ranked differently, indicating that toxicological research on cells in culture plates and living animals are not usually compatible. Despite this, Dr. Fujibuchi is optimistic that as more data is collected and advanced artificial intelligence methods are applied to moderate the differences in vitro as well as in vivo, scientists will understand how to minimize the number of animals used in drug trials using these cells (Center for iPS Cell Research and Application, Kyoto University 2021).

Both alternative approaches, as well as current research direction, point to an end to animal testing in the future. The US Environmental Protection Agency (EPA) announced in 2019 that it intends to significantly reduce animal testing, both those it commissions internally and those conducted by authorized companies, by 2025, and to phase it out completely by 2035 (Editor 2019). However, at the current level of technological development, it is still difficult to estimate for the moment what the scenario will be in 2035. Furthermore, there are currently many laboratories that only do animal testing, and many researchers and employees make their living doing animal testing. To stop animal testing altogether, it is not enough to consider only the development of technology, but also the social and economic impact.

After an overview of the major scientific and ethical arguments in the discussion on animal testing, this paper hereby proposes the question: Does it have to be proven that certain animals are sentient and be enshrined in law in order for researchers to have to treat them in an ethical manner? From a scientific perspective, since there are species that are newly found to be sentient beings, there is a chance that some of the 'non-sentient animals' that are being used in the laboratories are actually sentient beings. Furthermore, it is not difficult to conclude from the almost overwhelming arguments that animals also have moral worth and that humans need to treat them ethically. Thus, until the time when live animals are no longer needed in experiments, perhaps treating all animals in an ethical manner is the balance between science and ethics a researcher can achieve.

7 CONCLUSION

In this paper, different approaches to animal testing are studied and analyzed. The main conclusions can be summarized as follows: (1) from a scientific and practical perspective, the usage of animals in experiments is of great benefit to science and cannot yet be completely replaced in the current state of scientific advancement; (2) from an ethical and philosophical perspective, animals are sentient beings and should have basic rights. In terms of future work, the two promising alternatives, human organ-on-a-chip, and iPS cells should be developed to enhance the reduction of the use of living animals in experiments.

REFERENCES

About Animal Testing. Humane Society International. (2012). Retrieved 2 March 2022, https://www.hsi.org/news-media/about/.

Adolphe, M. (1995). Alternative methods to animal experimentation. Scientific and ethical problems. *Bulletin De L'academie Nationale De Medecine*, 179(6), 1009–16.

Animal experimentation up 73 percent, study says. (2015). Retrieved 29 April 2022, https://www.cbsnews.com/news/peta-study-finds-animal-testing-in-federal-labs-on-the-increase/.

Animal testing essential to medical progress but protocols could be improved. (2017). Retrieved 29 April 2022, https://www.sciencedaily.com/releases/2017/04/170426183030.htm

Cheluvappa, R., Scowen, P., & Eri, R. (2017). Ethics of animal research in human disease remediation, its institutional teaching; and alternatives to animal experimentation. *Pharmacology Research & Perspectives*, 5(4), e00332.

Editor, B. (2019). *Breaking news: EPA moves to end animal testing*. Retrieved 29 April 2022, https://blog.humanesociety.org/2019/09/breaking-news-epa-moves-to-end-animal-testing.html

Fernandes, M., & Pedroso, A. (2017). Animal experimentation: A look into ethics, welfare and alternative methods. *Revista Da Associação Médica Brasileira*, 63(11), 923–928.

Franco, R., & Cedazo-Minguez, A. (2014). Successful therapies for Alzheimer's disease: why so many in animal models and none in humans? *Frontiers In Pharmacology*, 5. https://doi.org/10.3389/fphar.2014.00146.

Franzen, N., van Harten, W., Retèl, V., Loskill, P., van den Eijnden-van Raaij, J., & IJzerman, M. (2019). Impact of organ-on-a-chip technology on pharmaceutical R&D costs. *Drug Discovery Today*, 24(9), 1720–1724.

Hubrecht, & Carter. (2019). The 3Rs and Humane Experimental Technique: Implementing Change. *Animals*, 9(10), 754.

Liebsch, M., Grune, B., Seiler, A., Butzke, D., Oelgeschläger, M., & Pirow, R. et al. (2011). Alternatives to animal testing: current status and future perspectives. *Archives of Toxicology*, 85(8), 841–858.

Paixão, R., & Schramm, F. (1999). *Ethics and animal experimentation: what is debated?* Cadernos De Saúde Pública, 15(suppl 1), S99-110.

Perel, P., Roberts, I., Sena, E., Wheble, P., Briscoe, C., Sandercock, P., Macleod, M., Mignini, L. E., Jayaram, P., & Khan, K. S. (2007). Comparison of treatment effects between animal experiments and clinical trials: a systematic review. *BMJ* (Clinical research ed.), 334(7586), 197. https://doi.org/10.1136/bmj.39048.407928.BE

Replacing animal experiments with stem cells | News and Events | CiRA | Center for iPS Cell Research and Application, Kyoto University. (2021). Retrieved 29 April 2022. https://www.cira.kyoto-u.ac.jp/e/pressrelease/news/210311-150000.html.

Singer, P. (1990). *Animal liberation.* New York, N.Y: New York Review of Books.

Stafl, H. (2019). The curtain on animal research at Queen's lifts an inch. *The Journal*. Retrieved 6 March 2022, https://www.queensjournal.ca/story/2019-02-13/investigations/the-curtain-on-animal-research-at-queens-lifts-an-inch/.

What are iPS cells? | For the Public | CiRA | Center for iPS Cell Research and Application, Kyoto University. Retrieved 29 April 2022, https://www.cira.kyoto-u.ac.jp/e/faq/faq_ips.html

Advances in Renewable Energy and Sustainable Development – Liang & Kasmani (Eds)
© 2023 Copyright the Author(s), ISBN: 978-1-032-39407-7

Screening of fungicides for mulberry sclerotinia in the field

Honglin Mou
The Chongqing Three Gorges Academy of Agricultural Sciences, Chongqing, China
Chongqing Three Gorges University, Chongqing, China

Minghai Zhang & Jiequn Ren*
The Chongqing Three Gorges Academy of Agricultural Sciences, Chongqing, China

Li Chen*
Extension Station for Plant Protection and Fruit Tree Technology of Wanzhou, China

Lixin Tan
The Chongqing Three Gorges Academy of Agricultural Sciences, Chongqing, China

Huaxian Yu & Mingjian Guo
Chongqing Silkworm Industry Management Station, Chongqing, China

Zhimin Fan
The Chongqing Three Gorges Academy of Agricultural Sciences, Chongqing, China
Chongqing Three Gorges University, Chongqing, China

Zhangyun Zheng & Yi Yang
The Chongqing Three Gorges Academy of Agricultural Sciences, Chongqing, China

ABSTRACT: In recent years, mulberry trees have been planted in large areas in China, and the Mulberry Sclerotinia sclerotiorum (*Sclerotinia sclerotiorum*) is a serious disease that endangers the quality of mulberry fruit. In the spring of 2016 ~ 2019, a variety of pesticides were selected for spraying control, and the effects of different medicament treatments on mulberry Sclerotinia sclerotiorum were observed and analyzed. Therefore, the problems of serious disease occurrence and poor prevention effect of medicament treatment are solved. The results showed that Tebuconazole, Pyrazolether, Myclobutanil, Prochloraz, Difenoconazole, and Azoxystrobin had better control effects on mulberry sclerotium disease, and the corrected control effect was up to 90%. However, because the high concentration of Tebuconazole and Prochloraz on mulberry leaves and mulberry fruits will produce different degrees of harm, it is more recommended to use the compound agent with good control effect and low degree of drug damage in production: such as Tebuconazole·Difenoconazole 2100 times liquid and Pyrazole·Pyracyl Bacteria 1200 times solution. It can not only delay the occurrence of drug resistance, but also prolong the service life of the medicament, which can provide technical methods for the green and high-quality production of mulberry in the Chongqing area.

1 INTRODUCTION

The fruit of the mulberry tree is called mulberries, consumers love them because they are rich in nutrients and bioactive ingredients, Mulberry is a kind of fruit that has been listed in the list

---

*Corresponding Authors: renjiequn@outlook.com and chenli420625@163.com

102   DOI 10.1201/9781003349648-16

of "homology of medicine and food" by the National Health Commission(Miao et al. 2020). In the past 10 years, the planting area of fruit mulberry has expanded rapidly in China. By 2020, the existing area of fruit mulberry in Chongqing has reached 22000 acres. The development of the fruit mulberry industry has achieved remarkable results, and has become an important development direction for the diversification transformation of the traditional sericulture industry (Ying 2009). However, due to the rapid development of the mulberry industry in recent years, the incidence rate of Sclerotinia sclerotiorum was more than 90%, which restricted the development of the mulberry industry (Wang 2015).

At present, Carbendazim and Thiophanate-Methyl have been used for a long time (E. R P 2005; Xiao 2019). However, with the increase in service life, the control effect decreased significantly (Bao 2011; Fan et al. 2015; Hu et al. 2011; Li & Zhang 2019; MO et al. 2017; Qin et al. 2017; Shi et al. 2014; Wei & Xu 2007; Wang 2019; Xiao 2019; Zhao et al. 2017). Unreasonable use of chemical agents is easy to cause pesticide residue beyond the standard, which not only causes environmental pollution, but also brings great safety risks to fruits and leaves (Xu 2020). So, it is urgent to screen new fungicides to prevent Mulberry Sclerotinia sclerotiorum. To select new fungicides for effective control of Sclerotinia sclerotiorum, a variety of single and compound fungicides were selected in the experiment from 2016 to 2019, it is expected that the screened fungicides can effectively prevent and control Sclerotinia sclerotiorum. This is particularly significant to reduce the economic losses caused by diseases, ensure the high-quality and high-yield mulberry, and achieve the increase farmers' income.

2 MATERIALS AND METHODS

2.1 *Experiment materials*

The mulberry variety used in this test was twelve-year-old "Big Ten without seeds". The test was listed in table 1. The applicator was a 3WBD-20 knapsack electric sprayer. The fungicides tested in this experiment were shown in Table 1.

2.2 *Experiment methods*

2.2.1 *Field trial design*
The experiment was carried out at the Ganning base of the Chongqing Three Gorges Academy of Agricultural Science, the experimental site is 325 m above sea level. Flat terrain, uniform fertility, and consistent field management were chosen for experiments. In the spring of 2016~2019, spraying control was carried out in the initial flowering stage, the full flowering stage, and the deflowering stage. Each fungicide has 5 concentrations, 3 trees per concentration, and 3 repetitions, and the control group was sprayed with clean water. The shape of the plot was square with basically the same area, around which guard rows were established. Pesticides were sprayed thoroughly and evenly on the mulberry trees by using electric sprayers after the dew dried up in the morning.

2.2.2 *Methods of investigation and calculation*
At 3 days after application, the damage to leaves and fruits of mulberry trees was investigated. The number of mulberry grains and diseased fruit grains was investigated from the beginning of mulberry discoloration. Mulberry fruits were picked once before ripening. Mulberry trees are picked every other day or every day after they mature. The number of mulberry fruit grains and diseased fruit grains in each group was recorded, and then the disease-bearing rate and corrective protection were calculated.

2.2.3 *Data analysis*
Statistical analysis of data was performed by using IBM SPSS 20.0 and Microsoft office 2016. One-Way ANOVA was used to conduct difference analysis, and the data were expressed $\bar{x} \pm s$, n = 3.

Table 1. Fungicides information.

Year	Fungicides	Manufacturer
2016	10% Difenoconazole WG	Qingdao Hansheng Biotechnology Co., Ltd.
	25% Tebuconazole EC	Qingdao Hansheng Biotechnology Co., Ltd.
	50% Procymidone WP	Dongguan Ruidefeng Biotechnology Co.,Ltd.
	250g/L Pyrazolether EC	BASF, Germany
	50% Dimethomorph WP	BASF, Germany
	240g/L Thifuroamide SC	Qingdao Hansheng Biotechnology Co., Ltd.
	75% Tricyclazole WG	Shanghai Heben Pharmaceutical Co., Ltd.
	45% Prochloraz EW	Hailier Pharmaceutical Group Co., Ltd.
	70% Azoxystrobin WP	Jiangxi Weiniu Crop Science Company
2017	70% Azoxystrobin WG	Jiangxi Weiniu Crop Science Co., Ltd.
	80% Captan WG	Hebei Guanlong Agrochemical Co., Ltd.
	10% Difenoconazole WG	Yifan Biotechnology Group Co., Ltd.
	40% Myclobutanil SC	Yifan Biotechnology Group Co., Ltd.
	25% Pyraclostrobine SC	Shandong Kangqiao Biotechnology Co., Ltd.
	430g/L Tebuconazole SC	Made by Mercer Technologies
	3% Imazalil GJ	Belgium Janssen Pharmaceutical Co., Ltd.
	45% Prochloraz ME	Shanxi Biaozheng Crop Science Co., Ltd.
2018	430g/L Tebuconazole SC	Tianjin Hanbang Plant Protection Co., Ltd.
	25% Pyraclostrobine SC	Shandong Kangqiao Biotechnology Co., Ltd.
	40% Myclobutanil SC	Yifan Biotechnology Group Co., Ltd.
	70% Azoxystrobin WG	Jiangxi Weiniu Crop Science Co., Ltd.
	450g/L Prochloraz EW	Hunan Dongyong Chemical Co., Ltd.
	10% Difenoconazole WG	Yifan Biotechnology Group Co., Ltd.
	46% Copper hydroxide WG	American Dupont Company
	50% Iprodione WP	Tianjin Hanbang Plant Protection Co., Ltd.
	50% Boscalid WG	BASF
	Tebuconazole : Difenoconazole =1 : 4	
	Prochloraz : Difenoconazole =1 : 5	
	25% Tebuconazole·Iprodione SC	Sichuan Woye Agrochemical Co., Ltd.
2019	40% Myclobutanil SC	Yifan Biotechnology Group Co., Ltd.
	25% Azoxystrobin WG	Jiangxi Weiniu Crop Science Co., Ltd.
	25% Pyraclostrobine SC	Yifan Biotechnology Group Co., Ltd.
	10% Difenoconazole WG	Yifan Biotechnology Group Co., Ltd.
	450g/L Prochloraz EW	Hunan Dongyong Chemical Co., Ltd.
	430g/L Tebuconazole SC	Tianjin Hanbang Plant Protection Co., Ltd.
	4:1 Imazalil·Fludioxonil SC	Yifan Biotechnology Group Co., Ltd.
	1:2 Trifloxystrobin·Tebuconazole	Shandong Rongbang Chemical Co., Ltd.
	1:10Metalaxyl-M·Chlorothalonil	Yifan Biotechnology Group Co., Ltd.
	1:2 Pyraclostrobine·Boscalid	Shanghai Yuelian Chemical Co., Ltd.
	Tebuconazole·Difenoconazole =1 : 4	
	1×10^{10}cfu/g Bacillus subtilis WP	Deqiang Biology Co., Ltd.

3 TEST RESULTS AND ANALYSIS

3.1 *Control effects of different fungicides on mulberry Sclerotinia sclerotiorum*

According to table 2, the most effective fungicides against mulberry Sclerotinia sclerotiorum were 25% Tebuconazole, 45% Prochloraz, 10% Difenoconazole, 250 g/L Pyraclostrobine, 40% Myclobutanil and 70% Azoxystrobin, and their effectiveness against more stable. In 2018, the control effect of three compound agents on mulberry Sclerotinia sclerotiorum was the best, and the control effect was above 99.09%. The effective fungicides against mulberry Sclerotinia sclerotiorum

104

were 50% Boscalid, 3% Imazalil, and 50% Procymidone, the control effect was 84.09%~94.51%. The control effect of 80% Captan was poor and only 71.93%. 46% Copper hydroxide, 240 g/L Thifluzamide, 50% Dimethomorph, and 75% Tricyclazole were the worst, only 42.20~59.46%, all of which were different from the 6 fungicides with the best control effect.

Table 2. Control effects of different fungicides on mulberry sclerotinia disease.

Year	Fungicides	Dilution multiple	Disease percentage/%	Control efficacy/%
2016	10% Difenoconazole	1500	7.24 ± 0.25 ab	92.72 ± 0.25 ab
	25% Tebuconazole	1800	0.14 ± 0.05 a	99.86 ± 0.05 a
	50% Procymidone	500	15.83 ± 0.58 b	84.09 ± 0.58 b
	250g/L Pyraclostrobine	1000	8.46 ± 0.49 ab	91.49 ± 0.49 ab
	50% Dimethomorph	900	53.04 ± 3.62 c	46.68 ± 3.64 c
	240g/LThifluzamide	1800	43.24 ± 0.60 d	56.54 ± 0.60 d
	75% Tricyclazole	1500	57.50 ± 7.81 c	42.20 ± 7.85 c
	45% Prochloraz	1200	0.88 ± 0.02 a	99.11 ± 0.02 a
	70% Azoxystrobin	2000	14.81 ± 0.58 b	85.11 ± 0.28 b
	Clear water (CK)		99.48 ± 0.44 e	0.00 e
2017	Azoxystrobin	900	7.71 ± 1.88 a	86.17 ± 3.37 b
	Captan	600	15.63 ± 0.47 b	71.93 ± 0.84 c
	Difenoconazole	900	2.53 ± 0.95 a	95.17 ± 0.47 a
	Myclobutanil	1200	1.32 ± 0.79 a	97.64 ± 1.42 a
	Pyraclostrobine	600	2.52 ± 0.77 a	95.47 ± 1.39 a
	Tebuconazole	2100	0.37 ± 0.65 a	99.33 ± 0.67 a
	Imazalil	600	3.06 ± 1.86 a	94.51 ± 3.33 a
	Prochloraz	1200	0.91 ± 0.46 a	98.37 ± 0.82 a
	Clear water (CK)		55.70 ± 5.97 c	0.00 d
2018	Tebuconazole	2400	0.07 ± 0.04 a	99.91 ± 0.05 a
	Pyraclostrobine	900	1.30 ± 0.15 ab	98.27 ± 0.20 ab
	Myclobutanil	600	0.21 ± 0.06 a	99.72 ± 0.08 a
	Azoxystrobin	1500	3.03 ± 0.44 ab	95.99 ± 0.59 ab
	Prochloraz	900	0.12 ± 0.05 a	99.84 ± 0.07 a
	Difenoconazole	1200	0.91 ± 0.19 ab	98.80 ± 0.25 a
	Copper hydroxide	600	30.63 ± 3.71 d	59.46 ± 4.92 d
	Iprodione	600	9.09 ± 0.22 c	87.96 ± 0.29 c
	Boscalid	900	4.82 ± 0.66 bc	93.62 ± 0.87 b
	Tebuconazole·Difenoconazole	900	0.29 ± 0.11 a	99.61 ± 0.15 a
	Prochloraz·Difenoconazole	900	0.69 ± 0.01 ab	99.09 ± 0.02 a
	Tebuconazole·Iprodione	900	0.23 ± 0.11 a	99.70 ± 0.14 a
	Clear water (CK)		75.55 ± 2.32 e	0.00 e
2019	Myclobutanil	1500	0.15 ± 0.13 a	99.82 ± 0.16 ab
	Azoxystrobin	600	9.69 ± 0.64 b	88.45 ± 0.76 c
	Pyraclostrobine	1200	0.66 ± 0.26 a	99.22 ± 0.31 b
	Difenoconazole	900	0.58 ± 0.31 a	99.30 ± 0.37 b
	Prochloraz	1500	0.00 ± 0.00 a	100.00 ± 0.00 a
	Tebuconazole	1500	0.00 + 0.00 a	100.00 ± 0.00 a
	Clear water (CK)		83.93 ± 3.88 c	0.00 d

Values followed by different letters in the same column indicate a significant difference ($P<$ 0.05), and those followed by the same letters indicate no significant difference ($P \geq 0.05$), the same as below.

Table 3.   Control effects of 9 fungicides with different dosages on mulberry sclerotinia disease in 2016.

Fungicides	Dilution multiple	Disease percentage/%	Control efficacy/%
10% Difenoconazole	1500	7.24 ± 0.25 abf	92.72 ± 0.25 abi
	1800	9.73 ± 0.44 abf	90.22 ± 0.44 abi
	2100	22.02 ± 0.97 cg	77.86 ± 0.97 cf
	2500	41.16 ± 2.97 d	58.62 ± 2.99 d
	3000	67.75 ± 1.82 eln	31.89 ± 1.83 eln
25% Tebuconazole	900	0.34 ± 0.13 a	99.66 ± 0.13 a
	1200	0.26 ± 0.10 a	99.74 ± 0.10 a
	1500	0.65 ± 0.15 a	99.34 ± 0.15 a
	1800	0.14 ± 0.05 a	99.86 ± 0.05 a
	2100	0.50 ± 0.09 a	99.50 ± 0.09 a
50% Procymidone	500	15.83 ± 0.58 cf	84.09 ± 0.58 fi
	800	23.35 ± 0.50 cg	76.53 ± 0.50 cf
	1000	29.42 ± 2.06 g	70.43 ± 2.07 c
	1200	44.26 ± 3.35 dhj	55.51 ± 3.37 dgj
	1500	57.53 ± 3.89 eik	42.17 ± 3.91 hkn
250g/L Pyraclostrobine	1000	8.46 ± 0.49 abf	91.49 ± 0.49 abi
	1500	13.07 ± 1.34 bcf	86.86 ± 1.35 bfi
	2000	22.75 ± 1.43 cg	77.13 ± 1.44 cf
	2500	28.78 ± 5.19 g	71.07 ± 5.21 c
	3000	27.64 ± 3.48 g	72.21 ± 3.50 c
50% Dimethomorph	900	53.04 ± 3.62 hij	46.68 ± 3.64 ghj
	1200	57.94 ± 1.16 eik	41.76 ± 1.16 hkn
	1500	55.29 ± 3.18 ijk	44.41 ± 3.19 hjk
	1800	70.62 ± 7.50 ln	29.01 ± 7.54 el
	2000	89.64 ± 1.26 mp	9.89 ± 1.27 mp
240g/LThifluzamide	1800	43.24 ± 0.60 dh	56.54 ± 0.60 dg
	2100	53.79 ± 2.66 hij	45.93 ± 2.67 ghj
	2400	52.82 ± 4.45 hij	46.91 ± 4.47 ghj
	2700	66.59 ± 7.66 ekln	33.06 ± 7.70 ekln
	3000	62.07 ± 6.22 eikl	37.61 ± 6.26 hkln
75% Tricyclazole	1500	57.50 ± 7.81 eik	42.20 ± 7.85 hkn
	1800	57.77 ± 9.11 eik	41.92 ± 9.16 hkn
	2100	82.13 ± 2.71 mo	17.44 ± 2.73 mo
	2400	75.75 ± 2.13 no	23.85 ± 2.14 eo
	2700	70.73 ± 11.26 ln	28.90 ± 11.31 el
45% Prochloraz	900	1.04 ± 0.23 a	98.95 ± 0.23 a
	1200	0.88 ± 0.02 a	99.11 ± 0.02 a
	1500	1.55 ± 0.17 ab	98.44 ± 0.18 ab
	1800	2.40 ± 0.40 ab	97.59 ± 0.40 ab
	2100	1.76 ± 0.56 ab	98.23 ± 0.56 ab
70% Azoxystrobin	2000	14.81 ± 0.58 cf	85.11 ± 0.28 fi
	3000	29.09 ± 0.88 g	70.76 ± 0.89 c
	4000	49.89 ± 2.11 dhij	49.85 ± 2.12 dghj
	5000	52.05 ± 4.25 dhij	47.68 ± 4.27 dghj
	6000	55.41 ± 2.73 ijk	44.30 ± 2.74 hjk
Clear water (CK)		99.48 ± 0.44 p	0.00 p

3.2   *Control effects of the same fungicide and different dosages on mulberry sclerotinia sclerotiorum*

The control effects of mulberry Sclerotinia sclerotiorum with different dosages in spring from 2016 to 2019 were studied (table 3 to table 6), there was a significant difference in the control effects

Table 4. Control effects of 8 fungicides with different dosages on mulberry sclerotinia disease in 2017.

Fungicides	Dilution multiple	Disease percentage/%	Control efficacy/%
Azoxystrobin	2100	10.33 ± 2.33 fghij	81.45 ± 4.18 cdef
	1800	14.76 ± 2.18 efg	73.50 ± 3.91 fgh
	1500	10.17 ± 2.25 fghij	81.74 ± 4.03 cdef
	1200	9.55 ± 1.72 fghijk	82.86 ± 3.08 bcdef
	900	7.71 ± 1.88 ghijklm	86.17 ± 3.37 abcdef
Captan	1800	26.78 ± 1.55 c	51.93 ± 2.78 ij
	1500	21.93 ± 2.77 cd	60.62 ± 4.97 hi
	1200	19.48 ± 1.98 de	65.03 ± 3.56 gh
	900	33.31 ± 10.37 b	40.21 ± 18.62 j
	600	15.63 ± 0.47 def	71.93 ± 0.84 fgh
Difenoconazole	2100	7.50 ± 2.21 ghjklm	86.54 ± 3.96 abcdef
	1800	7.81 ± 3.80 ghijklm	85.97 ± 6.81 abcdef
	1500	4.28 ± 0.24 ijklm	92.31 ± 0.43 abcd
	1200	4.65 ± 1.27 ijklm	91.65 ± 2.28 abcd
	900	2.53 ± 0.95 jklm	95.17 ± 0.47 abc
Myclobutanil	1800	6.34 ± 3.65 hijklm	88.62 ± 6.56 abcde
	1500	2.69 ± 0.26 jklm	95.17 ± 0.47 abc
	1200	1.32 ± 0.79 lm	97.64 ± 1.42 ab
	900	6.39 ± 1.24 hijklm	88.52 ± 2.23 abcdef
	600	5.21 ± 1.50 ijklm	90.64 ± 2.70 abcdef
Pyraclostrobine	1800	6.33 ± 0.39 hijklm	88.63 ± 0.70 abcdef
	1500	4.12 ± 1.78 ijklm	92.60 ± 3.19 abcd
	1200	3.03 ± 0.19 jklm	94.56 ± 0.34 abcd
	900	2.80 ± 0.93 jklm	94.98 ± 1.66 abcd
	600	2.52 ± 0.77 jklm	95.47 ± 1.39 abc
Tebuconazole	1200	1.48 ± 0.97 lm	97.34 ± 1.75 ab
	1500	0.85 ± 0.25 lm	98.48 ± 0.44 a
	1800	0.62 ± 0.24 lm	98.89 ± 0.43 a
	2100	0.37 ± 0.65 m	99.33 ± 0.67 a
	2400	2.35 ±1.90 klm	95.79 ± 3.41 abc
Imazalil	600	3.06 ± 1.86 jklm	94.51 ± 3.33 abcd
	900	5.06 ± 0.62 ijklm	90.91 ± 1.12 abcde
	1200	13.29 ± 2.71 efgh	76.15 ± 4.86 efg
	1500	8.34 ± 1.41 fghijkl	85.03 ± 2.54 abcdef
	1800	11.20 ± 0.30 fghi	79.89 ± 0.54 def
Prochloraz	2100	2.22 ± 0.41 klm	96.02 ± 0.74 abc
	1800	1.37 ± 0.42 lm	97.53 ± 0.75 ab
	1500	1.99 ± 0.75 klm	96.43 ± 1.35 abc
	1200	0.91 ± 0.46 lm	98.37 ± 0.82 a
	900	1.24 ± 0.64 lm	97.78 ± 1.16 ab
Clear water (CK)		55.70 ± 5.97 a	0.00 k

of different dosages, and the control effect increased with the increase of concentration of liquid medicine. The control effects of 25% Tebuconazole and 45% Prochloraz on mulberry Sclerotinia sclerotiorum were more than 97%; the control effect of 40% Myclobutanil was over 88%, and there was no significant difference in the control effects of these three fungicides on mulberry Sclerotinia sclerotiorum; the corrected efficacy of 10% Difenoconazole, 250 g/L Pyraclostrobine, 50% Procymidone, 50% Iprodione, 70% Azoxystrobin, 3% Imazalil and 80% Captan varied with the change of concentration gradient, and the control effects fluctuated greatly in different years.

Table 5. Control effects of 9 fungicides with different dosages on mulberry sclerotinia disease in 2018.

Fungicides	Dilution multiple	Disease percentage/%	Control efficacy/ %
Tebuconazole	1200	0.70 ± 0.54 abcd	99.08 ± 0.71 abc
	1500	0.45 ± 0.24 a	99.40 ± 0.32 abc
	1800	0.48 ± 0.35 a	99.36 ± 0.46 abc
	2100	0.23 ± 0.23 a	99.69 ± 0.31 ab
	2400	0.07 ± 0.04 a	99.91 ± 0.05 a
Pyraclostrobine	600	1.76 ± 0.24 abcd	97.68 ± 0.32 abcdef
	900	1.30 ± 0.15 abcd	98.27 ± 0.20 abcdef
	1200	2.32 ± 0.04 abcd	96.93 ± 0.05 abcdef
	1500	2.86 ± 0.41 abcd	96.22 ± 0.55 abcdef
	1800	4.75 ± 0.34 bcdef	93.72 ± 0.46 def
Myclobutanil	600	0.21 ± 0.06 a	99.72 ± 0.08 ab
	900	0.75 ± 0.11 abcd	99.01 ± 0.14 abc
	1200	2.95 ± 2.69 abcd	96.10 ± 3.56 abcdef
	1500	0.73 ± 0.56 abcd	99.03 ± 0.74 abc
	1800	1.01 ± 0.32 abcd	98.66 ± 0.42 abcde
Azoxystrobin	900	4.13 ± 0.71 abcd	94.53 ± 0.94 bcdef
	1200	4.21 ± 2.06 abcde	94.42 ± 2.72 cdef
	1500	3.03 ± 0.44 abcd	95.99 ± 0.59 abcdef
	1800	8.46 ± 0.30 efgh	88.80 ± 0.40 g
	2100	12.33 ± 0.20 hij	83.68 ± 0.27 hi
Prochloraz	900	0.12 ± 0.05 a	99.84 ± 0.07 a
	1200	0.12 ± 0.03 a	99.83 ± 0.03 a
	1500	0.14 ± 0.06 a	99.81 ± 0.08 a
	1800	0.59 ± 0.13 abc	99.22 ± 0.17 abc
	2100	1.10 ± 0.44 abcd	98.54 ± 0.59 abcde
Difenoconazole	900	1.32 ± 0.66 abcd	98.25 ± 0.88 abcdef
	1200	0.91 ± 0.19 abcd	98.80 ± 0.25 abcd
	1500	1.10 ± 0.05 abcd	98.54 ± 0.07 abcde
	1800	4.98 ± 0.48 defg	93.40 ± 0.63 f
	2100	9.72 ± 0.61 hi	87.14 ± 0.81 gh
Copper hydroxide	600	30.62 ± 3.71 l	59.46 ± 4.92 l
	900	32.80 ± 1.18 l	56.58 ± 1.56 lm
	1200	34.01 ± 2.38 lm	54.98 ± 3.15 m
	1500	37.27 ± 0.20 m	50.67 ± 0.26 n
	1800	45.15 ± 2.54 n	40.23 ± 3.36 o
Iprodione	600	9.09 ± 0.22 gh	87.96 ± 0.29 gh
	900	10.65 ± 1.21 hi	85.91 ± 1.60 ghi
	1200	13.94 ± 2.71	81.55 ± 3.58 ij
	1500	16.21 ± 1.48 j	78.54 ± 1.96 j
	1800	22.94 ± 1.26 k	69.64 ± 1.66 k
Boscalid	900	4.82 ± 0.66 cdefg	93.62 ± 0.87 ef
	1200	8.97 ± 0.29 fgh	88.13 ± 0.39 gh
	1500	9.71 ± 1.21 hi	87.15 ± 1.60 gh
	1800	11.34 ± 1.15 hi	84.99 ± 1.52 ghi
	2100	11.08 ± 0.92 hi	85.34 ± 1.22 ghi
Clear water (CK)		75.55 ± 2.32 o	0.00 p

3.3 Investigation of mulberry leaf damage and fruit damage caused by different fungicides

From 2016 to 2019, damages to leaves and fruits caused by several fungicides spraying were observed regularly, the results indicate that: there were 430 g/L Tebuconazole, 450 g/L Prochloraz,

Table 6. Control effects of 8 fungicides with different dosages on mulberry sclerotinia disease in 2010.

Fungicides	Dilution multiple	Disease percentage /%	Control efficacy/%
Myclobutanil	1500	0.15 ± 0.13 a	99.82 ± 0.16 ab
	2000	0.68 ± 0.28 abc	99.19 ± 0.33 abcd
	2500	2.18 ± 0.12 cdef	97.40 ± 0.15 fjh
	3000	1.78 ± 0.72 bcde	97.87 ± 0.86 efj
	3500	1.25 ± 0.13 abcd	98.51 ± 0.16 def
Azoxystrobin	600	9.69 ± 0.64 j	88.45 ± 0.76 i
	900	15.40 ± 0.39 h	81.65 ± 0.47 g
	1200	18.85 ± 0.49 i	77.55 ± 0.59 k
	1500	22.56 ± 0.98 g	73.12 ± 1.17 l
	1800	24.31 ± 0.97 k	71.03 ± 1.16 m
Pyraclostrobine	600	0.79 ± 0.20 abc	99.06 ± 0.24 bcde
	900	1.06 ± 0.22 abcd	98.74 ±0.26 bcde
	1200	0.66 ± 0.26 abc	99.22 ± 0.31 abcd
	1500	1.84 ± 1.13 bcde	97.80 ± 1.34 efj
	1800	2.86 ± 0.40 ef	96.59 ± 0.48 j
Difenoconazole	900	0.58 ± 0.31 abc	99.30 ± 0.37 abcd
	1200	1.11 ± 0.48 abcd	98.67 ± 0.58 cde
	1500	1.53 ± 0.47 abcde	98.18 ± 0.56 efj
	1800	3.65 ± 0.80 f	95.65 ± 0.96 h
	2100	2.61 ± 0.84 def	96.89 ± 0.10 fj
Prochloraz	1500	0.00 ± 0.00 a	100.00 ± 0.00 a
	2000	0.06 ± 0.10 a	99.93 ± 0.11 a
	2500	0.13 ± 0.15 a	99.85 ± 0.18 a
	3000	0.10 ± 0.10 a	99.88 ± 0.12 a
	3500	0.24 ± 0.13 ab	99.72 ± 0.16 abc
Tebuconazole	1500	0.00 ± 0.00 a	100.00 ± 0.00 a
	2000	0.00 ± 0.00 a	100.00 ± 0.00 a
	2500	0.00 ± 0.00 a	100.00 ± 0.00 a
	3000	0.00 ± 0.00 a	100.00 ± 0.00 a
	3500	0.00 ± 0.00 a	100.00 ± 0.00 a
Clear water (CK)		83.93 ± 3.88 l	0.00 n

and 40% Myclobutanil to mulberry leaves and fruits damage. In areas with high concentrations, the mulberry fruit was partially shed; Among them, 430 g/L Tebuconazole 1800 times solution, 40% Myclobutanil 900 times solution, and 450 g/L Prochloraz 1500 times solution resulted in yellowing and shriveling of tip tender leaves, the overall development of mulberry is normal; 25% Tebuconazole 1500 times solution, 430 g/L Tebuconazole 1200~1500 times solution, 40% Myclobutanil 600 times solution and 450 g/L Prochloraz 900~1200 times solution not only made mulberry leaves yellow, shrivel and even wither, but the dropping rate of mulberry fruit was also 5%~10%, and with the increase of the concentration, the drug damage gradually increased. After being treated with several other fungicides, the leaf shape and color of mulberry leaves were not obviously abnormal, and the shape and quality of mulberry had no significant effect, therefore, it can be preliminarily concluded that these fungicides are harmless to mulberry fruits and mulberry leaves.

As fungicides such as Tebuconazole and Prochloraz have been observed to cause certain harm to mulberry trees, therefore, compounding agents were prepared in an attempt to reduce the harm caused by the drug in 2018~2019 (table 7), the test results show that, in 2018, three kinds of compound preparation had good control effect on mulberry Sclerotinia sclerotiorum, reaching over 96%, the shape and quality of mulberry leaves and fruits were not abnormal. Tebuconazole

Table 7. Control effects of combination agent on mulberry sclerotinia.

Year	Fungicides	Dilution multiple	Disease percentage/%	Control efficacy/%
2018	Tebuconazole · Difenoconazole	900	0.29 ± 0.11	99.61 ± 0.15 a
		1200	0.42 ± 0.15	99.44 ± 0.20 ab
		1500	0.73 ± 0.17	99.03 ± 0.22 abc
	Prochloraz · Difenoconazole	900	0.69 ± 0.01	99.09 ± 0.02 abc
		1200	2.09 ± 0.50	97.23 ± 0.67 de
		1500	2.58 ± 0.52	96.59 ± 0.69 e
	Tebuconazole · Iprodione	900	0.23 ± 0.11	99.70 ± 0.14 a
		1200	1.51 ± 0.45	98.00 ± 0.60 cd
		1500	1.43 ± 0.48	98.10 ± 0.64 bcd
	Clear water (CK)		75.55 ± 2.32	0.00 f
2019	Imazalil · Fludioxonil	600	0.03 ± 0.05 a	99.97 ± 0.06 a
		900	0.43 ± 0.24 a	99.48 ± 0.28 a
		1200	0.69 ± 0.09 a	99.18 ± 0.11 a
		1500	0.98 ± 0.18 a	98.83 ± 0.22 a
		1800	1.67 ± 0.66 a	98.01 ± 0.78 a
	Trifloxystrobin · Tebuconazole	900	0.00 ± 0.00 a	100.00 ± 0.00 a
		1200	0.00 ± 0.00 a	100.00 ± 0.00 a
		1500	0.00 ± 0.00 a	100.00 ± 0.00 a
		1800	0.00 ± 0.00 a	100.00 ± 0.00 a
		2100	0.00 ± 0.00 a	100.00 ± 0.00 a
	Metalaxyl-M · Chlorothalonil	600	26.15 ± 5.74 b	68.85 ± 6.83 b
		900	37.79 ± 0.84 c	54.98 ± 5.77 c
		1200	46.58 ± 0.81 d	44.51 ± 0.96 d
		1500	50.41 ± 1.79 de	39.94 ± 2.13 de
		1800	52.12 ± 3.62 ef	37.91 ± 4.31 ef
	Pyraclostrobine · Boscalid	600	0.35 ± 0.16 a	99.58 ± 0.19 a
		900	0.48 ± 0.43 a	99.43 ± 0.51 a
		1200	0.23 ± 0.09 a	99.73 ± 0.11 a
		1500	0.86 ± 0.55 a	98.98 ± 0.65 a
		1800	1.30 ± 0.57 a	98.45 ± 0.68 a
	Tebuconazole · Difenoconazole	900	0.00 ± 0.00 a	100.00 ± 0.00 a
		1200	0.00 ± 0.00 a	100.00 ± 0.00 a
		1500	0.00 ± 0.00 a	100.00 ± 0.00 a
		1800	0.00 ± 0.00 a	100.00 ± 0.00 a
		2100	0.00 ± 0.00 a	100.00 ± 0.00 a
	Bacillus subtilis	600	52.64 ± 3.40 ef	37.28 ± 4.76 ef
		900	61.41 ± 7.11 j	26.83 ± 8.47 j
		1200	70.08 ± 6.34 h	16.51 ± 7.56 h
		1500	56.74 ± 2.12 fj	32.40 ± 2.53 fj
		1800	60.05 ± 9.79 j	28.46 ± 11.66 j
	Clear water (CK)		83.93 ± 3.88 i	0.00 i

combined with Difenoconazole had the best control effect, reaching over 99%. Therefore, it can be preliminarily considered that these 3 compounds are harmless to mulberry fruits and mulberry leaves. Based on the results of the 2018 trial, in 2019 the use of a combination of agents and biological agents, control results show that Imazalil·Fludioxonil, Trifloxystrobin·Tebuconazole, Pyraclostrobine·Boscalid, Tebuconazole·Difenoconazole have a good control effect on mulberry Sclerotinia sclerotiorum, which is more than 98%, among them, 900 times solution and 1200 times solution of Tebuconazole were harmful to mulberry leaves but had no significant effect on the shape and quality of mulberry. All things considered, it is recommended to use Tebuconazole or

Prochloraz mixed with Difenoconazole, Tebuconazole mixed with Iprodione, Imazalil mixed with Fludioxonil, Pyraclostrobine mixed with Boscalid to control mulberry Sclerotinia sclerotiorum.

4 DISCUSSION

Mulberry Sclerotinia sclerotiorum is a fungal disease. Because of its wide incidence and strong infectivity, Sclerotinia sclerotiorum is a major disease restricting the development of the mulberry industry (Kuai & Wu 2012; Yang et al. 2021), and the disease has brought huge economic loss to the fruit farmer. Our team concluded in a four-year field trial of fungicide screening, 430 g/L Tebuconazole, 25% Pyrazolether, 40% Myclobutanil, 450 g/L Prochloraz, 10% Difenoconazole, and 70% Azoxystrobin had a better control effect on mulberry Sclerotinia sclerotiorum, the effect of prevention is gradually increasing; the control effects of 430g/L Tebuconazole and 450 g/L Prochloraz on mulberry Sclerotinia sclerotiorum were not significant, showing the most stable. Because 430g/L Tebuconazole, 40% Myclobutanil, and 450 g/L Prochloraz caused different degrees of damage to mulberry leaves and fruits, the compound test was carried out to reduce the damage. According to some reports, Tebuconazole can be used as an excellent single agent to control mulberry Sclerotinia sclerotiorum. The preliminary test results of Linjian(Lin & Zhu 2015) and others using Tebuconazole SC to control mulberry Sclerotinia sclerotiorum were verified. However, excessive use of Tebuconazole can affect the quality of mulberry leaves and fruits, the negative effects could be alleviated to some extent if Pyrazolether was mixed with Tebuconazole (Ma 2009; Xue et al. 2017; Yang et al. 2021). In Yangmingfang's experiment, it was also pointed out that after Tebuconazole was mixed with Pyrazolether, the effect of correction and prevention was up to 100%[19], which is consistent with the conclusion of this experiment.

5 CONCLUSION

In this experiment, a variety of chemicals were selected to spray mulberry trees, the effects of different medicament treatments on the incidence, control effect, and phytotoxicity of Sclerotinia sclerotiorum were observed and analyzed. The main conclusions can be summarized as follows:

(1) From the point of view of the economy and environment protection, 25% Tebuconazole, 45% Prochloraz 2500~3500 times solution, 40% Myclobutanil 2500~3500 times solution, 10% Difenoconazole 1200~1500 times solution, 250 g/L Pyrazolether 1200~1800 times solution, 70% Azoxystrobin 900 times solution are recommended, with the same active ingredient, and we can use different dosage forms or different manufacturers;
(2) It is not recommended to use 46% Copper hydroxide, 50% Dimethomorph, 75% Tricyclazole, or 240 g/L Thifuroamide in production, for they were less than 60% effective.
(3) It is recommended to control mulberry Sclerotinia sclerotiorum with Tebuconazole · Difenoconazole, Prochloraz·Difenoconazole, Tebuconazole · Iprodione, Imazalil · Fludioxonil, Pyraclostrobine·Boscalid, it can not only enlarge the antibacterial spectrum and delay the drug resistance, but also prolong the service life of the medicament.

In future research work on the control of mulberry Sclerotinia sclerotiorum, according to the actual situation of local production, the agents with better control effect in this experiment were used alternately. It can not only prevent the pathogenic bacteria of Sclerotinia sclerotiorum from producing drug resistance, but also improve the control effect of Sclerotinia sclerotiorum. On the question of pesticide residues in mulberry fruit, the safety of the chemicals recommended in this study still needs to be further evaluated.

ACKNOWLEDGMENTS

These experiments were supported by the technical system of modern high-efficiency agriculture (sericulture) in Chongqing's modern mountainous areas (number 8-6).

REFERENCES

Bao, S.R. (2011) Experiment on the optimum control period of Sclerotinia sclerotiorum on Big Ten fruit mulberry. *Modern Agricultural Science and Technology*, (18): 186+189.

E. R P. (2005) A century of fungicide evolution. The Journal of Agricultural Science, 143(1).

Fan, J., Hu, X.M., Yu, C. (2015) Research progress on prevention and control technology of Mulberry Sclerotinia sclerotiorum. *China Sericulture*, 36(02): 11–14.

Hu, J.H., Cai, Y.X., Zhou, S.J. (2011) Preliminary Report on laboratory screening of fungicides against Sclerotinia sclerotiorum. *Shanghai Agricultural Science and Technology*, (03): 110–111+117.

Kuai, Y.Z., Wu, F.A. (2012) A Review on Pathogens of Mulberry Fruit Sclerotiniosis and Its Control Technology. *Acta Sericologica Sinica*, 38(06): 1099–1104.

Li, Y.Y., Zhang, Z.Z. (2019) Analysis of the incidence of Sclerotinia sclerotiorum and its control measures. *Agriculture and Technology*, 39(13): 49–50.

Lin, J., Zhu, X.T. (2015) Control effect of Tebuconazole SC on Mulberry Sclerotinia sclerotiorum. *China Sericulture*, 36(4): 35–37.

Ma, H.X. (2009) *Direction of the resistance in Sclerotinia sclerotiorum and controlling*. Nanjing Agricultural University.

Miao, Q., Hong, W.Y., Wu, Y.J. (2020) Different Combination Patterns of Fungicides Against Mulberry Sorosis Disease: Integrated Control Effect. *Journal of Agriculture*, 10(07): 39–44.

MO, C.Y., Chen, X.Q., Shi, M.N. (2017) Field trial of different drugs against Mulberry Sclerotinia sclerotiorum. *Guangxi Sericulture*, 54(04): 6–10.

Qin, H.S., Chen, W., Huang, Z.Q. (2017) Occurrence regularity and integrated control of Sclerotinia sclerotiorum on ten-fruit mulberry in North Guangxi. *Bulletin of Agricultural Science and Technology*, (12): 339–341.

Shi, X.P., Deng, Z.H., Peng, X.H. (2014) Occurrence and control of Sclerotinia sclerotiorum in different mulberry varieties. *Newsletter of Sericulture and Tea*, (06): 6–8.

Wang, R. (2019) Occurrence regularity and integrated control techniques of Sclerotinia sclerotiorum in Mulberry. *Anhui Agricultural Science Bulletin*, 25(10): 72–73.

Wang, Y. (2015) *Resistance management and synergistic interaction of SYP-7017 and BOSCALID in controlling sclerotinia stem rot*. Nanjing Agricultural University.

Wei, X.J., Xu, C.M. (2007) Occurrence and control of Mulberry Sclerotinia sclerotiorum. *Jiangsu Sericulture*, (04): 14–15.

Xiao, J.Q. (2019) Effect of different fungicides on Mulberry Sclerotinia sclerotiorum. *Agriculture and Technology*, 39(04): 16–17.

Xu, W.F. (2020) *Study on the Biological Control Effects and Mechanisms of Mulberry Endophytic Bacillus subtilis 7PJ-16 against Mulberry Fruit Sclerotiniosis*. Southwest University.

Xue, Z.M., Zhang, Y.J., Ma, X.B. (2017) Field Efficacy Trial of New Fungicide Tebuconazole to Prevent Mulberry Popcorn Disease. *North Sericulture*, 38(03): 20–23.

Yang, M.F., Sheng, S., Wu, F.A. (2021) Preliminary study on the control effect of several different fungicides on Mulberry Sclerotinia sclerotiorum. *China Sericulture*, 42(01): 13–16.

Ying, X. (2009) Main characteristics and cultivation techniques of fruit mulberry. *Journal of Zhejiang Agricultural Sciences*, (04): 832–833.

Zhao, A.C., Yu, M.D., Hu, W.L. (2017) Techniques for prevention and control of fruit mulberry sorosis disease in spring. *Newsletter of Sericultural Science*, 37(04): 20+39.

*Advances in Renewable Energy and Sustainable
Development – Liang & Kasmani (Eds)
© 2023 Copyright the Author(s), ISBN: 978-1-032-39407-7*

Effects of selenium application concentration on the content of secondary metabolites in dandelion

Zhiguo Zhao, Qi Lu*, Xianglong Meng & Xiangjun Lin
Northeast Forestry University, Harbin, China

ABSTRACT: At present, people have carried out in-depth metabolomics research on selenium stress in plants such as Salvia miltiorrhiza and tartary buckwheat, but the related research in dandelion is rarely reported. In order to further reveal the antibacterial and antiviral mechanism of dandelion, this experiment took dandelion as the research object and cultivated dandelion by foliar application of selenium. Through the detection of flavonoids, polyphenols, and other secondary metabolites in dandelion leaves and Analysis to obtain the effect of different selenium concentrations on plant metabolism. The results show that: (1) Generally speaking, a low concentration of exogenous selenium can promote the production of secondary metabolism in dandelion, while a high concentration of exogenous selenium can inhibit the production of secondary metabolism. (2) Taking the content of secondary metabolites as the response value, the optimum selenium application concentration of soil-cultivated dandelion was 200 μmol/L. (3) The contents of alkaloids, isoflavones, and flavonoids in leaves were significantly higher than those in roots, which indicated that selenium fertilization had a significant effect on the enrichment of secondary metabolites in dandelion. This study provides theoretical support for the research and application of secondary metabolites of Se-enriched dandelion.

1 INTRODUCTION

Plant secondary metabolites mainly include phenolic acids, alkaloids, flavonoids, isoflavones, and other substances. Among them, phenolic acids have allelopathic effects in plants, which are manifested as inhibiting cell division, elongation and submicroscopic structure, changing membrane permeability and inhibiting plant nutrient absorption, plant photosynthesis and respiration, various enzyme functions and activity, plant endogenous hormone synthesis and protein synthesis (Einhellig et al. 2004; Jacob & Sarada 2012). Phenolic acids also have biological activities such as antioxidants (Orsavová et al. 2019; Yadav et al. 2021), scavenging free radicals (Chen et al. 2021; Kong et al. 2021), and are widely used in health care products and medicine. Alkaloid compounds have antibacterial, antiviral, and antitumor effects (Chen et al. 2007; Guo et al. 2017). Isoflavones mainly perform two physiological functions in plants, one is a a phytoprotectant, which is used to resist pathogens and diseases to enhance plant resistance (Bennett et al. 2004). Another function is that of isoflavones as signaling molecules between legumes and rhizobia. Plant isoflavones have active functions such as lowering blood pressure, lowering blood cholesterol, improving the cognitive ability of the human brain, and relieving menopausal syndrome (Bennett et al. 2004). Flavonoids are a class of secondary metabolites formed by plants to resist adverse environments. It is a strong antioxidant, which can prevent the degeneration and aging of cells, prevent the occurrence of cancer, and also has the functions of improving immunity, lowering blood lipids, and lowering blood pressure. Su Min (Su 2017) and other studies found that selenium fertilization had a greater impact on the content of plant secondary metabolites, and also had a significant

---

*Corresponding Author: luqi42700473@126.com

DOI 10.1201/9781003349648-17

impact on plant growth and medicinal value. At present, there are few reports on the metabolism of dandelion selenium stress. In this study, dandelion was used as the research object to explore the effect of different selenium concentrations on the content of secondary metabolites in dandelion, and to provide a theoretical basis for the research and application of selenium-enriched dandelion in secondary metabolism.

2 EXPERIMENT

2.1 *Instruments and reagents*

LC-20A high-performance liquid chromatograph, Promosil C18 column (250 mm × 4.6 mm × 5 μm, Ageia Technologies), UV-1800 UV spectrophotometer, SB-5200DT ultrasonic cleaner, CP213 electronic balance (Ohaus Instruments) Company), TGL-20M high-speed refrigerated centrifuge, multiple test tube shaker (German Heidolph Company).

Drugs and reagents: Rutin standard, Beijing Soleibao Technology Co, Ltd; Gallic acid standard, Chengdu Kelong Chemical Plant; DPPH free radical, Tokyo Chemical Industry; $NaNO_2$, NaOH, $Al(NO_3)_3$, Folin-phenol, Na_2SeO_3, etc are all analytically pure drugs; the experimental water is distilled water; Northeast dandelion.

2.2 *Planting and selenium application of dandelion*

This research started on November 8, 2020, and was carried out in the greenhouse of the Yifu Building, Northeast Forestry University. The dandelion seeds with full grains were selected, and a 2:1 mixture of garden soil and sandy soil was used as the culture medium for pot planting and regular watering. The seeds began to emerge one after another on the 4th day (Leng & Zang 2021). After 53 days of sowing, the seedlings grew well and the body shape was moderate. 100 pots of dandelions with the same growth were selected and divided into 5 groups for selenium treatment. Sodium selenite was used as the selenium source(Yang & Liu 2012)and foliar spraying of selenium was used as the selenium application method(Liu & Liu 2012), and five treatment concentrations were set, namely 0, 50 μmol/L, 100μmol/L, 200 μmol/L, 400 μmol/L, 5 groups of dandelions were treated with selenium. After 18 days, there were obvious differences among the different treatment groups, so the treatment was stopped and sampling was carried out.

2.3 *Determination of total flavonoids*

2.3.1 *Configuration of the solution*

Standard solution: Weigh 20 mg of rutin standard substance, add 60% ethanol to it, dissolve it, and make up to a volume of 0.200 mg·mL $^{-1}$ of rutin standard solution (Li & Hu 2021).

Sample test solution: Weigh 2 g of dandelion dry sample, add 60% ethanol according to the ratio of material to liquid 1:20, reflux in a water bath for 1 h, and then filter; mL; take out 25 mL and make up to 50 mL with distilled water (Li & Hu 2021).

2.3.2 *Drawing of the standard curve*

Rutin standard curve: respectively take 3 mL, 4 mL, 5 mL, 6 mL, 7 mL, and 8 mL of the above standard solution, put them in a 25 mL volumetric flask, add 1 mL of 5% $NaNO_2$ solution to it, and shake up. Let them stand for 6 min; then add 1 mL of 10% $Al(NO_3)_3$ solution, shake well and let stand for 6 min; add 10 mL of NaOH solution to it, make up to the mark with 60% ethanol, mix well and let stand 15 min, measure the absorbance at 500 nm, take the test solution without rutin standard solution as the blank test solution; draw the standard curve with the rutin concentration (c) as the abscissa and the absorbance (A) as the ordinate, and the linearity is obtained. The regression equation is A=10.983c+0.0056, R^2=0.9996, and the linear range is 0.023~0.065 mg·mL $^{-1}$ (Li & Hu 2021).

2.4 Determination of total polyphenol content

2.4.1 Preparation of solution

Standard solution: Weigh 10 mg of gallic acid standard product, add distilled water to dissolve it and dilute to volume to prepare 0.100 mg $\cdot$mL^{-1} gallic acid standard solution.

Sample test solution: Weigh 2 g of dandelion dry sample, add 60% ethanol according to the ratio of material to liquid 1:20, reflux in a water bath for 1 h, and then filter; mL; take out 25 mL and make up to 50 mL with distilled water (Li & Hu 2021).

2.4.2 Drawing of the standard curve

Gallic acid standard curve: measure 0.2 mL, 0.3 mL, 0.4 mL, 0.5 mL, 0.6 mL, and 0.7 mL of the above gallic acid standard solution, put it in a 10 mL volumetric flask, and add 1.5 mL of 15% Na_2 CO to it 3 solutions, 0.5 mL of Folin-phenol reagent, supplemented with distilled water to the mark, water bath at 75°C for 15 min, and then measured the absorbance at 760 nm, the test solution without a gallic acid standard solution was used as the blank test solution, and the concentration of gallic acid was used as the blank test solution. (c)Using the abscissa and the absorbance (A)as the ordinate to draw the standard curve (Lv & Yang 2019), the linear regression equation is A=100.29 c+0.026, R^2=0.9994, and the linear range is 0.002~0.008 mg$\cdot$mL^{-1} (Li & Hu 2021).

2.5 Determination of secondary metabolite content

2.5.1 Pretreatment of samples

The collected dandelion samples were dried at 50°C, pulverized, accurately weighed 0.3 g (accurate to 0.001 g) of dry samples (Hua & Chen 2021), placed in a 10 mL test tube, and added 3 mL of 70% methanol solution(Cao & Yang 2021), Ultrasonic extraction was performed at 40°C for 45 min, centrifuged at 12000 r/min for 10 min, and extracted twice, and the supernatant was collected(Lv & Li 2020). Dry (Lv & Yang 2019), add 1 mL of 70% methanol solution for reconstitution, centrifuge at 12000 r/min for 10 min, and collect the supernatant for use.

2.5.2 Chromatographic conditions

Atlantis PREMIER BEH C18 AX column (100 mm×4.6 mm, 1.7 μm), the sample chamber is set to room temperature; mobile phase A is 10 mmol/L ammonium formate solution, mobile phase B is pure acetonitrile; injection volume is 10 μL, the flow rate was set to 0.3 mL/min, the column temperature was set to 30°C, and the gradient elution program was as follows: 0.0-1.0 min, 99% A; 1.0-3.0 min, 99%-88% A; 3.0-3.5 min, 88%~10% A; 3.5~4.5 min, 10% A; 4.5~5 min, 10%~99% A; 5.0~8.0 min, 99% A, 8 min, the program ends (Huo & Xie 2022).

2.5.3 Mass spectrometry conditions

Performed using an Electrospray Ion (Electronic Spray Ion, ESI) source and selecting the multiple reaction detection scan mode in the negative ion ionization mode. The nitrogen used is high-purity liquid nitrogen, the auxiliary gas is 0.27 MPa, the ion source temperature is 500°C, the ion spray voltage is -4.5 kV, the auxiliary gas pressure is 0.34 MPa, and the atomizing gas pressure is 0.34 Mpa (Huo & Xie 2022).

3 RESULTS AND DISCUSSION

3.1 The effect of different concentrations of selenium on the phenotype of dandelion

Figure 1 shows the effects of different selenium application concentrations on the morphology of dandelion seedlings. Compared with the blank group, when the selenium concentration was lower, the dandelion seedlings showed the phenomena of taller plants, denser leaves, and more developed roots. When the selenium concentration was 400 μmol/L, the dandelion plants showed the phenomenon of leaf shrinkage and dryness, which indicated that the 400 μmol/L selenium treatment had a serious inhibitory effect on the leaf growth of dandelion seedlings.

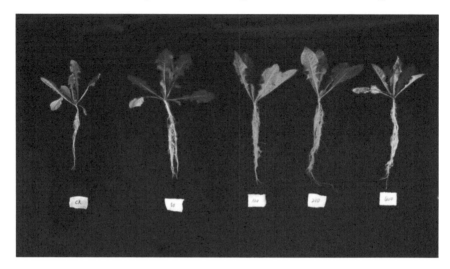

Figure 1. Effects of different selenium concentrations on dandelion phenotype.

3.2 *Analysis of total polyphenol content*

The changes in total polyphenol content in dandelion leaves and roots increased with the increase of selenium concentration and showed a trend of first increasing and then decreasing (Figure 2). The total polyphenol content of leaves in the blank control group was 75.60×10^{-2} mg/g, and the total polyphenol content in the root system was 66.75×10^{-2} mg/g. When the selenium concentration was 200 μmol/L, the total polyphenol content of leaves reached the peak value, and its value was 101.56×10^{-2} mg/g, an increase of 34.34% compared with the blank control group, and the total polyphenol content of the roots reached a peak value of 8 7.56×10^{-2} when the selenium application concentration was 100 μmol/L. mg/g, which was 31.18% higher than that of the blank control group, and the total polyphenol contents of leaves and roots were significantly different between high selenium application concentrations (P< 0.05).

3.3 *Analysis of total flavonoids*

Total flavonoid content of soil cultured dandelion leaves and roots showed a trend of first increasing and then decreasing with the increase of selenium concentration. In the blank control group, the total flavonoid content of leaves was 22.72 mg/g, and the total flavonoid content of roots was 5.477 mg/g. The total flavonoid content of each treatment group was higher than that of the control group. When the selenium application concentration was 50 μmol/L, the total flavonoid content of leaves reached a peak value of 39.36 mg/g, which was 1.732 times that of the blank control group. When the concentration was 100 μmol/L, the content of total flavonoids reached a peak value of 12.87 mg/g, which was 2.351 times that of the blank control group. There were significant differences in the total flavonoid content of roots among different selenium application concentrations (P<0.05). The content of total flavonoids in dandelion leaves was higher than that in roots.

3.4 *Analysis of dandelion secondary metabolites*

In dandelion samples of different treatment groups, 13 phenolic acids, 1 alkaloid, 5 isoflavones, and 14 flavonoids were detected as secondary metabolites. In order to gain an in-depth understanding of

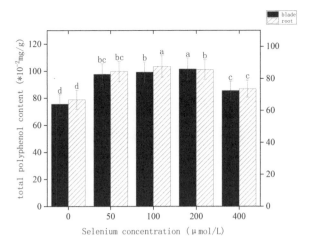

Figure 2. The effect of different selenium concentrations on the total polyphenol content of soil-cultivated dandelion (different letters indicate the correlation analysis of different concentrations in the same part).

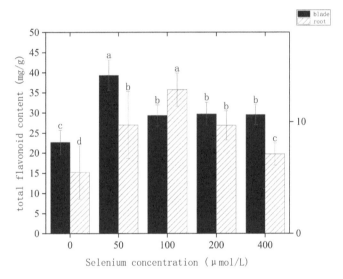

Figure 3. The effect of different selenium concentrations on the total flavonoid content of soil-cultivated dandelion (different letters represent the correlation analysis of different concentrations in the same part).

the metabolic changes caused by selenium fertilization, the OPLS-DA model was used to analyze 33 secondary metabolites of dandelion and screen out differential compounds. The results show that the model parameters of the above-ground part of dandelion $R^2Y=0.946$, $Q^2=0.895$; the model parameters of the underground part of dandelion $R^2Y=0.960$, $Q^2=0.884$, see Figure 4. R^2Y and Q^2 close to 1.0 indicate good accuracy and predictive power of the fitted model.

A total of 9 significantly different secondary metabolites were screened in the dandelion samples by the combined VIP value combined with a t-test (VIP>1, P<0.05) (Figure 5). Compared with the blank group, the contents of phenolic compounds such as ferulic acid, benzoic acid, and p-hydroxycinnamic acid in leaves and roots increased first and then decreased with the increase of selenium application concentration. The content of chlorogenic acid in leaves reached the highest value under the treatment of 200 μmol/L selenium application, and its value was 4.06×10^{-2}

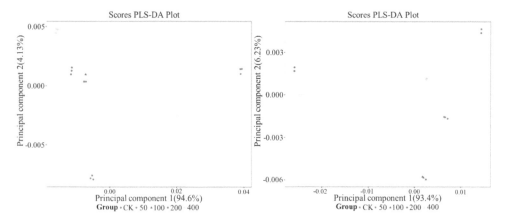

Figure 4. Fractional scatter of the secondary metabolite OPLS-DA in dandelion leaves and roots under different treatments under selenium stress.

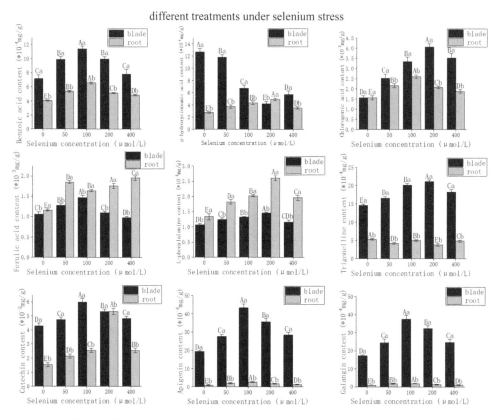

Figure 5. Effects of different selenium concentrations on the content of secondary metabolites in dandelion leaves and roots.

mg/g, the content of chlorogenic acid in roots under the same selenium application concentration was 2.61×10^{-2} mg/g. The contents of phenolic acids such as chlorogenic acid, benzoic acid, and p-hydroxycinnamic acid in leaves were significantly higher than those in roots, which indicated that selenium fertilization had a greater effect on the synthesis of phenolic acids in the aerial parts of dandelion. The content of trigonelline, an alkaloid compound in leaves, increased first and then decreased with the increase of selenium application concentration. The peak value was reached under the treatment with a selenium concentration of μmol/L, and its value was 2 1.10×10^{-4} mg/g. This indicated that the synthesis of alkaloids in the aerial parts of dandelion was greatly promoted under the selenium application concentration. In addition, the contents of alkaloids in leaves were significantly higher than those in roots under different selenium application concentrations. Compared with the blank group, the isoflavone catechin content in leaves and roots also increased first and then decreased with the increase of selenium application concentration. The value is 5.99×10^{-2} mg/g, 200 μmol/L selenium application concentration of chlorogenic acid in roots reached a peak value of 5.33×10^{-2} mg/g. The isoflavone content in leaves is higher than that in roots. The isoflavone content of each treatment group was higher than that of the blank group, indicating that the selenium application promoted the synthesis of isoflavones in dandelion. The contents of flavonoids apigenin and galangin in leaves and roots increased first and then decreased with the increase of selenium application concentration. The highest value was reached under the treatment with selenium concentrations of μmol/L, which were 4 3.3×10^{-4} mg/g and 37.4×10^{-4} mg/g. The contents of flavonoids in leaves were significantly higher than those in roots, indicating that flavonoids were mainly concentrated in dandelion leaves.

4 DISCUSSION

Chlorogenic acid and L-phenylalanine increased in dandelion after selenium fertilization. When the concentration of selenium was 200 μmol/L, the content increased to the highest value, which indicated that the anti-oxidative and anti-disease functions of dandelion could be significantly improved when the concentration of selenium was 200 μmol/L. The contents of alkaloids trigonelline in leaves and roots showed opposite trends, which may be because selenium affected the expression of related proteins in dandelion leaves and roots, and then affected the synthesis of alkaloids in different parts. The contents of isoflavones and flavonoids catechin, apigenin, and galangin in the selenium application were significantly higher than those in the blank group, indicating that the selenium application in dandelion significantly improved the efficacy of resisting pathogens and diseases and improving immunity. In this study, the flavonoids apigenin and galangin were mainly distributed in the leaves, indicating that the dandelion leaves treated with selenium had better functions in lowering blood pressure and blood cholesterol. In addition, selenium application could significantly increase the total polyphenol content of dandelion, and the total polyphenol content of dandelion reached the highest value under the condition of selenium application of 200 μmol/L. Under the selenium fertilization treatment, the total flavonoid content increased significantly, and the total flavonoid content in leaves was significantly higher than that in roots, indicating that selenium fertilization had a higher promoting effect on the accumulation of flavonoids in leaves.

5 CONCLUSION

In this paper, the content of secondary metabolites in dandelion under different selenium application concentrations was studied by liquid chromatography-mass spectrometry. The main conclusions can be summarized as follows: (1) The 50 μmol/L, 100 μmol/L, and 200 μmol/L selenium fertilization concentrations showed the effect of promoting the metabolism of dandelion, while the 400 μmol/L selenium application concentration showed the effect of inhibiting the metabolism of dandelion. (2) Taking the content of secondary metabolites as the response value, the optimum selenium application concentration of soil-cultivated dandelion was 200 μmol/L. (3) The contents of alkaloids, isoflavones, and flavonoids in leaves were significantly higher than those in roots, which indicated that selenium fertilization had a significant effect on the enrichment of secondary metabolites in dandelion. In future work, the research on the regulation mechanism of selenium in the secondary metabolism process should be carried out, and the research on the theory of selenium stress should be strengthened.

REFERENCES

Bennett Jo, Yu O, Heatherly Lg, et al. Accumulation of genistein and daidzein, soybean isoflavones implicated in promoting human health, is significantly elevated by irrigation[J]. *Journal of Agricultural and Food Chemistry*, 2004, 52(25): 7574–7579.

Cao Puqiong, Yang Shenying. Simultaneous determination of chlorogenic acid, ferulic acid, and violetine in Shujinhuoluo pills by high-performance liquid chromatography [J]. *Yunnan Chemical Industry*, 2021(48): 07.

Chen Miaojia, Ma Siming, Li Yang, et al. Synthesis of salvianolic acid A analogues with resveratrol as structural unit and determination of their antioxidant activity in vitro [J]. *Zhongnan Journal of Medical Sciences*, 2021, 49(4): 402–405. DOI: 10.15972/j.cnki.43-1509/r.2021.04.007.

Chen Sibao, Wang Liwei, Yang Junshan, et al. The chemical constituents and resource utilization of medicinal plants in Aquilegia [J]. *Chinese Herbal Medicine*, 2001, 32(11): 3–5.

Einhellig FA, Galindo JCG, Molinillo JMG, et al. Mode of allelochemical action of phenolic compounds. In: Allelopathy: Chemistry and Mode of Action of Allelochemicals [M]. CRC Press: Boca Raton, FL, USA. 2004: 217–238.

Guo CH, Zhu XF, Duan YX, et al. Suppression of different soybean isoflavones on Heterodera glycines. *Chinese Journal of Oil Crop Sciences*, 2017, 39(4): 540–545.

Guo Chunhong, Zhu Xiaofeng, Duan Yuxi, etc. Study on isoflavone species inhibiting soybean cyst nematode. *Chinese Journal of Oil Crops*, 2017, 39(4): 540–545.

Guo, Zhu, Duan, et al, 2017 Gong Hongdong. Aquilegia in Gansu Research on medicinal plant resources[J]. *Gansu Science and Technology*, 2007, 23(7):217–218.

Hua Zhenggang, Chen Xi. Rapid determination of four toxic components in Tripterygium wilfordii poisoning samples by ultra-high performance liquid chromatography-tandem mass spectrometry [J]. *Chemical Analysis and Metrology*, 2021, (30): 10

Huo Yingyi, Xie Muxiding·Mairepati. Simultaneous and rapid determination of four pyridine nucleotide coenzymes in cells by ultra-high performance liquid chromatography-mass spectrometry [J]. *Applied Chemistry*, 2022(39):02.

JACOB J, SARADA S. Role of phenolics in allelopathic interactions[J]. *Allelopathy Journal*, 2012(29): 215–230.

Kong J, Xie Y, Yu H, et al. Synergistic antifungal mechanism of thymol and salicylic acid on Fusarium solani[J]. *LWT- Food Science and Technology*, 2021, 140: 110787. DOI: 10.1016/j.lwt .2020.110787.

Leng Kai, Zang Chuanjun. Biological characteristics and planting technology of dandelion in Shuangliao City [J]. *Modern Agricultural Science and Technology*, 2021(05): 78–79.

Li Na, Hu Yueyue. Determination of total flavonoids and total polyphenols in sea buckthorn from different origins and their antioxidant activities [J]. *Chemical and Biological Engineering*, 2021(18): 08.

Liu Chunju, Liu Fuguo. Effects of selenium foliar spraying on its accumulation in fresh corn[J]. *Jiangsu Journal of Agricultural Sciences*, 2012(4): 713–716.

Lv Guanghui, Li Xingyu. Simultaneous determination of chlorogenic acid, isochlorogenic acid A and luteolin in Chinese cabbage by high-performance liquid chromatography [J]. *Medicine Herald*, 2020(39): 07.

Lv Ming, Yang Lingxiao. Simultaneous determination of chlorogenic acid and caffeic acid in Escherichia coli by high-performance liquid chromatography [J]. *Zhongnan Pharmacy*, 2019(17): 07.

Orsavová J, Hlaváčov Á I, Mlček J, et al. Contribution of phenolic compounds, ascorbic acid, and vitamin E to the antioxidant activity of currant (Ribes L.) and gooseberry (Ribes uva-crispa L.) fruits[J]. *Food Chemistry*, 2019, 284: 323–333. DOI: 10.1016/j.foodchem.2019.01.072.

Su Min. Effects of selenium enrichment on the content of secondary metabolites in Salvia miltiorrhiza and research on polysaccharide extraction [D]. Chengdu University of Technology, 2017.

Yadav Mp, Kaur A, Singh B, et al. Extraction and characterization of lipids and phenolic compounds from the brans of different wheat varieties[J]. *Food Hydrocolloids*, 2021, 117: 106734. DOI: 10.1016/j.foodhyd.2021.106734.

Yang Bo, Liu Zhikui. Effects of cultivation modes on the performance of selenium-enriched Candida utilis[J]. *Chinese Journal of Bioprocess Engineering*, 2012(4): 7-11

Advances in Renewable Energy and Sustainable Development – Liang & Kasmani (Eds)
© 2023 Copyright the Author(s), ISBN: 978-1-032-39407-7

Improvement of EFB palm fiber pelletizing with castor meal

Jer-Yuan Shiu*
Hubei Key Laboratory of Mine Environmental Pollution Control & Remediation, Huangshi, Hubei, China
Hubei Polytechnic University, School of Environmental Science & Engineering, Huangshi, Hubei, China

Xiang Wang
College of Environment and Safety Engineering, Fuzhou University, Fuzhou, Fujian, China

Chih-Hung Wu
School of Resource & Chemical Engineering, Sanming University, Sanming, Fujian, China
Cleaner Production Technology Engineering Research Center of Fujian Universities, Fujian, Sanming, Fujian, China

ABSTRACT: One of the most promising alternatives for fossil fuel replacement is biofuels derived from different types of biomass. Biofuels are usually molded under heat and pressure with the pelletizer to form dense pellets for industrial and residential applications. The needed temperature and pressure of pelletizing process are the two factors for the pellet quality and economic efficiency and are determined by the density of the raw biomass material used. High-density materials, such as the Empty Fruit Bunch (EFB) palm fiber, require more heat and pressure to form the pellets due to the lacking of lignin content; low-density materials, such as starches and proteins, are difficult to form durable pellets because they do not have enough fiber content to support the structure. However, if these two kinds of biomass are properly blended, they can create complementary effects, which can enhance energy-efficiency of the pelletizing process, and produce premium EFB palm fiber pellets with excellent quality. In this study, low-density castor meal is added to high-density EFB palm fiber to verify the effects in the pelletizing process. With the protein richness and high oil content of castor meal, the resistance in the pelletizer is reduced which also increases the production rate, moreover, the EFB pellets produced are durable and shiny. The data indicate that adding 15% castor meal to raw palm fiber materials can reduce energy consumption by up to 25%, while the mechanical properties of pellets (including durability, ash content, bulk density, and calorific value) meet all the requirements of the commercial specification.

1 INTRODUCTION

To pursue the goal of carbon neutrality by 2060, the replacement of fossil fuels with other renewable fuels is undergoing recently. Among all carbon-neutral fuels available today, biofuel is one of the most promising alternatives (Decicco J et al. 2016). Biofuels are a sustainable form of energy derived from the harvesting and processing of different types of biomass, including organic wastes, charcoal, wood, fishery, and agricultural products. Burning biofuels can have a net-zero carbon impact on the environment (Mathews 2008).

Moreover, because of the high surface per unit volume of fine particles, surface forces like Van der Waals' play an important role in creating bonds between particles. The formed bonds and charged particles lead to blockage of storage and handling equipment. As a result, microalgae particles need to be agglomerated to larger particles before processing. Pelleting improves fuel

---

*Corresponding Author: frankshiu@qq.com

DOI 10.1201/9781003349648-18

quality by increasing the density and creating agglomerated structures (Kenny & Opalicki 1996). The high calorific value of pellets compared to raw biomass and their homogenous structure make pellets valuable fuel to be used in industrial and residential applications. Their elevated mechanical stiffness will reduce handling, storage, and transportation costs. Pelletization is carried out by forcing the particle toward each other via applying a mechanical force to produce homogeneous pellets (Hosseinizand et al. 2018).

The pelletizing process is a very important key when converting biomass feedstock into biofuel resources. The traditional pelletizing process can be divided into three conditions: (A) normal temperature and high pressure; (B) high-temperature medium pressure; (C) low pressure with adhesive. Pelletizing process under high pressure usually requires large energy consumption and has low economic efficiency. By adding the appropriate amount of adhesive, the raw material can be well bonded with higher compressive strength and cohesion (Grover & Asia 1996; Silva lora et al. 2011).

For premium EFB pellets, not only the qualities of the raw material quality (such as calorific value, ash content, etc.) must reach a certain level, but also the pellets must meet all the commercial specifications, such as durability, ash content, bulk density, combustion performance, etc. The main function of the additive is to enhance the pellet properties of the biomass fuel, it can be an adhesive that bonds the raw materials together and produces smooth and luster pellets without cracking as in Figure 1. In addition to those quality adjustments, the additive can also be a lubricant that allows the pelletizing process to be carried out more smoothly and improves the efficiency of the pelletizing process (Yao et al. 1997). At present, thermoplastic is often used as pelletizing binder for both mechanical and economical reasons which are not acceptable from the environmental point of view. The premium pellets must be consistent with pure natural biomass only, which is a big challenge for the current pelletizing industry (Bellatrache et al. 2020; Marwah et al. 2016).

Currently, palm husk is already an important biomass fuel for industrial boilers, and its demand is increasing, which causes the purchase cost to increase year by year; besides this valuable byproduct, the majority of biomass in the palm oil industry, such as Empty Fruit Brunch (EFB) and residue, has been discarded. If these biomass resources can also be utilized as biofuel, it will be greatly helpful for both energy conservation and environmental benefits. The data from the real trial and experimental analysis shows that the pellets of EFB palm fiber have a calorific value of about 2600~4000 kcal/kg (Nasrin et al. 2015). However, in the pelletizing process of EFB pellets, more heat and pressure are needed than other regular raw materials, for example, wood and straw, serving as the EFB raw material with insufficient lignin (~15%) will be more energy intensive and uneconomical. In order to improve such disadvantages, it is suggested that oil-rich substances, such as castor meal, may be added to the palm fiber raw material to smoothen the pelletizing process and the qualities of outcomes (Yang 2018).

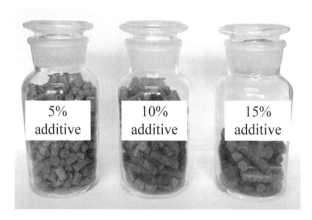

Figure 1. EFB pellets with different amounts of the addition of castor meal.

Castor has received a lot of attention among all other raw materials that can be converted into biomass because of its vitality, its ability to grow on barren land, and its excellent ability to conserve water and soil. Castor ripens in 6 months and can be harvested two to three times a year. Castor is currently one of the world's top ten oil crops, its main distribution areas include Africa, South America, Asia, Europe, India, and China, and Brazil castor seed production accounted for 81.2% of the global total. Castor seeds are oval in shape and divided into two parts: the core and the shell. Grains account for about 70%~75% of the total weight of seeds. Castor seeds are rich in oil and have thus been referred to as "green renewable petroleum." Castor oil can be extracted under pressure, with or without heating, or by using solvents (Ogunniyi 2006).

Castor oil is an important industrial raw material, and the international market is mainly concentrated in North America, Europe, and China. The residue left behind after extruding oil from castor seeds is called castor meal, which is a rich biomass raw material. Castor oil is usually obtained by squeezing castor beans, a process that can be done by cold pressing or hot pressing. By cold pressing, castor oil can be obtained without steam heating. In this way, the resulting castor oil is not exposed to high temperatures, so its medicinal properties are preserved. In contrast, by hot pressing, castor oil is obtained by steam heating. A slow heat treatment process reduces the viscosity of residual castor oil, allowing the extraction of oil that cannot be obtained by cold pressing alone. However, conventional squeezing methods have some limitations, therefore castor meal retains a considerable amount of oil (Chen et al. 2017).

Castor meal is the residue of castor oil production. It is rich in biomass up to 35%, which is three times that of general food crops. It is a high-quality high-protein feed. Untreated cakes, containing 7.59% total nitrogen, 2.85% total phosphorus, and 0.79% total potassium, are good fertilizers and fuels, causing castor to be used as fertilizer for a long time. According to modern nutrition analysis, Castor meal is a high-protein food containing about 35% of high-quality protein, 1.2% of fat, 14% of sugar, crude fiber, and calcium and phosphorus. Because of less than 20% lignin content, it is not easy to bond raw palm fiber into pellets. The addition of castor meal can increase the protein and oil content in feedstock, which will promote the bonding strength within the raw material, and improve the operation of its pelletizing process; The lipid content in castor meal acts as a natural lubricant to reduce friction during the expulsion of pellets from the die, and the protein content can be used as a secondary binder to improve the mechanical properties of co-pelletized pellets. (Wang 2019).

However, during the composition adjustment, partial high-density EFB palm fiber was replaced by low-density castor meal, this could cause the density of the pellet to decrease and moisture content to increase slightly, but on the contrary, pellet durability is improved due to the extra binding effect of castor meal. Based on Pellet Fuels Institutes (PFI) Standards, premium pellets shall meet the several commercial standards as in following Table 1:

Table 1. PFI defined fuel grade requirements.

Parameter	Range (unit)
Bulk Density	≥ 0.77 kg/m^3
Diameter	5.84–7.25 mm
Durability	≥ 96.5
Fines	$\leq 0.5\%$
Ash Content	$\leq 1\%$
Length	$\leq 1\% > 1.5$in
Moisture	$\leq 8.0\%$
Chlorides	≤ 300 ppm

The amount of additives is related to the performance of EFB palm fiber pellets, which refers to the density of the particles and the quality of the pellet formation. The denser the particles, the stronger the structure (Harun et al. 2016). The advantage is that the particles are more able to withstand transportation and work in the pellet burner. When squeezed out from the pelletizer under heat and pressure, the premium pellets should have a smooth surface with minor cracks on the surface (Monica regina et al. 2018). If the pellets are easily broken, it means that the quality needs improvement (Blok et al. 2018).

When pure EFB palm fiber is used for energy applications, it could be pelletized in order to increase the bulk density, under high pressure and high temperature. Theoretically, adding oily biomass can improve pellets' properties (e.g., density and durability) and also decrease pelletizing energy. This can improve the economics of palm fiber pelletization. Hosseinizand et al.'s research used dried fine powder microalgae Chlorella Vulgaris, as a natural binding agent for pine sawdust to produce solid fuels. The results of compressing using a mounting press indicated that discs containing > 20% microalgae had better strength compared to pine sawdust discs (Hosseinizand et al. 2018). The objective of this research was to understand the effects of co-pelletization of EFB palm fiber and castor meal and to evaluate the properties of the pelleted fuel. In this study, the effects of different ratios of castor meal on EFB palm fiber pellets and the operating parameters of pelletizing process were investigated to find the optimal operation process of premium EFB pellets, which might be used as a reference for large-scale pelletizing production.

2 MATERIALS AND METHODS

2.1 *EFB*

Palm fiber was provided by Dingpu Industrial Co., Ltd. in Taiwan. Palm fiber was air-dried and shredded into 20~25 mm in length using a knife mill and was stored in sealed plastic bags to maintain a stable moisture content of around 7.0%. Castor meal was provided by Changfeng Organic Agricultural Fertilizer Co., Ltd. in Taiwan, castor meal was in powder, ingredients: total nitrogen 7.0%, total phosphoric anhydride 2.5%, total potassium oxide 1.5%, organic matter 80.0%. The working procedure of this experiment is shown in Figure 2.

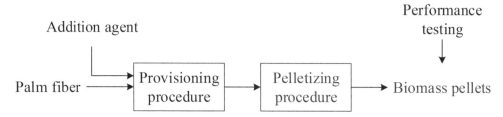

Figure 2. Experimental flowchart for EFB pelletizing process improvement.

2.2 *Feedstock preparation*

The palm fiber after pretreatment (shredding /sieving /drying) was mixed with different amounts of castor meal to prepare feedstock for pelletizing procedure. The experiments were conducted with different ratio (0%, 5%, 10%, 15%, and 20% w/w) of castor meal adding into EFB palm fiber. Since the consistency of the raw materials feedstock is critical to the stability and efficiency of the pelletizing process. These two materials were well mixed with a batch mixer to get homogeneous feedstock for pelleting process next.

2.3 *Pelletizing procedure*

The prepared feedstock was then conveyed into the pelletizing process. A ring die pelletizer was used in this study, and the raw material was taken from the sump bin via a regulator above the pelletizer, the raw material is fed into the inside of the ring mold by a screw auger. The theory of ring die pelletizer mainly are internal roller fixing and ring die driving. The operation mode is similar to that of the drum-type washing machine. Once the raw material is put into place, the inner roller will press the raw material by the ring die (Luo 2017). This study mainly explores the effects of different addition amounts of castor meal on the energy consumption of the pelletizer, the molding temperature, and the properties of EFB pellets(Xiao & He 2017).

Ideally, the horsepower required to produce the particles is the energy required to compress the biomass plus the energy required to extrude the particles from the mold, plus heat loss which is absorbed by the pellets. In an estimated manner, the heat loss is neglected here. Thus, specific energy is the total input energy (=horsepower x time) divided by the mass of the particle. If the operating time and production rate (1 ton per hour) are both constants, then the energy consumption is proportional to the horsepower required.

2.4 *Analysis methods*

The analysis of the EFB pellet's performance includes Submerging (Density) test, Buck Density test, Durability test, and Combustion test. In order to obtain the accuracy and credibility of the data, each set of experiments was repeated 3 times, and the average line was drawn as a representative:

(A) Submerging (Density) Test – Based on past experience, the easiest way to test the quality of EFB pellets is to place the particles in a glass of water. If the particles sink to the bottom of the cup, it has a high density, which is formed under sufficient pressure. If the particles float, it means that the quality is poor because of their lower density, lower mechanical durability, and more prone to chipping and fines.

(B) Bulk Density – Fill a liter container with the pellets and measure their net weight. In the case of high-quality pellets, the result should be between 600 and 700 g/L. This value can also be called the bulk density of the EFB pellets and is a key indicator of the pellets produced under the correct pressure. Generally, those pellets with a specific gravity below 0.6 can be easily broken or pulverized and produce excessive fines, which is not acceptable in a real application.

(C) Durability - Simulates the fragmentation of the pellets caused by crushing during shipping. The durability means that the integrity of the pellets is easily maintained during transportation, and the fragmentation loss is smaller. The test method is based on ASAES 269.4. The mass of particles remaining on the sieve was recorded, and the pellet durability was calculated from Eq. 1. (Hosseinizand et al. 2018).

$$\text{Durability (\%)} = \frac{\text{Mass left on the screen}}{\text{Initial Mass}} \times 100 \tag{1}$$

(D) Combustion test - the most important specification of commercially available EFB pellets: calorific value and ash, which are the basis for judging whether the performance is superior or not, and of course also related to economic benefits. The test methods are based on ASTM D3174-12 (heat value) and ASTM D5865-13 (ash) (Olivier 2015).

The ring dies pelletizer used in this test has a horsepower of 150 Hp and a production capacity of up to 1 ton per hour. The feed is made of palm fiber and added with castor meal as raw material, and the ratios added are 0%, 5%, 10%, 15%, and 20% w/w respectively. The EFB pellets produced by the ring die pelletizer have a uniform size of $\phi 8$ mm and 2 cm long. The length of the pelletizer is not the focus of production, but if the particles are too long, for example, more than 2.5 cm, it may cause obstruction or damage to the auger in the pellet burner (Li & Zhang 2017).

3 RESULTS AND DISCUSSIONS

3.1 *Energy consumption*

The main function of the additive used in this experiment is to increase the smoothness of pelletizing in order to reduce the energy consumption during the pelletizing process. The effect of castor meal on the energy consumption of pelletizer under different addition amounts (0%, 5%, 10%, 15%, 20%), the results are shown in Figure 3. From the results, it was found that the addition of castor meal can reduce the energy consumption during the pelletizing process; when the amount of castor meal added is 5%, there is a significant decrease.

When the amount increased to 10%, the energy consumption is only 92% of the blank (without additive), and the output power from 120 Hp drops to 105 Hp; as the amount of additive continuously increased to 15%, the energy consumption gradually decreases to 75% of the blank, output power down to 90 Hp, the change is significant. Thereafter, as the addition gradually increases to 20%, the decrease of output power becomes less and less.

The data indicates that using castor meal as a lubricant can be energy-saving during pelletizing by deducing the friction, which might also cause the properties of EFB pellets to deteriorate. To fulfill energy saving and all commercial quality requirements, the optimum amount of additives need to be determined through the following experimental results.

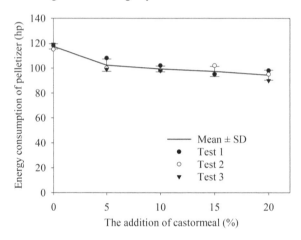

Figure 3. Energy consumption of pelletizer V.S. The addition of castor meal.

3.2 *Pelletizing temperature*

In addition, during the pelletizing process, the temperature rises due to the continuous extrusion of the raw materials by the pelletizer, and the temperature can be as high as 80°C or higher. In general, the extreme temperature has an adverse effect on pelletizing. According to the wood pellet processing reference, 80°C is the preferred pelletizing temperature.

Therefore, palm fiber used in this experiment is pelletized at around 60°C to 80°C when no additives are added, and the temperature is reduced to about 55°C when a 15% castor meal is added. The results similarly indicate the optimum value is reached at 15 (Figure 4).

3.3 *Properties of EFB pellets*

Following illustrate the effects on various physical properties of solidified EFB pellets under different amounts of castor meal added. First is the effect on the density change of EFB pellets. From the test results, it is found that the EFB pellets prepared under different addition amounts all have a density of more than 1, about 1.1, indicating that the addition of Castor meal has an impact on EFB pellets.

Next, the influence on the bulk-specific gravity of the EFB pellets was studied. It was found that the addition of castor meal did not significantly affect the bulk specific gravity of the molded EFB pellets, and the value was about 550 g/l. Similarly, the measured fastness was all around 98% or more regardless of the amount of addition, which indicates the product quality was very good.

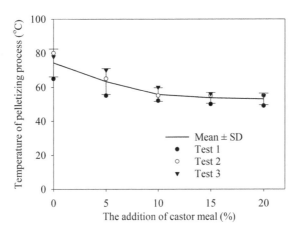

Figure 4. The temperature of pelletizing process V.S. The addition of castor meal.

Finally, the effect on the calorific value of the EFB pellets was analyzed. The pure biomass pellets by EFB with different amount of castor meal was found to have only a slight change in calorific value, about 3800 kcal/kg. The reason might be that the caloric value of EFB and castor meal are very close, so the overall calorific value of final products shows no significant difference.

Moisture and ash are also factors for evaluating the performance of EFB pellets. From the moisture test results, it was found that the palm fiber pellets obtained by increasing ratios (0%~20%) of castor meal have water content all between 9%~10%, slightly increase but not significantly. Although it is necessary to maintain a certain proportion of water during the pelletizing process (more than 12%), the less water in the finished pellets, the more energy the burner can use, because the water will absorb a lot of heat when vaporized. The heat of vaporization (539 cal/g) reduces its calorific value. Therefore, the moisture content of the finished pellets should be less than 10%, otherwise, the energy efficiency will be greatly reduced.

For ash content, the so-called "Premium pellets" in the general market mean that the pellets contain a very low percentage of ash after incineration, basically, the lower ash content the better pellet quality. As in Table 2, the ash content of the blank palm fiber pellets used in this experiment is less than 3%, and those pellets with different ratios of castor meal added are all less than 3%. The reason is due to the ash content of both raw biomass materials (palm fiber and castor meal) all being less than 3%, and no other non-biomass materials are added during the pelletizing process, this is one of the major advantages of pure biomass biofuel.

Table 2. The calorific value of pellets with different addition of castor meal.

Addition of castor meal (%)	Calorific value (Kcal/Kg)
0	3794.3 ± 88.79
5	3781.3 ± 99.25
10	3772.0 ± 80.73
15	3767.7 ± 55.81
20	3658.0 ± 78.37

Calorific value is also the major parameter of premium EFB pellets, the higher the calorific value the better the energy value. The pure palm fiber pellet has a calorific value of ~3800 Kcal/kg which is better than most biomass pellets, but the energy-intensive pelletizing process also reduces the total energy efficiency; thus, the castor meal was added under controlled ratios for improvement in this study.

Based on the TGA results, Table 2, the pellets obtained by increasing ratios (0%~15%) of castor meal have calorific values slightly decrease from 3795 Kcal/kg to 3768 Kcal/kg, but not obvious; as the adding ratio increases to 20%, the calorific value is reduced more significantly. Therefore, considering both pellets' quality and pelletization mechanism, the most appropriate adding ratio of castor meal with palm fiber is 15%, which will reduce energy consumption up to 25% (the output power from 120 Hp drops to 105 Hp), while the mechanical properties of pellets (include durability, ash content, bulk density, calorific value, etc.) meet all the requirements of the commercial specification.

The results of this study can be used as a reference for large-scale co-pelletization of EFB palm fiber and castor powder. In industrial production, raw biomass composition with an appropriate mixing ratio of high-density material (palm fiber) and low-density material (castor meal) has a great impact on the pelletizing process economics. The results of experiments demonstrated that adding 15% castor meal to palm fiber can fulfill the commercial standers and improve energy efficiency. Furthermore, adding castor meal to EFB palm fiber in large-scale operations causes pelletization to occur at lower applied horsepower and temperatures compared to pure palm fiber biomass. Moreover, castor meal can be used as auxiliary biomass to approach the standards of commercial pellets (durability and density). Because of the high calorific value of castor meals, less biomass is needed to be used for energy applications, which will lead to fewer environmental impacts.

4 CONCLUSIONS

In this study, the effects of castor meal addition on EFB palm fiber pellets quality and pelletization mechanism are investigated. From the experimental results, it is found that the adding castor meal to the pure palm fiber pelletizing process has certain tradeoff effects between pellet quality and energy efficiency; as the amount of castor meal added increases, the energy efficiency did get improved (lower applied horsepower and lower operating temperatures), but certain qualities of pellets (such as calorific value and bulk density) also gradually decreasing. However, by controlling castor meals at an appropriate ratio, the energy efficiency can reach significant improvement and commercial criteria of premium pellets can also be fulfilled. The following outcomes are obtained:

1. Two abundant biowastes with high-density EFB palm fiber and low-density castor meal, are difficult to be reused as energy resources individually; however, after mixed in proper ratio and be pelletized, premium RDF pellets can be produced, which not only meet all the commercial specifications but also have high Calorific value around 3800 kcal/kg.
2. The energy efficiency of the palm fiber pelletizing process is using lipid in castor meal as a natural lubricant to reduce friction during the expulsion of pellets from the die, and the protein content can be used as a secondary binder to improve the mechanical properties of co-pelletized pellets. EFB palm fiber is generally pelletized at 80°C and the output power at 120 Hp when no additives are added, but when 15% castor meal is added, the working temperature is reduced to about 55°C and output power drops to 105 Hp, and the energy consumption of pelletizing process will reduce up to 25%. The improvement in energy efficiency is significant.
3. During the composition adjustment, partial high-density EFB palm fiber was replaced by low-density castor meal, this could cause the density of the pellet to decrease and moisture content to increase slightly, but on the contrary, pellet durability is improved due to the extra binding effect of castor meal.
4. The ash content of the blank palm fiber pellets used in this experiment is less than 3%, and those pellets with different ratios of castor meal added are all less than 3%. The reason is that the ash content of both raw biomass materials (palm fiber and castor meal) is less than 3%, and

no other non-biomass materials are added during the pelletizing process, this is one of the major advantages of pure biomass biofuel.

5. Moisture and ash are also factors for evaluating the performance of EFB pellets. From the moisture test results, it was found that the palm fiber pellets obtained by increasing ratios (0%~20%) of castor meal have water content all between 9%~10%, slightly increase but not significantly.

ACKNOWLEDGMENTS

Financial support for this project by the Introduced High-level Talent Research Start-up funds of Hubei Polytechnic University (22xjz17R) is gratefully acknowledged.

REFERENCES

Bellatrache Y, Ziyani L, Dony A, et al. Effects of the addition of date palm fibers on the physical, rheological and thermal properties of bitumen [J]. *Construction and Building Materials*, 2020, 239: 117808.

Blok L G, Longana M L, Yu H, et al. An investigation into 3D printing of fiber reinforced thermoplastic composites [J]. *Additive Manufacturing*, 2018, 22: 176–86.

Chen G-B, Li Y-H, Chen G-L, et al. Effects of catalysts on pyrolysis of castor meal [J]. *Energy*, 2017, 119: 1–9.

Decicco J M, Liu D Y, Heo J, et al. Carbon balance effects of U.S. biofuel production and use [J]. *Climatic Change*, 2016, 138(3): 667–80.

Mathews J A. Carbon-negative biofuels [J]. *Energ Policy*, 2008, 36(3): 940–5.

Grover P D, Asia R W E D P I. Biomass Briquetting: Technology and Practices [M]. *Food and Agriculture Organization of the United Nations*, 1996.

Harun M Y, Che Yunus M A, Ismail M H S, et al. A comparative investigation on the effect of thermal treatments on the mechanical properties of oil palm fruitlet components [J]. *J Taiwan Inst Chem Eng*, 2016, 60: 582–7.

Hosseinizand H, Sokhansanj S, Lim C J. Co-pelletization of microalgae Chlorella Vulgaris and pine sawdust to produce solid fuels [J]. *Fuel Process Technol*, 2018, 177: 129–39.

Kenny J M, Opalicki M. Processing of short fiber/thermosetting matrix composites [J]. *Composites Part A: Applied Science and Manufacturing*, 1996, 27(3): 229–40.

Li T Y, Zhang Z. Mod. Manuf. Technol. Equip., 1 (2017) 155–156. [J]. *Mod Manuf Technol Equip*, 2017: 155–6.

Luo W. Analysis on the selection of extrusion granulator [J]. *Mod Manuf Technol Equip*, 2017, 10: 114–6.

Marwah O M F, Halim N F A, Shukri M S, et al. A study on palm fiber reinforces as a filament in portable FDM [J]. *ARPN Journal of Engineering and Applied Sciences*, 2016, 11(12): 7828–34.

Monica Regina G, Antonio Shigueaki T, Marilia B. <b> Production of self-healing asphalt with steel short fibers and microwave heating: a pilot study [J]. *Acta Scientiarum Technology*, 2018, 40(1).

Nasrin A B, Choo Y M, Joseph L, et al. *Improved process for the production of low-ash empty fruit bunch pellet* [M]. MPOB Information Series. 2015: 1511–7871.

Ogunniyi D S. Castor oil: A vital industrial raw material [J]. *Bioresour Technol*, 2006, 97(9): 1086–91.

Olivier P. Analysis of the technical obstacles related to the production and utilization of fuel pellets made from agricultural residues [J]. *Pellets for Europe project*, EUBIA, 2015.

Silva Lora E E, Escobar Palacio J C, Rocha M H, et al. Issues to consider, existing tools and constraints in biofuels sustainability assessments [J]. *Energy*, 2011, 36(4): 2097–110.

Wang P F. Extrusion pelletizing process to produce special compound fertilizer [J]. *Yunnan Chem Ind*, 2019, 46(3): 26–31.

Xiao D X, He L X. Structural characteristics and quality control of large-scale twin-screw extrusion granulator [J]. *Chem Equip Pipeline*, 2017, 54(3): 33–64.

Yang Y. *Research on heat transfer and temperature control of extrusion granulator barrel* [D]; Dalian University of Technology, 2018.

Yao K Y, Zhao C Y, Wang Z M, et al. Study on humic acid pelletizing binder [J]. *J Harbin Univ Sci Technol*, 1997, 2: 124–6.

Advances in Renewable Energy and Sustainable
Development – Liang & Kasmani (Eds)
© 2023 Copyright the Author(s), ISBN: 978-1-032-39407-7

Research on the sublimation of distiller's grain quantum technology peptide sublimation and the application of livestock feed technology and market management and sales

Huai Lu*

Department of Industrial Management, Yang-En University, Quanzhou, China

ABSTRACT: Kinmen Gaoliang Wine Co., Ltd., with an annual turnover of more than 10 billion yuan, has begun to devour the ecology of the surrounding waters of Kinmen because of the residue and sewage after traditional brewing. There are two main sources of environmental pollution caused by enterprises, one is production wastewater, and the other is distiller grains. Disasters brought about by global climate change in recent years have also greatly affected the existing interdependence between man and land. Therefore, establishing a new production system such as recycling is a key measure to avoid exceeding the carrying capacity of the earth's environment and move towards sustainable development. Green supply chains are another important development in tackling the environmental problems caused by waste. The purpose of this experiment is to study the transformation of waste distiller grains in the production of quantum technology peptide feed to produce various probiotics, to study the effect of intestinal bacteria on the growth performance and immunity of livestock, and to explore bacteria or substances. The purpose of this study is to address the issue of waste distiller's grains produced by wineries, to address the problem of secondary environmental pollution, and to respond to the government's circular economy policy in achieving the goal of waste reduction, while also promoting economic effects to address economic growth and ecological environmental protection.

1 INTRODUCTION

In the early days, the Kinmen Sorghum Company dumped all the lees into the sea. Under the pressure of public opinion, the daily output of sorghum distiller grains in Kinmen Winery is as high as 300 tons. The troughs taken by local farmers in Kinmen are mostly used for livestock feed, and some are used as agricultural fertilizers. In recent years, Taiwanese wineries have also produced a large number of distiller grains for market use. In the case of oversupply, the price of distiller's grains has declined, resulting in relatively high cleaning and transportation costs of distiller's grains, lower willingness to bid by manufacturers, and slower processing efficiency of distiller's grains than in the past, making gin and distiller's grains "invisible bombs" that pollute Kinmen. Farmers in the Kinmen area bring distiller grains back to compost. They don't use closed composting methods. They found a place in their farmland, using distiller grains as the base material, and then mixing chicken manure, vegetable leaves, kitchen waste, and other things into it. The "smell" emanating from the fermentation process can be carried around by the wind, and pollution can also enter the air.

---

*Corresponding Author: ttpadc@gmail.com

2 EXPERIMENTAL WORK

2.1 *Research technology theory*

In addition to the four forces of the basic matter of the universe: electromagnetic force, gravitational force, weak nuclear force, and strong nuclear force, under the principle of superstring: quantum wave, particle duality, virtual particle interdependence, and alternately discover quantum consciousness.

The state of quantum ghosts and quantum entanglement was produced, and the void force of the universe was discovered, which is the origin of life and matter in the universe, and the identity of the universe, space-time, and matter. The existence of dust particles, void forces, and mesogens promotes matter and life continues to perform its essential function of present existence, which further confirms Planck's theory of the non-existence of matter.

$$E = h \cdot v \tag{1}$$

Where E refers to energy, v refers to the vibrational frequency, and h refers to quantum constant; h = (6.6260693(11) *10^-34 J.S).

The universe has produced the Xunzi super-light curtain, the light body world and the multi-universe life code have occurred, and the function and value of matter and maintaining the existence of life in the universe have been created, and with this theoretical framework, the method of breaking the net has been developed without high pressure, high heat and temperature. It is a major investment in equipment; with a drop of water and a little active agent, the omentum produced by binding substances can be broken, and the material debonding refiner can be activated and deodorized, such as reservoir sludge, oil sludge, kitchen waste, distillers grains, vegetables, peels, chicken manure, pig manure, cow manure, and other major environmental pollution problems, and these wastes can be recycled to regenerate organic matter or renewable energy to facilitate the recycling of environmental resources.

2.2 *Theoretical basis of quantum technology*

(1) Quantum technology is used to modify soil, the nutrient structure of soil aggregates, restore the soil force, and use it as the basis for crop production and nutrition.
(2) Mercury and heavy metals should be relieved, as well as Dioxin and bromide.
(3) Activate the soil with beneficial bacteria to eliminate the presence of bad bacteria.

2.3 *Quantum technology application process*

(1) Materials
  Materials include a quantum active agent, water, and distiller's grains (wet sorghum distiller's grains. The waste from the production of red sorghum (Sorghum bicolor Lin) after distillation is sealed and packaged in plastic bags on the same day, delivered to the laboratory, and stored at 4°C.
(2) Instructions
  The process is as follows:
  (a) Distiller grains and water (active agent plus water)
  (b) Stir the three evenly (stir three times).
  (c) Apply the diluted mixture and the distiller's grains and stir evenly.
  (d) Add active powder and stir evenly (special research and development).
  (e) Stir evenly after adding the diluted mixture and distiller grains.
  (f) Pour the secondary quantum mixture into the distiller's grains and stir well.
(3) Experimental equipment:
  (a) Rapid grinder (RT08 25,000 rpm)
  (b) pH meter (Jenco Instruments Model 6071 pH meter)

(c) 4-digit electronic balance (Sartorius TE214S)

(d) Electronic scale BBX-3000, Circulator Oven DO45

(e) 4° C refrigerator (DAYTIME TV-6200)

(f) Ultrasonic oscillator (DELTA DC400H)

(g) Constant temperature & humidity incubator (HCS LTI-500)

(h) Vacuum Concentrator (EYELA Rotary Evaporation N1000)

(4) Determination of oil on the surface of distiller's grains:

Weigh 1 g of wet sorghum distiller's grains and 1 gram of washed sorghum distiller's grains respectively, then add 20 ml of N-hexane, stir and extract with a magnet for two hours, filter with filter paper and take the clear liquid, add 50 ml of 0.88% KCl solution, shake well, then put the solution into a separatory funnel, and then add 2 to 3 gram of anhydrous sodium flow. Finally, the N-hexane layer was taken out into a weighed round bottom flask and concentrated to dryness under reduced pressure. The surface oil weight of the sample can be obtained by subtracting the weight of the empty flask from the weight of the round-bottomed flask after concentration.

(5) Determination of pH value:

Referring to the pH measurement method in the Fertilizer Inspection Technology and Management Process Description of the Bureau of Standards and Inspection of the Ministry of Economic Affairs, the freshly collected samples were dried at 65–70°C to constant weight, ground with a pulverizer, and passed through a 35-mesh screen. Then dry at 65–70°C to constant weight. Take 10 g of the treated sample into a 100 mL beaker, add 50 mL of deionized water, stir evenly, and let stand for 60 minutes (stir occasionally). The pH is directly measured with a pH meter.

Step 1: Add the active agent to the quantum stock solution;

Step 2: Dilution of Quantum Mixture;

Step 3: Pour the quantum mixture into the lees;

Step 4: Mix and stir;

Step 5: Add active powder;

Step 6: Add the quantum stock solution to the diluted mixture;

Step 7: Pour the secondary quantum mixture into the lees;

Step 8: Mix again.

3 EXPERIMENTAL RESEARCH AND DISCUSSION

3.1 *Experimental results*

The discarded distiller's grains of Kinmen Winery were added with the quantum active agent in this experiment, stirred evenly by active powder, and then mixed with the diluted mixture and distiller's grains. During the stirring process, quantum and secondary quantum mixtures were fermented again. The status of the finished Quantum Technology raw titanium feed is shown in Figure 1.

3.2 *Analysis of quantum technology peptide feed*

(1) Generally, the traditionally concentrated feed is powdery, which melts when it meets water, causing waste, and is not conducive to the chewing and digestion of ruminants. The quantum technology peptide feed is easy to store and feed, which improves the utilization rate of the feed.

(2) Quantum Technology Peptide Feed is simple to use and does not need to be operated in a pre-mixed mode according to the traditional proportional principle, saving labor and no management errors.

(3) This feed can strengthen beneficial and dominant bacteria in the digestive system of ruminants, maintain an acidified environment, antagonize viruses, produce various digestive enzymes,

Figure 1. Status of quantum mixture fermented distillers' grains.

Table 1. Composition analysis of technical products.

Ingredient	Ratio	Value	Unit
Protein	6.53 %	max	
Moisture	67.74 %	min	
Crude fat	1.63 %	max	
Crude fiber	3.34 %		
Thermal energy		1551	Kcal/kg
Calcium		203.00	ppm
Phosphorus	0.17 %		
Ash	1.44 %		
Protein digestibility	73.4 %		

The weight left by the deduction method can be regarded as carbohydrates.

vitamin B group, and unknown growth factor (UGF) (Huang 2018), and increase the number of beneficial bacteria in the rumen. In the rumen, the buffer effect helps to prevent bulging and hyperacidity, improve digestibility, maintain normal metabolism, increase disease resistance, reduce the probability of infectious diseases such as cattle fever and foot-and-mouth disease, promote improved breeding health, and prevent bulging.

(4) It can reduce the incidence of cow mastitis milk (CMT milk), thereby improving milk quality, increasing milk fat rate, and reducing somatic cytokines in milk.

(5) In addition, feeding with Quantum Technology peptide feed can significantly enhance the visual perception and flavor of ruminant meat products as well as improve the feed conversion rate, and promote the improvement of the quality of farmed products, including the refined meat content and the flavor of milk and meat (Hsiao 2016).

(6) This kind of feed has good palatability, is easy to absorb and digest, increases feed intake, and shortens the feeding period. The improvement of feed conversion rate, digestion, and absorption rate and the increase of milk yield, milk fat rate, and meat production rate have improved the breeding capacity and the economic benefits of operators based on reducing the feeding cost.

(7) Without any harmful residues such as drugs and antibiotics, it conforms to the green standards of HACCP, promotes the safety improvement of breeding, and improves the market competitiveness of products. Quantum technology peptide feed is used for feeding, and animal excrement has low ammonia nitrogen, no hydrogen sulfide odor, and no residue of any drug antibiotics, which can greatly improve the environmental protection effect of breeding.

(8) The technical research and development strength of "high-tech quantum technology peptide" is to make the cattle (Hu 2017), horses, and sheep grow rapidly and healthily after use, increase the proportion of oil flowers, increase the milk, and make the meat fresher and sweeter.

Table 2. Commonly used types of probiotics.

Strain (alternative designations)	Brand name
Bifidobacterium animalis DN 173 010	Activia
Bifidobacterium animalis subsp. lactis BB-12	
Bifidobacterium breve Yakult	Bifiene
Bifidobacterium infantis 35624	Align
Bifidobacterium lactis HN019 (DR10)	Howaru Bifido
Bifidobacterium longum BB536	
Enterococcus LAB SF 68	Bioflorin
Escherichia coli Nissle 1917	Mutaflor
Lactobacillus acidophilus LA-5	
Lactobacillus acidophilus NCFM	
Lactobacillus casei DN-114 001	Actimel
Lactobacillus casei CRL431	
Lactobacillus casei F19	Cultura
Lactobacillus casei Shirota	Yakult
Lactobacillus johnsonii La1 (Lj1)	LC1
Lactococcus lactis L1A	
Lactobacillus plantarum 299V	GoodBelly
Lactobacillus reuteri ATTC 55730	Reuteri
Lactobacillus rhamnosus ATCC 53013 (LGG)	Vifit and others
Lactobacillus rhamnosus LB21	Verum
Lactobacillus salivarius UCC118	
Saccharomyces cerevisiae (boulardii) lyo	DiarSafe, ltralevure, and others
Lactobacillus acidophilus CL1285 &	
Lactobacillus casei Lbc80r*	Bio K+
Lactobacillus rhamnosus GR-1 &	FemDophilus
Lactobacillus reuteri RC-14*	
VSL#3 (mixture of 1 strain of Streptococcus thermophilus, four	VSL#3
Lactobacillus spp., and three Bifidobacterium spp. strains*	
Lactobacillus acidophilus CUL60 and	
Bifidobacterium bifidum CUL 20*	
Lactobacillus helveticus R0052 and	Abiotica and others
Lactobacillus rhamnosus R0011*	
Bacillus clausii strains O/C, NR, SIN, and T*	Enterogermina

3.3 Analysis of the property market related to distiller's grains

(1) Domestic market

The annual output of Taiwan's feed is about 9 million metric tons. 90% of the feed in Taiwan is imported, and the cost of feed accounts for 65–75% of the production cost.

(2) Fishery market

The nutrients extracted from the lees can be used to produce aquatic products such as mullets and shrimp. The shrimp raised is four times the normal weight. Breeding mullets can tolerate low temperatures. Usually, mullets do not grow in winter, as they do not hibernate. However, mullets that eat lees will continue to grow during the winter (Chou 2018).

(3) Beauty Market

Due to the large daily output of distiller's grains, there is no other application except that the winery uses the distiller's grains to make masks and sells them. A large amount of distiller's grains is also prone to pollution problems.

(4) Food market

a. The extracted distiller's grains powder is used as a food additive and added to bakery products to make distiller's grains toast, distiller's grains biscuits, distiller's grains roselle candied fruit, etc.

b. The meat quality of "Taiwan Distillers' Grains Chicken" is firmer than that of ordinary feed chickens, and there is no traditional meaty smell. The price is similar to that of ordinary chickens.

c. As for the healthy soy sauce of distiller's grains powder, the soybeans were steamed first, followed by the fried and crushed wheat, then added the bacteria, then they were placed into an urn combined with the distiller's grains and salt water. The residue was filtered after fermentation for six months. The raw soy sauce was seasoned and then cooked into soy sauce. This natural healthy soy sauce is delicious and nutritious.

3.4 *Quantum Technology Raw Titanium Feed Cost and Economic Analysis*

(1) Taking fattening pigs as an example, it can replace more than 50% of the full-price feed. The market retail price of full-price feeds for fattening pigs is around 5.8-6.3 NT$/ kilograms, while the raw titanium feed of Quantum Technology is 1.4-1.8 NT$/ kilograms. The cost per kilogram is more than 4 NT$, and a fattening pig uses 500-600 kilograms of full-price feed in a fattening period. In this way, a pig can save about 1,100 NT$ in feed costs, and it does not affect the slaughtering period. It also improves the quality of pork. The sales price of feed for livestock farms in Taiwan is about 23 NT$/ kilograms; our team sells each ton of feed at 15 NT$/ kilograms, providing a replacement for the original supply and demand market.

(2) Trial cost

a. The purchase price is 300 NT$/ton, the output is 400 kilograms/ton, the selling price is 15 NT$/kilograms, and the sales price is 6000 NT$/ton.

b. The daily production is 50 tons, and the daily profit is 285,000 NT$.

c. The monthly production is 1,100 tons, and the profit is 6.27 million NT$/month.

d. The annual production is 13,200 tons, and the profit is 80.4 million NT$/year.

e. After deducting 25% of administrative sales and 5% tax rate, the net profit is about 56 million NT$.

f. The investment amount is 25 million NT$ (including plant rental, machinery, equipment, water and electricity, administrative management, and sales).

4 CONCLUSIONS

In this paper, quantum technology is used to study the quantum technology peptide feed converted from waste distiller's grains. The main conclusions can be summarized as follows:

(1) Quantum Technology Peptide Feed uses the principles of chemistry and microbiology to degrade distiller's grains rich in lignin, cellulose, and hemicellulose into biological protein feed rich in bacterial protein, vitamins, and other components. The feed is highly palatable, the fiber degradation rate can reach 20–35%, and the discarded distiller's grains can increase agricultural production and efficiency. To increase the economic benefits and added value of products, a long-term mechanism for comprehensive utilization of distiller's grains must be developed through multichannel, multi-level rational utilization (Hsu 2017).

(2) Its components are rapidly decomposed into small molecules by rapidly decomposing the crude fiber, starch, crude fat, crude protein, and other macromolecules in the distiller's grains. Small molecules decomposed into amino acids and glucose, live bacteria fermentation of nutritional bacterial microorganisms (cellular protein) added to minerals and trace elements (silicon, calcium, iron, magnesium, manganese, selenium, zinc, potassium, sodium, phosphorus, titanium, copper and more than 20 kinds). The use of this feed for breeding greatly increases the cost of breeding and does not affect the slaughter period of meat and poultry and the egg production rate of egg-laying poultry.

(3) Quantum Technology Peptide Feed is beyond the quality of the current livestock feeding, which can make the feed more palatable, make the gastrointestinal tract of livestock healthier, and

significantly improve the visual appearance and flavor of ruminant meat products, and promote the quality of aquaculture products. To reduce the investment, waste, and emissions of resources, it is more important to solve the problem of waste distiller's grains produced by wineries, solve the problem of secondary environmental pollution, and implement the government's circular economy policy. Promoting economic effects while considering ecological environmental protection and economic development is an important step towards sustainable development.

REFERENCES

Ching-Ching Hsiao, *Bacterial cellulose production using sorghum distiller's grains as substrate and application of improved stationary culture system*, Chung Hsing University, 2016.

Chou, Ching-Yun, *Yunzhi fermented sorghum distiller's grains as an alternative source of protein for grouper feed*, Taiwan Ocean University, 2018.

Hu, Wei-Cheng, *The conditions of solid-state fermentation of sorghum distiller's grains with Yunzhi and the changes of functional components of the product*, Yilan University, 2017.

Huang, Zi-Hsin, *The Effect of Sorghum Distillers Grains/ Cow Manure Mixed Compost on Earthworm Growth and Its Properties Analysis*, Tunghai University, Quemoy University, 2018.

Hao-Yuan Chao, *Evaluation of vermicomposting of sorghum distillers grains as organic fertilizer*, Yilan University, 2019.

Ming-An Hsu, *Using sorghum distiller's grains as a matrix to improve the yield of bacterial cellulose production using a static culture system*, Chung Hsing University, 2017.

Wang, Sheng-Chi, Study on Production of Enzymatic Feed by Using Aspergillus Niger 2039 Fermented Rice Distiller's Grains, Pingtung University of Science and Technology, 2016.

Yi-Jyun Cai, Bacterial cellulose growth model of acetic acid bacteria Komagataeibacter rhaeticus NCHU R-1 using sorghum distiller's grains water extract as the main matrix and its application in immobilized probiotics, Chung Hsing University, 2018.

Advances in Renewable Energy and Sustainable
Development – Liang & Kasmani (Eds)
© 2023 Copyright the Author(s), ISBN: 978-1-032-39407-7

Effects of rumen fluid treatment on nutritional components and feeding value of sweet sorghum

Baoyu Yang, Liangyao Bai, Feng Chen, Jiao Wang, Kai Zhang & SuJiang Zhang*
College of Animal Science and Technology, Tarim University, Alar, Xinjiang, China
Key Laboratory of Tarim Animal Husbandry Science and Technology, Xinjiang Production & Construction Corps, Alar, Xinjiang, China

ABSTRACT: In this experiment, sweet sorghum straw was selected as raw material to study the effects of rumen fluid pre-treatment and anaerobic solid-state fermentation time on the nutritional components and feeding value of sweet sorghum, to provide data support for the rational development and utilization of forage resources of herbivorous livestock. One control group (CK, 0% rumen fluid, 0 mL/100g) and three rumen fluid groups (R1, 10% rumen fluid, 10 mL/100g; R2, 20% rumen fluid, 20 mL/100g; R3, 30% rumen fluid, 30 mL/100g) were set up. The mixture of sterile distilled water and rumen fluid was evenly added to the straw, and the moisture content was about 75%. After continuous anaerobic fermentation for 45 days, the samples (3-time points) were tested at 0, 15, and 45 days after fermentation × 4 processes × 3 replicates (36 bags in total) for sampling and analysis. The results showed that: 1) with the extension of anaerobic fermentation time and the increase of rumen fluid, the dry matter (DM), ether extract (EE), neutral detergent fiber (NDF), and acid detergent fiber (ADF) of sweet sorghum decreased as a whole, and the difference of crude protein (CP) and crude ash (Ash) was significant ($P < 0.05$). 2) With the extension of anaerobic fermentation time and the increase of rumen fluid, the dry matter intake (DMI), digestible dry matter (DDM), total digestible nutrients (TDN), and relative feeding value (RFV) of sweet sorghum showed an upward trend ($P < 0.05$). In conclusion, with the extension of fermentation time and the increase of rumen fluid, the fiber content of sweet sorghum can be significantly reduced, the nutritional composition can be improved, and the feeding value and effective energy value can be improved. Under the same anaerobic fermentation time and conditions, the addition ratio of rumen fluid in the R3 group (30% rumen fluid, 30 mL/100g) has the best overall fermentation effect on sweet sorghum, which can provide a theoretical and experimental basis for the rational development and utilization of sweet sorghum and other crop straw.

1 INTRODUCTION

With the adjustment of China's agricultural industrial structure and the change in people's consumption concept, the rapid development of animal husbandry has been promoted, and the demand of farmers for forage is also increasing year by year. If the crop straw with high cellulose content can be fully processed and utilized, it can not only reduce the problems of resource waste and environmental pollution but also solve the problem of insufficient supply of forage for herbivores in southern Xinjiang. At the same time, it can reduce the breeding cost and improve the economic benefit.

As a crop straw with high photosynthetic capacity and high reducing sugar, sweet sorghum is a C4 energy crop with great potential (Amar et al. 2020). It is also one of the forage crops that are often promoted and planted in animal husbandry to solve the shortage of roughage resources

---

*Corresponding Author: zsjdky@126.com

DOI 10.1201/9781003349648-20

in southern Xinjiang. As a forage crop, sweet sorghum has relatively low crude protein and high crude fiber, especially the refractory characteristics of lignocellulose, which limits the resource utilization of sweet sorghum and cannot meet the needs of an animal body. Moreover, its harvest is seasonal and cannot meet the demand of annual utilization (Wang et al. 2021; Zhang et al. 2014).

At present, the green sweet sorghum straw mostly adopts anaerobic fermentation methods such as silage (Hou et al. 2018), mixed silage (Wang et al. 2021), silage with green fluid fermentation broth (Lin et al. 2012) and silage with a bacterial agent (Yu et al. 2018) to reduce the cellulose of sweet sorghum straw and increase the palatability. In terms of fermentation time, adding microbial agents can save more time. However, due to the high screening cost of microbial agents, if ordinary farmers use many microbial agents in silage for a long time, they will not be able to support this capital economically for a long time.

Therefore, it is imperative to develop efficient and cheap microbial agents. Rumen microorganisms of ruminants are important places to degrade fiber. Zhang et al. (2016) showed that treating rice straw with rumen fluid can significantly reduce the content of fiber components. Hu et al. (2008) showed that using rumen fluid for straw anaerobic fermentation also found that rumen microorganisms can improve the rate of degradation of fiber components. Although rumen fluid is a relatively cheap microbial agent, there are relatively few studies on the fermentation quality and microbial diversity of straw fermented by rumen fluid.

Based on the principle of anaerobic fermentation, the solid-state anaerobic fermentation of sweet sorghum was carried out by adding rumen fluid to study the effects of different amounts of rumen fluid and different fermentation times on the nutritional characteristics of sweet sorghum. The nutritional composition, fiber composition, feeding value, and effective energy value were analyzed to provide a theoretical basis for improving the dimensional degradation rate of sweet sorghum straw and the development and utilization of roughage resources.

2 MATERIALS AND METHODS

2.1 *Test material*

The sweet sorghum variety used in this experiment is Hercules. It is planted in the animal experiment station of the school of animal science and technology of Tarim University. In October 2020, sweet sorghum straw will be cut with a moisture content of about 70%, placed in a cool and dry place for air drying, and crushed to 2-3 cm for standby.

The rumen fluid used in this experiment was taken from the Xinjiang Aksu Tarim cattle and sheep slaughterhouse in April 2021. The contents of the front, middle and rear parts of the rumen of eight adult sheep were collected, filtered with four layers of gauze, discarded the residue, placed in a thermos, quickly brought back to the laboratory, and placed in a $-80°C$ refrigerator for standby.

2.2 *Preparation of sweet sorghum straw by fermentation*

In this experiment, after the sweet sorghum straw crushed 2-3 cm in the early stage is fully mixed, the sampling is carried out according to the quartering method, and 200 g of sweet sorghum straw is weighed (the water content after air drying is about 5%). After thawing the rumen fluid, measure different volumes of rumen fluid, dilute them in an appropriate amount of sterile distilled water, and evenly spray them on the weighed sweet sorghum straw (the control group is sprayed with the same volume of sterile distilled water), adjust the water content of sweet sorghum straw to 75%, mix them evenly, and put them into polyethylene plastic bags (25.5 cm × 35.0 cm), after vacuum sealing with a vacuum packaging machine, weigh its mass (about 750 g). Then, the fermentation bag was subjected to solid-state anaerobic fermentation at room temperature (about 25°C).

2.3 Experimental design

This experiment adopts a two-factor completely random design, i.e., 3 × 4 (fermentation days × rumen fluid addition) factor treatment structure. One control group (CK, with an equal volume of sterile distilled water added to the straw, i.e., 0 mL/100g) and three rumen fluid groups (R1, with 10% rumen fluid added to the straw, i.e., 10 mL/100g; R2, with 20% rumen fluid added to the straw, i.e., 20 mL/100g; R3, with 30% rumen fluid added to the straw, i.e., 30 mL/100g) were set up, and three repeated tests were set up in each test group. After continuous anaerobic fermentation for 45 days, the samples (3-time points) were tested at 0, 15, and 45 days after fermentation × four processes × three replicates (36 bags in total) were sampled for analysis of nutritional components, feeding value, and other related indicators.

2.4 Measurement index and method

2.4.1 Determination of nutrient content
Dry matter (DM) was determined by the oven drying method (GB/T 6435-2014) (GB/T 6435-2014, 2014). Crude protein (CP) was determined by the Kjeldahl method (GB/T 6432-2018) (GB/T 6432-2018,2018). Ether extract (EE) was determined by Soxhlet extraction method (GB/T 6433-2006) (GB/T 6433-2006, 2016). Determination of crude ash (Ash) by high-temperature burning method (GB/T 6438-2007) (GB/T 6438-2007, 2007.). Neutral detergent fiber (NDF) and acid detergent fiber (ADF) were determined by the method of Van Soest et al. (Van Soest et al. 1991).

2.4.2 Feeding value evaluation
Relative feeding value (RFV), dry matter intake (DMI), and relative forage quality (RFQ) were assessed using the method of Rohweder et al. (1978). Total digestible nutrients (TDN) were calculated by the method of Lithourgidis et al. (2006). The relative forage quality (RFQ) was calculated by the method of Moore et al. (2002).

2.5 Data statistics and analysis

WPS Office 2021 was used to preliminarily sort out the data and draw charts. IBM SPSS statistics 26.0 software was used for two-way ANOVA, and Duncan's multiple comparison method was used to compare and analyze each group. Results were expressed by "mean value and SEM" (when $P < 0.05$, it shows a significant difference; when $P < 0.01$, it shows a high significant difference).

3 RESULTS AND DISCUSSION

3.1 Effects of rumen fluid addition amount and fermentation time on the nutrient content of sweet sorghum

It can be seen from Table 1 that when the fermentation time is the main effect, with the extension of anaerobic fermentation time, in the same treatment group, DM, EE, NDF, and ADF of sweet sorghum straw show a downward trend, and CP and Ash show an upward trend. There were significant differences in DM, CP, EE, NDF, ADF, and Ash in each group on the 0th, 15th, and 45th days of fermentation ($P < 0.05$); the CP of the CK group and R1 group was lower than that of the 15th day of fermentation ($P > 0.05$); the EE of CK group, R1 group, and R3 group was higher on the 15th day of fermentation than that on the 45th day of fermentation, and the difference was not significant ($P > 0.05$).

When the rumen fluid treatment group was the main effect, with the increase in the amount of rumen fluid, the DM, EE, NDF, and ADF of sweet sorghum straw showed a downward trend, while CP and Ash showed an upward trend at the same fermentation time. On the 0th day of fermentation, there was no significant difference in DM, EE, NDF, ADF, and Ash in each group ($P > 0.05$); the CP of the R3 group was higher than that of the CK group, R1 group, and R2 group, and there was no significant difference with CK group and R1 group ($P > 0.05$). On the 15th and 45th days of fermentation, DM, EE, NDF, and ADF in the R2 group and R3 group were significantly lower than those in the CK group and R1 group ($P < 0.05$); CP and Ash in the R2 group and R3 group were significantly higher than those in CK group and R1 group ($P < 0.05$); On the 15th day of fermentation, CP, and EE in R1 group were lower than those in R2 group, and the difference was not significant ($P > 0.05$).

Table 1. Effects of rumen fluid addition and fermentation time on the nutrient content of sweet sorghum (DM basis) %.

Items	Day	Treatment				SEM
		CK	R1	R2	R3	
DM	0	24.93^{Aa}	24.95^{Aa}	24.95^{Aa}	24.90^{Aa}	0.04
	15	24.74^{Ba}	24.60^{Ba}	24.23^{Bb}	24.14^{Bb}	
	45	23.71^{Ca}	23.51^{Cb}	23.14^{Cc}	22.86^{Cd}	
CP	0	3.71^{Bb}	3.77^{Bb}	3.83^{Cab}	3.94^{Ca}	0.11
	15	3.84^{Bc}	4.24^{Bb}	4.53^{Bab}	4.75^{Ba}	
	45	4.56^{Ac}	5.30^{Ab}	6.04^{Aa}	6.47^{Aa}	
EE	0	3.22^{Aa}	3.19^{Aa}	3.20^{Aa}	3.22^{Aa}	0.03
	15	2.70^{Ba}	2.54^{Bb}	2.58^{Bb}	2.33^{Bc}	
	45	2.67^{Ba}	2.48^{Bb}	2.39^{Cc}	2.22^{Bd}	
NDF	0	58.51^{Aa}	58.47^{Aa}	58.81^{Aa}	58.44^{Aa}	0.26
	15	55.42^{Ba}	54.70^{Ba}	51.49^{Bb}	48.66^{Bc}	
	45	48.54^{Ca}	46.82^{Cb}	45.08^{Cc}	43.20^{Cd}	
ADF	0	36.24^{Aa}	36.42^{Aa}	36.07^{Aa}	36.46^{Aa}	0.22
	15	34.42^{Ba}	32.54^{Bb}	29.84^{Bc}	28.22^{Bd}	
	45	27.73^{Ca}	26.60^{Cb}	25.33^{Cc}	24.01^{Cd}	
Ash	0	7.11^{Ca}	7.10^{Ca}	7.10^{Ca}	7.12^{Ca}	0.03
	15	7.71^{Bc}	7.94^{Bb}	8.36^{Ba}	8.31^{Ba}	
	45	8.39^{Ab}	8.42^{Ab}	8.52^{Aa}	8.54^{Aa}	

Note: 1) Different capital letters in the same column showed significant differences ($P < 0.05$), and different lowercase letters in the same row showed significant differences ($P < 0.05$).

The content of nutrients is the most basic index to directly measure the fermentation quality of forage, in which DM is a relatively important index in the fermentation process. In this experiment, the DM content of sweet sorghum straw decreased significantly with the increase of rumen fluid and fermentation time. On the 45th day of fermentation, the content of DM in the R3 group reached the lowest value, 22.86%. This experiment shows that in the process of anaerobic fermentation, the participation of rumen microorganisms enables the rapid fermentation of sweet sorghum straw. Dai et al. (2019) used alfalfa and sweet sorghum for mixed storage, and its DM decreased in varying degrees before and after fermentation. When Lu et al. (2021) studied the effect of fermentation time on alfalfa silage, the DML before and after silage unsealing was 10%. The test results are consistent with the above research results. CP is the general name of all nitrogen-containing substances in forage and contains various essential amino acids. It is an important index to measure the feeding value of forage (Ma et al. 2019). Qin et al. (2012) showed that although CP was decomposed into amino acids during silage, the total content of nitrogen decreased and the total amount of microbial synthesis increased. Bilal (2009) research shows that bacterial microorganisms composed of protein cannot grow and reproduce for a long time in the fermentation process, and become a part of the feed, thus increasing CP. In this experiment, CP of sweet sorghum straw increased with the increase of rumen fluid and fermentation time; This may be that replacing additives with rumen fluid is equivalent to providing foreign microorganisms to the fermentation environment to increase CP.

with the extension of fermentation time, microorganisms can produce part of bacterial protein, to increase CP. The test results are consistent with the above research results. EE is a substance that can provide energy for the animal body and is an important index to evaluate the nutritional value of forage (Zhao 2019). An et al. (2014) showed that the use of mixed bacteria fermented feed consumed sugars in the substrate, resulting in a relative increase in EE, and the other part of EE was transformed during microbial fermentation. This is consistent with the research results of Wang et al. (2020). In this experiment, the EE content of sweet sorghum straw is relatively small. With the gradual increase of fermentation time and rumen addition, there can be significant differences in EE. This experiment shows that the addition of rumen fluid can produce fat soluble nutrients under the interaction of enzymes secreted by its microbial flora, to improve the nutritional value and fermentation quality of silage forage. Cellulose is an important part of the plant cell wall, and the levels of NDF and ADF in the feed are important indicators for the quality evaluation of fermented feed. The level of fiber is very important for the growth and development of animals, rumen fermentation of ruminants, and the digestion and absorption of nutrients; The lower the ADF, the stronger the ability to degrade to WSC, provide more substrates for lactic acid bacteria fermentation, and the higher the feed value of its feed (Baba et al. 2019; Kou et al. 2021; Liu et al. 2021; Lu et al. 2021; Ran et al. 2020; Ren et al. 2021; Shi et al. 2021; Sun et al. 2022; Wang & Wu 2018; Xiong et al. 2018; Yang et al. 2022; Zhang et al. 2021; Zhou et al. 2020). In this experiment, the fiber components (NDF and ADF) of sweet sorghum straw decreased significantly with the extension of anaerobic fermentation time and the addition of rumen fluid. Ren et al. (2021) and Baba et al. (2019) also proved that adding rumen fluid and cellulase can effectively degrade sweet sorghum straw fiber. The test results are consistent with the above research results. This experiment shows that the cellulose macromolecular chain of the cell wall is decomposed to a certain extent under the action of rumen microorganisms and related enzymes in rumen fluid; the degree of polymerization decreased and more holocellulose was released. The result showed that the fiber content decreased and the WSC in the fermentation quality was improved. NDF and ADF are closely related to ruminant DMI. It further shows that the microbial flora in the rumen can make full use of the sugars and other substances transformed from cellulose to provide the energy source required by the ruminant body. Ash content can reflect the content of mineral elements in forage and its habitat conditions (Xiong et al. 2018). Ash in forage is mostly inorganic substances such as inorganic salts and oxides, which can provide relevant minerals for the animal body and ensure animal health. In this experiment, ash increased gradually with the increase of fermentation time and rumen addition. This experiment shows that during the fermentation process of sweet sorghum straw, physical and chemical changes will occur. Finally, most of the organic matter will be utilized by microorganisms, a small part will be volatilized, and the residual ash will gradually increase. Sun et al. (2022) studied that adding different additives had the same trend on the fermentation feed ash of distiller's grains.

3.2 *Effects of rumen fluid addition and fermentation time on the feeding value of sweet sorghum*

It can be seen from Table 2 that when the fermentation time is the main effect, the DMI, DDM, TDN, RFV, and RFQ of sweet sorghum straw show an overall upward trend with the extension of anaerobic fermentation time in the same treatment group. There were significant differences in DMI, DDM, TDN, RFV, and RFQ among the treatment groups on the 0th, 15th, and 45th days of fermentation ($P < 0.05$).

When the rumen fluid treatment group was the main effect, with the increase in the amount of rumen fluid, the DMI, DDM, TDN, RFV, and RFQ of sweet sorghum straw showed an upward trend at the same fermentation time. On the 0th day of fermentation, there was no significant difference in DMI, DDM, TDN, RFV, and RFQ among the groups ($P > 0.05$). On the 15th and 45th days of fermentation, the DMI, DDM, TDN, RFV, and RFQ of the R2 group and R3 group were significantly higher than those of the CK group and R1 group ($P < 0.05$); DMI, DDM, TDN, RFV and RFQ in R3 group were significantly higher than those in R2 group ($P < 0.05$).

Table 2. Effects of rumen fluid addition and fermentation time on the feeding value of sweet sorghum (DM basis) %.

Items	Day	CK	R1	R2	R3	SEM
DMI/ %BW	0	2.05^{Ca}	2.05^{Ca}	2.04^{Ca}	2.05^{Ca}	0.01
	15	2.17^{Bc}	2.19^{Bc}	2.33^{Bb}	2.47^{Ba}	
	45	2.47^{Ad}	2.56^{Ac}	2.66^{Ab}	2.78^{Aa}	
DDM	0	54.67^{Ca}	54.53^{Ca}	54.48^{Ca}	54.50^{Ca}	0.17
	15	56.08^{Bd}	57.55^{Bc}	59.66^{Bb}	60.92^{Ba}	
	45	61.30^{Ad}	62.18^{Ac}	63.17^{Ab}	64.20^{Aa}	
TDN	0	54.61^{Ca}	54.37^{Ca}	54.82^{Ca}	54.32^{Ca}	0.29
	15	59.94^{Bd}	59.37^{Bc}	62.86^{Bb}	64.94^{Ba}	
	45	65.58^{Ad}	67.03^{Ac}	68.68^{Ab}	70.38^{Aa}	
RFV	0	86.93^{Ca}	86.77^{Ca}	86.69^{Ca}	86.76^{Ca}	0.67
	15	94.14^{Bd}	97.87^{Bc}	107.79^{Bb}	116.47^{Ba}	
	45	117.49^{Ad}	123.54^{Abc}	130.35^{Ab}	138.24^{Aa}	
RFQ	0	91.07^{Ca}	90.73^{Ca}	90.95^{Ca}	90.69^{Ca}	0.87
	15	100.24^{Bd}	105.89^{Bc}	119.12^{Bb}	130.24^{Ba}	
	45	131.83^{Ad}	139.69^{Ac}	148.95^{Ab}	158.95^{Aa}	

DMI, DDM, TDN, RFV, and RFQ are all comprehensive evaluation indexes of feed quality. The higher the value, the higher the nutritional value of silage, which is a comprehensive reflection of ADF content and NDF content in the feed (Ran et al. 2020; Wang & Wu 2018; Zhou et al. 2020). In this experiment, ADF and NDF decreased significantly with the time of anaerobic fermentation, and the amount of rumen fluid, and DMI, DDM, TDN, RFV, and RFQ increased significantly. Because these indexes were negatively correlated with ADF and NDF, the R3 group reached the maximum on the 45th day of fermentation, which was (2.47% BW, 60.92% DM, 64.94% DM, 138.24% DM, and 158.95% DM). This experiment showed that the fiber of sweet sorghum straw was significantly degraded by cellulose degrading microorganisms and related enzymes in rumen fluid, and the nutritional value and feeding value of sweet sorghum straw were improved. Kou et al. (2021) showed that the addition of lignin enzyme can significantly reduce the ADF of Huangzhu grass silage, improve DDM and TDN, and then improve RFV and RFQ. Liu et al. (2021) showed that different microbial additives decreased NDF and ADF, and RFV and RFQ increased in whole plant corn silage. Shi et al. (2021) showed that adding ammonium chloride (9 g/kg) to whole plant silage corn can significantly improve its TDN, effective energy value, and nutritional value. The test results are consistent with the above research results.

4 CONCLUSION

With the extension of fermentation time and the increase of rumen fluid, the nutritional components, degraded fiber components, and feeding value of sweet sorghum could be significantly improved. After 45 days of anaerobic fermentation, the fermentation effect of the R3 group (30% rumen fluid, 30 mL/100g) was significantly better than that of the CK group, R1 group, and R2 group. The interaction effect of sweet sorghum with the amount of rumen fluid and fermentation time significantly affected the nutritional components, fiber components, and feeding value of sweet sorghum.

ACKNOWLEDGMENTS

This paper is an open project of the Key Laboratory of Tarim Animal Husbandry Science and Technology Corps—demonstration and promotion of forage sweet sorghum of high quality, and condensed results of research (Grant No. HS202101). This paper is also the autonomous region graduate education innovation plan project in 2021—effects of rumen fluid pretreatment on fermentation quality and aerobic stability of sweet sorghum straw (Grant No. XJ2021G299).

REFERENCES

Amar, A., Wang, J., Zhang, S.J., 2020. Comparison of gas production and rumen degradability on sweet sorghum silages of different cultivars in vitro. *Journal of Gansu Agricultural University*. 55, 7–12+19.

An, X.P., Wang, Z.Q, Qi, J.W, Yu, C.Q, Tong, B.S, 2014. Effect of mixed bacteria solid state fermentation on nutritional characteristics of soybean meal. *Feed Research*. 66–70.

Baba, Y., Matsuki, Y., Kakizawa, S., Suyama, Y., Tada, C., Fukuda, Y., Saito, M., Nakai, Y., 2019. Pretreatment of lignocellulosic biomass by cattle rumen fluid for methane production: fate of added rumen microbes and indigenous microbes of methane seed sludge. *Microbes and Environments*. 34, 421–428.

Bilal M.Q., 2009. Effect of molasses and corn as silage additives on the characteristics of Mott Dwarf elephant grass silage at different fermentation periods. *Pakistan Veterinary Journal*. 29, 19–23.

Dai, S., Wang, F., Dong, X., Hao, J., 2019. Effects of mixing ratio of alfalfa and sweet sorghum on nutritional quality and aerobic stability of total mixed ration silage. *Chinese Journal of Animal Nutrition*. 32, 2306–2315.

GB/T 6432-2018,2018. *Determination of crude protein in the feeds-Kjeldahl method*. Standardization Administration of China, Beijing, China.

GB/T 6433-2006, 2016. *Determination of crude fat in feeds*. Standardization Administration of China, Beijing, China.

GB/T 6435-2014, 2014. *Determination of moisture in feedstuffs*. Standardization Administration of China, Beijing, China.

GB/T 6438-2007, 2007. *Animal feeding stuffs-determination of crude ash*. Standardization Administration of China, Beijing, China.

Hou, M.J., Zhou, E.G., Fu, X.Y., Shang Z.H., Wang, H.C., 2018. Effects of sweet sorghum silage on the blood routine and the concentration of Lipopolysaccharide in sheep. *Chinese Journal of Veterinary Medicine*. 54, 36–39+44.

Hu, Z.H., Liu, S.Y., Yue, Z.B., Yan, L.F., Yang, M.T., Yu, H.Q., 2008. Microscale analysis of in vitro anaerobic degradation of lignocellulosic wastes by rumen microorganisms. *Environmental Science and Technology*. 42, 276–281.

Kou, J.T., Lei, J.X., Zhang, H.B., Guo, D.S., Meng, J.Q., Liu, X., Li, L., Chen, Y., 2021. Effects of different additives on silage quality of yellow bamboo grass. *Chinese Journal of Animal Science*. 57, 178–182.

Lin, S.X., Gao, C.F., Li, W.Y., Liu, Y., Zhang, X.P., Dong, X.N., 2012. Effects of adding fermented green juice and for a form on the fermentation quality of forage sorghum silage. *Journal of Domestic Animal Ecology*. 33, 69–72.

Lithourgidis, A.S., Vasilakoglou, I.B., Dhima, K.V., Dordas, C.A., Yiakoulaki, M.D., 2006. Forage yield and quality of common vetch mixtures with oat and triticale in two seeding ratios. *Field Crops Research*. 99, 106–113.

Liu, G.K., Wang, S.W., Liu, S.X., Li, K.Y., Guo, W.T., Wang, Y.T., Wang, K., 2021. Effects of different microbial additive combinations on silage quality of whole plant maize. *Chinese Journal of Animal Science*. 57, 215–218+223.

Lu, G.C., Xu, H., Yu, Y.X., Jiang, C.D., 2021. Effects of different additives and densities on silage quality of hybrid Pennisetum. *Pratacultural Science*. 38, 2191–2199.

Lu, Q., Sun, L., Ren, Z.H., Sa, D.W., Du, S., Li, J.F., Yuan, N., Jia, Y.S., 2021. Dynamic analysis of nutritional quality and microbial community of alfalfa silage. *Chinese Journal of Grassland*. 43, 111–117.

Ma, X.N., Zhu, F.H., Ge, W., Chen, M., Liu, Z.Z., Liu, Y.T., 2019. Effects of moisture ratio and fermentation time on nutrients of fermented total mixed ration based on whole plant corn. *Chinese Journal of Animal Nutrition*. 31, 2367–2377.

Mooer, J.E., Undersander, D.J., 2002. *Relative forage quality: An alternative to relative feed value and quality index*. Florida: Proceedings 13th Annual Florida Ruminant Nutrition Symposium, USA. 16–32.

Qin, M.Z., Shen, Y.X., 2012. Effects of maturity stage on fermentation quality of whole crop wheat silage. *Scientia Agricultura Sinica*. 45, 1661–1666.

Ran, F., Jiao, T., Lei, Z.M., Gao, X.M., Zhao, S.G., 2020. Effects of different steam explosion conditions on the feeding value of corn straw. *Pratacultural Science*. 37, 2133–2141.

Ren, H.W., Zhao, Y., Liu, Y.L., Feng, Y.P., Zhang, B.Y., Li, Z.Z., 2021. Comparison of the improvement efficacies for ensiling quality and biodegradation performance of sweet sorghum silage by different additives. *Transactions of the Chinese Society of Agricultural Engineering*. 33, 2191–2199.

Rohweder, D.A., Barnes, R.F., Jorgensen, N., 1978. Proposed hay grading standards based on laboratory analyses for evaluating quality. Journal of Animal Science. 43, 747–759.

Shi, Q.T., Jiang, F.G., Cheng, H.J., Wei, C., Wang, X.M., Song, E.L., Xu, Z.X., 2021. Effects of ammonium chloride on nutritional value and fermentation quality of whole-plant corn silage. *Chinese Journal of Animal Nutrition*. 33, 2063–2072.

Sun, Y.N., Li, M., Lu, L.R., Ren, Y.W., Li, R.C., Pan, L.C., Ren, H.W., 2022. Effect of additives on the nutritional value of feed made from distillers grains. *Liquor-Making Science & Technology*. 17–23.

Van soest, P.J., Roberton, J.B., Lewis, B.A., 1991. Methods for dietary fiber, neutral detergent fiber, and nonstarch polysaccharides in relation to animal nutrition. *Journal of Dairy Science*. 74, 3583–3597.

Wang J., Yang B.Y., Zhang S.J., Amar A., Chaudhry A.S., Cheng L., Abbasi I.H.R., Al-Mamun M., Guo X.F., Shan A.S., 2021. Using mixed silages of sweet sorghum and alfalfa in total mixed rations to improve growth performance, nutrient digestibility, carcass traits and meat quality of sheep. *Animal*. 15, 100246.

Wang, H.W., Wu, D.Q., 2018. Study on the nutrient composition and feeding value of phragmites before and after ensiling. *Cereal & Feed Industry*. 59–61.

Wang, J., Yang, B.Y., Amar, A., Zhang, S.J., 2021. Effects of sweet sorghum and alfalfa silage mixtures on digestive tissue morphology in meat sheep. *Feed Industry*. 42, 54–60.

Wang, X.P., Wang, X.Q., Li, B., Wang, L., Duan, Z.Y., 2020. Testing the performance of compound microbial additives in silage maize nutrients and the effect of feeding Tan sheep. *Chinese Journal of Eco-Agriculture*. 28, 1258–1264.

Xiong, Y., Xu, Q., Yu, Z., Ji, G., Ou, X., Ma, L.Y., Lang, Q., Shi, Y., Li, J.L., 2018. Evaluation of nutritional components and feeding value of different alfalfa hay. *Acta Agrestia Sinica*. 26, 1262–1266.

Yang, J., Feng, F., Gao, Q.X., Wang, J., Xin, G.S. 2022. Effect of feed rape on the quality of fermented total mixed diet. *Chinese Journal of Animal Science*. 58, 192–197.

Yu, F.Y., Hua, J.L., Guo, L., Fu, J.W., 2018. Effects of different silage additives on quality of sweet sorghum silage. *Journal of Anhui Science and Technology University*. 32, 1–6.

Zhang, H.B., Zhang, P.Y., Ye, J., Wu, Y., Fang, W., Gou, X.Y., Zeng, G.M., 2016. Improvement of methane production from rice straw with rumen fluid pretreatment: A feasibility study. *International Biodeterioration and Biodegradation*. 113, 9–16.

Zhang, J., Shi, W.J., Han, X.L., Zhang, L., Li, S.T., Feng, Q.X., Li, Y., Yang, F.L., Zhou, J., 2021. Effects of malic acid and lactobacillus acidophilus on quality and microbial fiversity of chamaecrista rotundifolia silage. *Chinese Journal of Animal Nutrition*. 33, 6941–6952.

Zhang, S.J., Osman, A., Xue, X.Z., Zhang, X., Guo, X.F., Chen, L.Q., 2014. Quality analysis of different sweet sorghum silages in Southern Xinjiang compared with corn silage. *Acta Prataculturae Sinica*. 23, 232–240.

Zhao, H.B., 2019. Effects of silage on nutritional components of corn straw. *Today Animal Husbandry and Veterinary*. 35, 71.

Zhou, E.G., Wang, H.C., Shang, Z.H., 2020. Nutritional value of forage sweet sorghum and its gas production performance evaluated using incubation with sheep rumen fluid in vitro. *Acta Prataculturae Sinica*. 29, 43–49.

Advances in Renewable Energy and Sustainable Development – Liang & Kasmani (Eds)
© 2023 Copyright the Author(s), ISBN: 978-1-032-39407-7

Evaluation of the performance characteristics of colored melting ice-snow asphalt concrete

Weidong Ji*
Ningxia Communications Construction Co., Ltd., Yinchuan, Ningxia, China

Benju Zhang
Jingbo Middle School, Helan County, Ningxia, China

Zhitao Zhang, Guangyu Men & Yuchuan Feng
Ningxia Communications Construction Co., Ltd., Yinchuan, Ningxia, China

ABSTRACT: Tunnel entrances and exits can easily lead to traffic accidents due to driving fatigue and easy icing of road surfaces in winter. To reduce traffic accidents, this study combines the advantages of colored asphalt pavement and low freezing point (LFP) asphalt pavement to research the performance of colored LFP asphalt concrete with strong exploratory and practicality. First, the amount of pigment was determined by a colorimeter. Secondly, based on the indoor test, the mixing ratio design, pavement performance, and melting ice-snow performance tests of colored LFP asphalt mixture were carried out. The performance differences between the colored asphalt mixture and the colored LFP asphalt mixture were compared and analyzed. To check whether the road performance meets the specification requirements and to verify the active ice and snow melting effect of the colored LFP asphalt concrete. The results show that: the color pigment content is 2.5%–3.5%, and the color rendering effect of colored asphalt concrete was the best. When the LFP filler content is 4%–5%, the pavement performance of the colored LFP asphalt concrete can meet the requirements of the specification, and its active melting ice-snow effect was good. The colored LFP asphalt concrete proposed in this study is a new type of green road paving material that has great application value.

1 INTRODUCTION

With the rapid development of the social economy, China's infrastructure construction has made remarkable achievements. The rapid growth of modern transportation networks has enhanced people's awareness of road environment aesthetics and safety (Tang et al. 2018; Wang et al. 2019). Especially for tunnel safety issues, research and investigations show that tunnels are prone to traffic accidents and are prone to frequent traffic accidents. The traffic accidents and fatalities are more severe than in ordinary road sections (Ye et al. 2022).

Since the color of the road at the entrance and exit of the tunnel is the gray-black of the asphalt concrete itself, gray or black has easily made human nervous fatigue and other problems (Yuan et al. 2021). Therefore, colored asphalt pavements, which play a specific role in inducing traffic, relieving driving fatigue, and improving driving safety and comfort, have received increased attention. The accident rate can be reduced by 85% to 90% (Ye et al. 2022). At present, colored pavements are mostly used in tunnel entrance and exit sections, which can remind, induce, decelerate, and improve pavement skid resistance (Autelitano & Giuliani 2019). At the same time, due to the

---

*Corresponding Author: wedoji@163.com

DOI 10.1201/9781003349648-21

pressure difference between the entrance and the exit of the tunnel, the air outside the tunnel will enter the tunnel under the action of the pressure difference, thereby changing the temperature of the tunnel road surface (Xiwang et al. 2021). Especially in the rainy and snowy weather in winter in the cold region, the inertia of the wheel motion carries the rain and snow into the tunnel. The cold air has intense heat exchange with the temperature in the tunnel, which changes the temperature field of the road surface at the entrance and exit of the tunnel, and causes the road at the entry and exit to freeze quickly. It is straightforward to cause traffic accidents, resulting in many casualties and property losses (Zhong et al. 2019). Therefore, it is necessary to research the problems mentioned above by combining the color of pavement and the technology of melting ice and snow.

Colored asphalt refers to decolorized asphalt or artificially prepared light-colored binder materials which are mixed with various colors of pigments, additives, and other materials. Colored asphalt concrete can be formed after paving and rolling. Colored pavement has the advantages of coordinating with the surrounding environment, beautifying the city, alleviating driving fatigue, promoting driving safety, reducing road noise, and alleviating the "heat island effect" (Petrukhina et al. 2021; Yang et al. 2022). The LFP asphalt pavement can reduce the freezing temperature from $-3°C$ to $-10°C$, increase the exposure rate of the road surface, reduce the incidence of traffic accidents in winter snow and ice weather, improve the peel ability and snow removal efficiency of the road surface and ice, and effectively protect the road surface (Yiqiu et al. 2019). Therefore, this study combined the advantages of colored asphalt pavement and LFP asphalt pavement, pigments and LFP fillers were selected to replace mineral powder, and they were mixed into colored asphalt concrete in different proportions. The mixing ratio of the colored LFP asphalt mixture was designed. The influence of pigment content on the color rendering effect of colored pavement was analyzed. The effects of LFP filler content on road performance and ice melting performance were discussed. The optimum dosages of pigments and LFP filters are proposed.

2 MATERIAL PROPERTIES AND EXPERIMENT METHODS

2.1 *Material properties*

2.1.1 *Decolorized binder*
This study selected the decolorized binder produced in Yinchuan, Ningxia. The test results of its main technical indicators are shown in Table 1.

Table 1. Technical indicators of the decolorized binder.

Parameters	Units	Requirements	Results
Penetration	0.1 mm	80–100	86
Ductility (10°C)	cm	≥ 45	78
Softening point	°C	≥ 50	54

2.1.2 *Pigment*
The pigment used in this paper is iron red, which is the most stable compound in iron oxides (Izadi et al. 2022). The effect of iron-red on light is very durable, and it can strongly absorb the ultraviolet part, and the impact on water and atmosphere is stable. The technical requirements and test results of iron oxide red pigment are shown in Table 2.

2.1.3 *Aggregate*
The aggregate in this paper is red sandstone, which is produced in Enhe, Ningxia. The aggregate is suitable for the material required for the upper layer of high-grade highways. The color is reddish-brown, which can be better combined with the iron red pigment to make the color of the asphalt pavement studied more prominent. The specific technical requirements and test results of red sandstone are shown in Table 3.

Table 2. Technical index of iron oxide red pigment.

Parameters	Units	Requirements	Results
Iron content	%	≥ 95	97
Soluble content	0.1	≤ 1.0	0.1
105°C Volatile	mm	≤ 1.5	0.4
Tinting strength	%	98–102	100

Table 3. Technical indicators of red sandstone.

Parameters	Units	Requirements	Results
Crush value	%	≤ 22	14.3
Los Angeles attrition loss	%	≤ 25	15
Needle-like content	%	≤ 15	2.9
Apparent relative density	–	≥ 2.6	2.69
Water absorption	%	≤ 2.0	1.2
Solidity	%	≤ 8	2
< 0.075 mm content	%	≤ 1.0	0.4
Adhesion	grade	≥ 4	4

2.1.4 *LFP filler*

In this paper, the LFP filler is from Harbin, which can play the role of actively melting ice and snow. The specific technical indicators are shown in Table 4.

Table 4. Technical indicators of LFP fillers.

Parameters	Units	Requirements	Results
Exterior	–	Uniform color	white powder
PH value	–	7.0–9.0	8.0
Relative density	–	≥ 1.7	2.2
Water content	%	≤ 2	0.3
Freezing point	°C	≤ -5	−7

2.2 *Mixture design*

Colored LFP asphalt pavement is mainly used for the upper layer of asphalt pavement. The AC-13 type is mainly used to gradate the upper layer of asphalt pavement in China (Fu et al. 2021), so the gradation of the mixture in this study is tested by the AC-13 type. According to the particle size sieving test results, the mix proportion design is carried out regarding the JTG E20-2011. In the test, the gradation curve of different contents only affects the aggregate below 0.6 mm and has little effect on the overall gradation. The test gradation is shown in Figure 1.

According to JTG E20-2011, the Marshall sample was formed, and its porosity, stability, and flow value were tested. Combined with the JTG F40-2004 hot mix asphalt mixture proportion design method, the optimal oil-stone ratio of colored LFP asphalt mixture is 5.0%.

Studies in the literature have shown that the addition of anti-stripping agents can improve the adhesion between asphalt and aggregates and improve the road performance of asphalt mixtures (Amir & Ali 2013). To improve the overall road performance of the colored LFP asphalt mixture, slaked lime was selected as the anti-stripping agent in the test process, and the dosage was 1%.

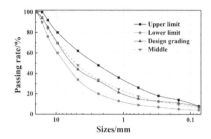

Figure 1. AC-13 grading curve.

2.3 Experiment methods

2.3.1 Color rendering performance

The color difference between the colored asphalt mixture with pigment and the asphalt mixture without pigment was defined as the total color difference. The amount of pigment in colored asphalt concrete was determined based on the change of color difference, and the colorimetric parameters of colored asphalt concrete were measured using a CM-26d colorimeter. The chromaticity parameter is calculated according to Eq. (1).

$$\Delta E^* = \sqrt{(\Delta L)^2 + (\Delta a)^2 + (\Delta b)^2} \qquad (1)$$

Where ΔE^* is the chromaticity parameter, ΔL is the lightness index difference, Δa is the red-green axis chromaticity index difference, and Δb is the yellow-blue axis chromaticity index difference. ΔL, Δa, and Δb can all be measured by a CM-26d colorimeter.

2.3.2 Pavement performance

Since there is no separate performance index for colored asphalt mixture at present, the colored asphalt mixture is evaluated regarding the evaluation parameters of petroleum asphalt mixture. According to the relevant test methods in JTG E20-2011, high-temperature rutting, low-temperature bending, immersion Marshall, and freeze-thaw splitting tests were performed on colored LFP asphalt mixture and colored asphalt mixture, respectively. We evaluate the high-temperature stability, low-temperature crack resistance, and water stability of colored LFPasphalt mixtures to verify whether they meet the specifications.

2.3.3 Melting ice-snow performance

The freezing point test mainly tests the performance of active ice and snow melting. In this method, the freezing point of the antifreeze asphalt mixture is evaluated by the sponge drawing method.

3 ANALYSIS OF RESULTS

3.1 Analysis of the effect of pigment dosage on color rendering

The content of the pigment is the most critical factor affecting the color rendering effect of colored asphalt mixture. Therefore, in this study, the color difference change was used as the evaluation index, and 1.5%, 2.5%, 3.5%, and 4.5% of the pigment were used to replace the mineral powder to explore the optimal pigment content colored asphalt mixture. The results are shown in Figure 2. It can be seen from the figure that with the increase of the pigment content, the color difference of the colored asphalt mixture also increases. When the pigment content is 2.5%, the color difference is 27.7, which increases by 42.1% compared with the colored asphalt mixture with a pigment content of 1.5%. When the pigment content is greater than 2.5%, the color difference of the colored asphalt mixture is relatively tiny. Therefore, considering the economic cost and color rendering effect, this study determined the pigment content to be 2.5%–3.5%.

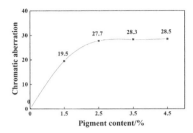

Figure 2. The effect of pigment content on the color difference of the mixture.

3.2 *Analysis of LFP fillers content on pavement performance*

As a pavement surface material, colored LFP asphalt concrete will be affected by vehicle loads and various environmental factors such as light, rain, and water. Therefore, colored LFP asphalt mixtures should have good high-temperature stability, low-temperature crack resistance, water stability, and other technical properties to achieve the purpose of long-term use of the pavement surface. Based on determining the optimal pigment content, this study replaces mineral powder with LFP fillers in different proportions to prepare colored LFP asphalt concrete. The substitution rates of LFP fillers were 2%, 4%, 5%, and 6% of the total aggregate, respectively. The effects of different LFP filler contents on pavement performance and ice and snow melting performance of colored LFP asphalt mixture were analyzed. The test results are shown in Figures 3 to 6.

It can be seen from the figure that the high-temperature stability, low-temperature crack resistance, and water stability of the colored asphalt mixture mixed with LFP fillers are lower than those of the ordinary-colored asphalt mixture. And the decreasing range increases with the increase of LFP filler content. The colored LFP asphalt mixture with the LFP filler content of 5% is compared with the ordinary-colored asphalt mixture. The dynamic stability decreased by 34.2%, the low-temperature bending strain decreased by 36.4%, the Marshall residual stability decreased by 8.4%, and the freeze-thaw splitting tensile strength ration (TSR) decreased by 7.1%. However, the technical indicators of the colored low-freezing-point asphalt mixture with a LFP filler content of 5% all meet the specification requirements.

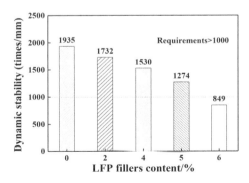

Figure 3. Rutting test results.

3.3 *Analysis of LFP fillers content on melting ice-snow performance*

Through the freezing point test of color LFP Marshall specimens with different contents (Figure 7), the test results show that the Marshall specimens with LFP contents of 4%, 5%, and 6% are at

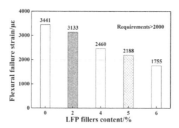

Figure 4. Low temperature bending test results.

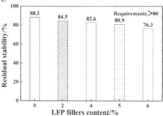

Figure 5. Marshall immersion in water test results.

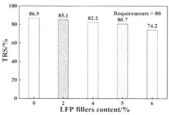

Figure 6. Freeze-thaw tensile split test results.

−5°C, the surface of the sample and the sponge is non-adhesive. The surface of the test piece and the sponge are easily peeled off at −10°C. The Marshall test piece of ordinary colored asphalt and the Marshall test piece with a LFP content of 2% is difficult to peel off from the surface of the test piece and the sponge at −5°C. Therefore, combined with the analysis of pavement performance and active melting ice-snow performance, the LFP content determined in this study should be 4%–5%.

Figure 7. Freezing point pull test.

4 CONCLUSIONS

In this paper, we combined the advantages of colored asphalt pavement and LFP asphalt pavement, and evaluation of the performance characteristics of colored melting ice-snow asphalt concrete. The main conclusions can be summarized as follows:

(1) The pigment content increase, and the color difference of the colored asphalt mixture also increases.

(2) Compared with ordinary colored asphalt concrete, the indicators of dynamic stability, bending failure strain, Marshall residual stability after immersion, and freeze-thaw splitting tensile strength ration of colored LFP asphalt concrete have decreased to varying degrees.

(3) To ensure that the colored LFP asphalt mixture has an excellent structural frame, color rendering effect, stability, better pavement performance, and melting ice-snow performance, the pigment content to replace the mineral powder should be 2.5%–3.5%, and the LFP fillers dosage should be 4%–5%.

In terms of future work, relying on the project, a test pavement will be built to verify the application effect of active melting ice-snow and safety warning on colored LFP asphalt concrete pavement.

ACKNOWLEDGMENT

This paper was financed by the Natural Science Foundation of Ningxia Hui Autonomous Region (Grant No. 2022AAC03759). We appreciate the reviewers for their comments that have notably helped us improve the manuscript.

REFERENCES

Amir M, Ali K. A review of state of the art on stripping phenomenon in asphalt concrete[J]. *Construction and Building Materials*,2013,38(1): 423-442.

Autelitano F, Giuliani F. Daytime and nighttime color appearance of pigmented asphalt surface treatments[J]. *Construction and Building Materials*, 2019,207:98-107.

Fu X, Chen Y, Sun M, et al. Effects of different colorants on service performance for colored asphalt pavement in cold regions[J]. *Pigment & Resin Technology*, 2021,32(2):1-8.

Izadi A, Zalnezhad M, Bozorgi Makerani P, et al. Mix design and performance evaluation of coloured slurry seal mixture containing natural iron oxide red pigments[J]. *Road materials and pavement design*, 2022,23(4):907-924.

Petrukhina N N, Bezrukov N P, Antonov S V. Preparation and Use of Materials for Color Road Pavement and Marking[J]. *Russian Journal of Applied Chemistry*, 2021,94(3):265-283.

Tang P, Mo L, Pan C, et al. Investigation of rheological properties of light-colored synthetic asphalt binders containing different polymer modifiers[J]. *Construction and Building Materials*, 2018,161:175-185.

Wang C, Xiao X, Lu Y, et al. Utilization and properties of modified epoxy resin for colorful anti-slip pavements[J]. Construction and Building Materials, 2019,227:116801.

Xiwang Z, Qinguo M, Haiqiang J, et al. Study on the Temperature Field and Anti-freezing Length of High-speed Railway Tunnel in Cold Regions under Natural Wind Conditions[J]. *Railway Standard Design*, 2021,65(09):140-147.

Yang W, Zhang K, Yuan J, et al. Tire-track resistance performance of acrylic resin emulsion coatings for colored asphalt pavements[J]. *Road materials and pavement design*, 2022,23(4):874-889.

Ye F, Su E, Liang X, et al. Review and Thinking on Landscape Design of Highway Tunnel[J]. *China Journal of Highway and Transport*, 2022,35(01):23-37.

Yiqiu T, Chi Z, Huining X, et al. Review on Snow-melting and Deicing Characteristics and Pavement Performance of Active Deicing and Snow Melting Pavement[J]. *China Journal of Highway and Transport*, 2019,32(04):1-17.

Yuan Jingyu, Liu Xiaojian, Yao Sheng, et al. Visual Characteristics of ColoredPavementat Entrance and Exit of Expressway Tunnel[J]. *Journal of Chongqing Jiaotong University*(Natural Science), 2021,40(09):60-67.

Zhong Y, Du K, Zhao D, et al. Real time early warning of ice and snow and visibility at entrance and exit of tunnel[J]. *Highway*, 2019,64(04):191-196.

Green energy security and urban sustainable development

Advances in Renewable Energy and Sustainable
Development – Liang & Kasmani (Eds)
© 2023 Copyright the Author(s), ISBN: 978-1-032-39407-7

Research on site selection of prefabricated component factory based on P-median model

Ziji Liu*

The Department of Business Administration, Shanghai University of Engineering and Technology, Shanghai, China

ABSTRACT: Investing in the construction of prefabricated component factories is a very important aspect to be considered during the transformation of building industrialization; however, there is a problem of unreasonable site selection for prefabricated component factories. To help enterprises select a site scientifically and effectively, this study analyzes the current problems in the development of prefabricated buildings, improves the standard P-median model based on its characteristics, and comprehensively considers the two factors of transportation cost and construction cost. The greedy takeaway algorithm finds the optimal solution. This paper analyzes this method for a practical case.

1 INTRODUCTION

Promoting sustainable development, the development of the prefabricated construction industry, and the transformation of construction industrialization has been the general trend in the development of China's construction industry. Prefabricated buildings are inseparable from prefabricated component factories. With the government's introduction of a series of policy support, the market demand for prefabricated building components has been stimulated. More funds have been invested into the construction of prefabricated component factories. However, blind investment and construction can easily lead to a waste of capital and production capacity, which is not conducive to the overall development of the prefabricated industry. The issue of location selection is the primary consideration for investing in prefabricated components factories. Therefore, how to scientifically and effectively help the prefabricated component factory in site selection planning is crucial.

2 RESEARCH STATUS

Aiming at the general location problem, Hakimi established a mathematical model for the P-median value to study the optimal location of the "switching center" in the communication network. Jorge H. Jaramillo analyzed and proved that the genetic algorithm is feasible in solving the location problem. According to the characteristics of the e-commerce distribution center, Bizheng Liu proposed an integrated model of distribution center location and vehicle route and designed a combination of genetic algorithm and simulated annealing algorithm to solve the problem.

In response to the site selection of prefabricated component factories, the "Guidelines for the Construction of Prefabricated Concrete Component Factories for Prefabricated Buildings" issued by Shanghai provides guidance from seven aspects, including service areas, transportation conditions, and enterprise collaboration. Qianrong Luo, using the analytic hierarchy process, established the location index system of the concrete precast component factory and considered infrastructure, economy, society, policy, and other factors as the main influencing factors. Wei Song constructed

---

*Corresponding Author: 1374793836@qq.com

DOI 10.1201/9781003349648-22

the objective function with the total cost as the best one and used it as a greedy takeaway heuristic algorithm to solve the objective function to obtain the optimal solution to the prefabricated component factory location problem.

Scholars at home and abroad have already established a certain research foundation for the site selection of prefabricated component factories, but they still need to be further improved. Based on the summary of the existing research, this paper conducts a quantitative study on the location selection of the prefabricated component factory and provides a reference method for the location selection of the prefabricated concrete component factory in the future.

3 QUESTION FORMULATION AND MODEL BUILDING

This paper is based on the current problems in the development of prefabricated buildings: 1) The transportation cost of components is relatively high, while the transportation cost of prefabricated components is proportional to the transportation distance. 2) Due to the particularity of the product, its transportation distance is controlled by the service radius. 3) The cost of prefabricated components is high, and the fixed construction cost of the prefabricated construction plant greatly affects the unit price of the product. Therefore, it is very meaningful to consider the location of the prefabricated component factory from the aspects of transportation cost and construction cost.

In the site selection study of this paper, there are M candidate prefabricated component factory addresses and N prefabricated component assembly sites in demand; among them, the location of the prefabricated component factory can be selected, and the address and demand of the assembly site are fixed. By minimizing the transportation and construction costs of the prefabricated component factory, the supply and demand relationship between the prefabricated component factory and the assembly site is obtained, and P points are determined from the M candidate points to ensure the efficient and smooth transportation of the prefabricated components. The site with the lowest total cost is selected.

3.1 *Standard P-median model*

The objective function of the P-median problem is

$$\min Z = \sum_{i \in N} \sum_{j \in M} d_i c_{ij} y_{ij}$$

Restrictions:

$$\sum_{j \in M} y_{ij} = 1, i \in N$$

$$\sum_{j \in M} x_j = p$$

$$y_{ij} \leq x_j, i \in N, j \in M$$

$$x_i \in \{0, 1\}, j \in M$$

$$y_{ij} \in \{0, 1\}, i \in N, j \in M$$

In the formula: n is the collection of the demand point in the area, $N = \{1, 2, ..., n\}$; M is a collection of candidate facilities in the area, $M = \{1, 2, ..., m\}$; d_i is the i-s one. The demand point is required; c_{ij} is the unit transportation cost from the demand point i to the facilities point j; P allows the number of facilities to be constructed, $P < m$; x_j is 0-1 variable, x_j means Otherwise, it is 0; y_{ij} means that the demand point i is provided by the facility j; otherwise, it is 0.

The standard P medium value model refers to the collection of a given quantity and location demand collection and a collection of a candidate facility. The lowest transportation costs between facilities and demand points are reached. The P-median model is generally applied to the location of the location or warehouse, but this model only considers the transportation distance and user needs in the site selection. The prefabricated component factories have the characteristics of large area, high construction costs, and high operating costs. Therefore, the middle-value model of the original P is not suitable for the selection of prefabricated component plants, and we need to improve it.

3.2 *Improved P-medium-value model*

In the modeling of the minimum cost in the modeling, this article mainly considers the transportation cost and the construction cost of the facility point to the demand point. The target value is the sum of the annual depreciation of annual transportation costs and construction costs. The improved P-median site selection model has the following basic assumptions:

(1) A demand point is only supplied by one facility point.
(2) The production volume of each facility point can meet multiple demand points.
(3) The unit transportation cost is proportional to the transportation distance.
(4) The demand of the point of each need must be known.
(5) The construction costs of different facilities are different and known.

$$\min Z = \sum_{i \in N} \sum_{j \in M} d_i c_{ij} y_{ij} + \sum_{j \in M} \frac{f_j}{g_j} x_j$$

Restrictions:

$$\sum_{i \in N} d_i \leq q$$

In the formula: f_j is the construction cost of the jth candidate facility; g_j is the estimated service life of the jth candidate facility.

3.3 *Greedy takeaway heuristic algorithm*

In the greedy takeaway heuristic algorithm, we must first clearly solve the target and formulate a feasible greedy criterion and then take the current basis as the optimal choice, and gradually search for a better choice based on the current basis until the target value that satisfies the problem is found. The basic calculation steps for a prefabricated component plant are as follows:

(1) Initialization, let the number of cycles $k = m$, select all m prefabricated component factory candidate sites, and then assign each construction project to the nearest prefabricated component factory candidate site to find the total cost.
(2) Select and take a point from the m candidate sites of prefabricated component factories, which satisfies the following conditions: If it is taken away, and the construction projects allocated by it are re-assigned to other candidate sites of prefabricated component factories, the new total cost increment h is generated under the assignment. If increment h is the smallest, then delete the candidate site of the prefabricated component factory from the candidate, let $k = k-1$.
(3) When $k = p$, end the calculation; otherwise, return to (2).

4 CASE ANALYSIS

A company plans to build two prefabricated construction factories in a certain place to meet the construction projects ($I_1, I_2, I_3, I_4, I_5, I_6$ and I_7) that need to order prefabricated components around the region. After inspections on the aspects of local industrial policies, transportation conditions, venue environment, and residents' wishes, five candidate buildings (J_1, J_2, J_3, J_4, and J_5) were now determined. The candidate address of the prefabricated component factory for the construction project, the transportation cost C_{ij} and the demand of each construction project D_i are shown in

Table 1. Transportation cost and the demand for each construction project.

$C^3_{ij(yuan/m)}$	I_1	I_2	I_3	I_4	I_5	I_6	I_7
J_1	744	651	465	744	1395	1860	1674
J_2	930	837	372	558	520.8	837	1023
J_3	855.6	1674	744	1209	558	465	1860
J_4	2046	1599.6	1581	669.6	744	1767	372
J_5	2101.8	1860	1525.2	1395	372	930	558
$Di/10000m^3$	5.4	4.8	6.5	4.6	5.3	5.7	4.2

Table 2. Construction cost and expected service life.

	J_1	J_2	J_3	J_4	J_5
$F_j/10000$yuan	16,000	11,000	12,000	11,500	13,000
$G_j/$year	12	8	10	9	10

Table 1. The construction cost of the prefabricated component factories F_j and the expected service life G_j are shown in Table 2.

Use greedy takeaway heuristic analysis and solution.

Step 1: When $p = 5$, assign the prefabricated component demand points (I_1, I_2, I_3, I_4, I_5, I_6, and I_7) according to the principle of minimizing transportation costs. (J_1, J_1, J_2 J_2, J_5 J_3, and J_4). At this time, the annual transportation cost is 183.117 million yuan, the annual fixed construction cost is 91.9444 million yuan, and the total cost is 275.0614 million yuan.

Step 2: Take the candidate points J_1, J_2, J_3, J_4, and J_5 of the prefabricated component factory, respectively, and re-assign them according to the principle of minimizing transportation costs. First, take I1, the prefabricated component factory candidate points (J_3, J_2, J_2, J_2, J_5 J_3 and J_4) assigned by the prefabricated component demand points (I_1, I_2, I_3, I_4, I_5, I_6, and I_7), respectively, at this time the total cost is 289.0992 million yuan, and the total cost increment is 14.0377 million yuan. Similarly, when J_2 is taken away, the total cost increment is 9.7897 million yuan; when J_3 is removed, the total cost increase is 22.954 million yuan. When the J_4 is taken away, the total cost increment is RMB 8,034,200. When the J_5 is taken away, the total fee increment is 8.858 million yuan. Since the cost increment when taking J_4 is the smallest, the candidate point J_4 is taken for the first time. At this time, $p = 4$, the total cost is 283.0957 million yuan.

Step 3: Repeat step 2, take out the candidate points J_1, J_2 J_3, and J_5 of the prefabricated component factory, respectively, and re-assign them according to the principle of minimizing the transportation cost. When J_1, J_2, J_3, and J_5 are taken away, respectively, the increments of the total cost are 19.8053 million yuan, 13.7677 million yuan, 22.954 million yuan, and 28.9164 million yuan. Since the cost increment when taking J_2 is the smallest, the candidate point I2 is taken for the second time. At this time, $p = 3$, the total cost is 296.8633 million yuan.

Step 4: Repeat Step 3, take out the candidate points J_1, J_3, and J_5 of the prefabricated component factory, respectively, and re-assign them according to the principle of minimizing transportation costs. When J_1, J_3, and J_5 are taken away, respectively, the increment of the total cost is 9.2005 million yuan, 27.505 million yuan, and 56.0633 million yuan. Since the cost increment when taking J_1 is the smallest, the candidate point J_1 is taken for the second time. At this time, p=2, and the total cost is 306.0638 million yuan. To sum up, the final decision result of the candidate point of the prefab factory is J_3 and J_5.

5 CONCLUSION

The site selection of the prefabricated component factory is an important early stage of investment and construction of the prefabricated component factory, and it is an important part of promoting the development of prefabricated buildings, which directly affects the development quality of prefabricated buildings in my country. Starting from the two cost factors that need to be considered in the site selection of prefabricated buildings, this paper improves the original P-median model, and comprehensively considers the transportation cost and construction cost of investing in the construction of a prefabricated construction plant. It uses the greedy takeaway algorithm to optimize the target continuously, and the cost is minimized. Factors such as cost, local policies, transportation conditions, and market demand should be comprehensively considered in the site selection of a prefabricated component factory. However, this paper only conducts quantitative analysis from the perspective of cost, which has certain limitations. The model and algorithm proposed in this paper have clear steps and simple methods, which are easy to be accurately grasped and reasonably used by planning decision-makers, and have certain practical significance and reference value.

REFERENCES

Bizheng Liu, Chao Mao Distribution center location problem and its optimization under E-commerce [J] *Systems Engineering*, 2008, 26 (10): 17–21.

Hakimi. Optimum Locations of Switching Centers and the Absolute Centers and Medians of a Graph[J]. *Operations Research*, 1964, 12(3): 450–459.

Jorge H. Jaramillo and Joy Bhadury and Rajan Batta. On the use of genetic algorithms to solve location problems[J]. *Computers and Operations Research*, 2002, 29(6): 761–779.

Qianrong Luo, Xiyue Dong, Deheng Zeng Location of PC component factory of prefabricated building based on fuzzy analytic hierarchy process [J] *Journal of Civil Engineering and Management*, 2018, 35 (03): 111–117 + 123 DOI:10.13579/j.cnki. 2095-0985.2018.03.018.

Wei Song, Yikun Su Research on the location of PC component factory based on greedy take heuristic algorithm [J] *Architectural Technology*, 2019, 50 (05): 615–618.

Advances in Renewable Energy and Sustainable
Development – Liang & Kasmani (Eds)
© 2023 Copyright the Author(s), ISBN: 978-1-032-39407-7

Traditional ecological wisdom of Guilin Longji terraced field landscape from the perspective of water adaptability

Luyao Hu*
Guilin Tourism University, Guilin, Guangxi, China

ABSTRACT: Based on the field research of terraced agricultural landscape, this paper from the perspective of traditional ecological wisdom, through on-site interpretation, participant observation, in-depth interviews and other methods, in-depth and comprehensive understanding of local natural knowledge, traditional agricultural practices and social mechanisms, combined with existing Literature, local chronicles, genealogy, and digital archives of village landscapes are studied, and from the perspective of water adaptability, the traditional ecological wisdom of Guilin Longji terraced rice landscape is deeply excavated, and the core value of its ecological wisdom is summarized and extracted, in order to contribute to the construction of modern rural landscapes.

1   INTRODUCTION

During the long-term water resources management and the struggle against various water disasters, ancient human beings have accumulated simple agricultural production experiences with ecological value. Traditional agricultural production and life are highly dependent on irrigation technology and settlement planning technology. Adaptive Landscape (Chen & Yu 2015; Yu & Chen 2014). Water adaptability is an important part of ecological adaptability (Li et al. 2014), which is the result of long-term natural selection and artificial adaptation (Zhang 2020). These simple ecological experiences, as unique cultural heritage, are important for inheriting national cultural traditions and solving contemporary water problems. Dilemma problems, building an ecologically harmonious society, etc., still have reference significance (Huang 2014).

The Longji terraced field system in Guilin is a living specimen of ancient ancestors adapting to and using nature (Yang 2010; Zheng et al. 2019). Longji Ancient Zhuang Village is located in Longsheng County, Guangxi, at the southwestern foot of the Yuechengling Mountains in northern Guangxi. The history of terrace farming has been more than 500 years. Longji is a typical mountainous landform with very little flat land. The Longji Mountains and the Jinzhu Mountains form a watershed here. The Jinjiang River winds through the valley from the northeast to the southwest. The ancestors of the Zhuang people migrated here. Taking advantage of the local landform, climate, and vegetation characteristics, following the spatial variation law of the ecological environment, choosing terraced agriculture as a means of livelihood and developing a terraced agricultural culture in hundreds of years of production practice, forming a "Forest Resource View-Village Settlement-Terrace Agriculture" landscape pattern, constructing a terraced rice farming ecological landscape system that adapts to natural conditions, making the terraced landscape and the ecosystem integrated.

For a long time, the Longji Zhuang people have lived in harmony with nature, forming an ecological concept of harmony and unity between man and nature. The landscape of Longji terraced fields is a reflection of the fundamental relationship between man and land. As a typical "rice farming

---

*Corresponding Author: 1092407145@qq.com

160   DOI 10.1201/9781003349648-23

nation," the Zhuang people obtain rice from the land to meet the needs of survival and development through the reclamation and maintenance of terraced fields. And customs also continue to develop and mature. Longji terraced fields are the product of the harmonious coexistence of man and nature. In the long-term relationship between man and land, a balanced and stable ecosystem has been formed, where nature and humanities have formed a highly coordinated relationship, which is constantly circulating.

2 WATER RESOURCE UTILIZATION TECHNOLOGY OF LONGJI TERRACED FIELDS

Water is the lifeblood of terraced fields, and the formation of the cultural landscape of terraced fields in Longji is closely related to water. The success of agro-ecosystems lies in effective water resources management (Figure 1). There is a popular saying in Guangxi that "the Zhuang nationality lives in the water head." The Zhuang nationality is the oldest ethnic group in the Lingnan area, and its farming culture has a strong appeal and dependence on water. The system construction and ecological civilization concept of water resources management of Longji terraced fields are full of scientific truth and ecological wisdom and have a complete set of terrace water resources utilization and management systems for drinking water, irrigation and fire prevention. Under the background of increasingly prominent problems, it has important reference value and practical significance for coping with climate change and promoting sustainable agricultural development.

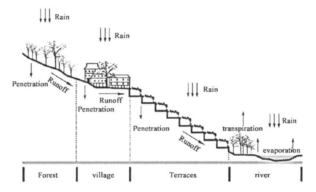

Figure 1. Landscape of Longji terraced fields.

2.1 *Village location layout*

The Longji Ancient Zhuangzhai is named after the dragon's back that hovers down from the top of the mountain, and the Longji Village is located on it. The location of the village is determined by fully considering the local topography and natural water system. The Longji Mountain and the opposite Jinzhu Mountain overlap each other. The place where the two mountains intersect forms a "V"-shaped river valley. The bottom end is the Jinjiang River flowing from the northeast to the southwest. The river and the mountain block each other. They are all hidden by the mountain peaks, which can better hide the wind, get water, and take advantage of the vitality. The maximization of the location of the village combines the advantages of the environment to provide suitable production and living space for the villagers, so that they can fully enjoy and rationally use the natural resources (Figure 2).

2.2 *Water source development*

The irrigation system of Longji Ancient Zhuang Village is mainly formed by diverting mountain spring water to terraced fields through canals and diverters at all levels. Mountain spring water is

Figure 2. Longji rice terraced landscape.

distributed on the mountainside or in the village. There are hundreds of canals at all levels, and they are scattered in all directions of the terraced fields. They form an irrigation water system from top to bottom. In this way, the layers are filled, and the excess water flows into the Jinjiang River.

2.2.1 *Canal*
The ancestors of Longji built hundreds of canals on the mountainside that can be connected to terraced fields. In order to prevent the sand and gravel from entering the terraced fields with the water flow, causing the sand-alkalinization of the terraced fields, the ancestors dug a deep pit for the sedimentation of sand and gravel at the place where the water channel entered the terraced fields. Through this ancient method, the soil of the terraced fields can keep its fertility forever.

2.2.2 *Well*
There are several wells in Longji Village, which are composed of well bottom, well rail, well platform, well cover and drainage ditch. Longquan Well is the earliest one. There are two types of wells in Longji: house wells and pavilion wells. Among them, the pavilion-style well is built into a pavilion where the spring water spouts and the water is drained to the water storage tank by stone troughs or bamboo sticks. People can wash their hair and clothes and have social activities such as relaxing and chatting here. Longquan Pavilion is a public activity place in Liaojiazhai, ancient Zhuangzhai. The water here is not only used for the villagers' domestic use but the remaining water is directly introduced into the terraced fields below the village through ditches as an irrigation water source. The water volume from another nearby well, "Qingquan Pavilion," remains the same throughout the year, and the water temperature is warm in winter and cool in summer.

2.3 *Water diversion technology*

2.3.1 *Water diversion tool*
Bamboo tree: People cut down the fast-growing local bamboo, cut it in half, hollowed out the inner knot of the bamboo, and then connected it end to end as a water diversion facility for several kilometers, providing people's life and irrigation, reflecting the simple ecology and wisdom of the people of Longji. Although bamboo saplings are now replaced by PVC pipes or steel pipes in some places, the ecological properties of bamboo saplings are more suitable for application and promotion here (Figure 3).

Connecting the tube: "With the bamboo in the hollow, the hundreds and ten are connected, and the river can be drawn even if it is three or four miles away." Compared with the continuous tube,

Figure 3. Water diversion tool.

the advantage of the bamboo stalk is that it saves half of the material, but the continuous tube can ensure the water quality is clean and free from pollution. The advantage of the two materials is that they can drain water overhead and cross complex terrain. Generally, the spring water drained by bamboo tubes is used for drinking, and the water drained by bamboo sap is used to irrigate terraced fields.

Mushu: The raw materials of Mushu are fir and pine trees that are abundant in local high-altitude forests. The wood is made into grooves, and the spring water can be drained by connecting end to end. The service life of the wooden arbor is longer than that of the bamboo arbor, but the manufacturing cost of the wooden arbor is higher, and the curved wood needs to be found at the turning point.

2.3.2 *Water diversion method*

Generally, a canal is opened at the bottom of the forest for irrigation. The water source in the forest is first collected, and then the canal is used to divert it to the terraced field, and then the small ditches or bamboo stalks are used to divert the water to the terraced fields. In order to prevent the water from infiltrating and not flowing into the next terraced fields, the villagers often repair the field walls or directly use bamboo to drain the water, effectively saving water resources. In natural canals, to retain water and raise the water level to facilitate water intake and water distribution, simple "dams" composed of large stones are often seen, similar to the combination of stone and water features in classical gardens. It also reflects the original ecological wisdom and constitutes a unique regional agricultural landscape.

2.4 *Terraced fields*

In order to allow each terraced field to evenly distribute water, the ancestors of Longji used the method of ditching canals for irrigation. If there are many fields in a plot that need to divide water, install a "water divider" at the appropriate location of the mainstream or tributary. The water divider is made of polished stones or wood blocks, usually the locally abundant turquoise stone, and the polished slate is buried deep for reinforcement. It needs to be divided into several streams, and a few gaps are cut on it so that the section is in the shape of "concave and concave" of equal depth, and the corresponding water output is controlled by controlling the width of each groove, so as to be divided according to the specific terrace area. Water, making water distribution scientific and reasonable, saving water sources, and avoiding unnecessary disputes among villagers.

3 WATER RESOURCES MANAGEMENT SYSTEM OF LONGJI TERRACED FIELDS

3.1 *Township rules and regulations*

The ancestors of Longji set a "township agreement" for water use to maintain the normal agricultural production order and coordinate the village's production and domestic water use. As stipulated by the local customary law, the irrigation water for terraced fields should first meet the water consumption of the rice fields opened first and the rice fields irrigated by the branch canals dug first (Fu 2010). In addition, according to the number of irrigated fields, wood or stone should be used to chisel "concave" openings in the main canal or branch canal for water diversion. These township regulations ensure the order of irrigation and the fairness of irrigation water to avoid unnecessary water disputes (Fu 2010).

3.2 *Institutional customs*

Longji Zhuang ancestors have formed a complete set of irrigation water methods in the long-term cultivation process to ensure the optimal benefit of rice farming. The local village old organization formulated the terms of the water use agreement to resolve the water resources contradiction and solve the water dispute: "In the drought years, the fields and canals will still take water as usual, and it is not allowed to exchange for new ones privately, and take water by force..." In times of drought, the mountains and fields fetch water according to the ancients, and they dare not destroy the old and start the new; if they do not obey, they will be reported to the officials and sent to officials for the treatment..." (Fu 2010). This "village contract" system, which villagers consciously manage and follow, maintains the long-term operation of the terraced irrigation system and the healthy and sustainable agricultural production, and creates a terraced agricultural landscape full of vitality and harmonious coexistence (Fu 2010).

4 THE TRADITIONAL CONCEPT OF HUMAN AND WATER DEPENDENCE ON LONGJI TERRACED FIELDS

Water is the foundation for the survival of Longji terraced fields, and water occupies an extremely important position in the minds of the Longji Zhuang people. Under the influence of the concept of animism, the Longji Zhuang people regard water as their god, and their worship of water has profoundly affected the society and culture of the Longji Zhuang people (Wei 2007). Water resources are indispensable in the process of rice growth, and whether the rainfall is adjusted or not also affects the rice production in terraced fields. In the process of coexisting with nature for a long time, the Longji Zhuang people gradually realized the interactive relationship between "forest-water-field-food-people." They believed that "there is water only when there is forest, and there is the field when there is water. Only when there are fields can there be food, and when there is food, there is life." Under the guidance of this simple natural ecological view, they regard water and forest resources as the support of life and form the consciousness and habits to protect water sources and forests. (Wei 2007).

5 CONCLUSION

The agricultural landscape is an integral part of the rural landscape and contains the ecological wisdom of the interaction and relationship between man and nature. The terraced rice farming landscape of Longji Ancient Zhuangzhai is the product of Zhuang compatriots interacting with nature in the long-term production and life practice. It contains rich ecological wisdom, which is of great value for the protection of the ecological environment and the sustainable utilization of resources. Water is an indispensable foundation for local people's life and production. In the interaction between people and land for thousands of years, the ancestors of Longji have gradually formed a set of ecological wisdom systems and used this to guide production and life practice, forming splendid terraced agriculture. It is composed of water resource utilization technology, water resource management technology and the traditional concept of human and water interdependence. The Longji terraced rice farming landscape is based on the ecological cognition concept of "harmony between man and nature" and the landscape construction concept of "adopting measures to local conditions."

ACKNOWLEDGMENTS

This work is supported by the Middle-aged and Young Teachers' Basic Ability Promotion Project of Guangxi (Grant No. 2021KY0810) and the Scientific research project of Guilin Tourism University (Grant No. 2020C02).

REFERENCES

Chen Yiyong, Yu Kongjian. Ancient "Sponge City" Thought: Experience and Enlightenment of Water Adaptable Landscape [J]. *China Water Resources*, 2015(17): 19–22.

Fu Guanghua. Climate catastrophes and local response: the traditional ecological knowledge of Longji Zhuang people [J]. *Guangxi Ethnic Research*, 2010(02):84–92.

Fu Guanghua. The traditional water culture of the Zhuang nationality and the construction of contemporary ecological civilization [J]. *Guangxi Ethnic Studies*, 2010(03):86–94.

Huang Longguang. An Introduction to Minority Water Culture [J]. *Journal of Yunnan Normal University* (Philosophy and Social Sciences Edition), 2014, 46(03): 147–156.

Li Liang, Shang Hongchi, Xu Xi. Research on the ecological wisdom of water adaptive environment creation in ancient villages in western Beijing [J]. *Architecture and Culture*, 2014(12):85–87.

Wei Yichun. *Investigation and Research on Water Conservancy of Terraced Fields in Liaojia Ancient Zhuangzhai Village in Longji, Guangxi* [D]. Guangxi University for Nationalities, 2007.

Yang Zhuquan. Research on ecological wisdom in ethnic minority terrace culture in the "Yuechengling" area: Taking Longsheng Longji as an example [J]. *Agricultural Archaeology*, 2010(06):397–399.

Yu Kongjian, Chen Yiyong. Foreign traditional agricultural water adaptation experience and water adaptation landscape [J]. *China Water Resources*, 2014(03):13–16.

Zhang Jin. Discussion on the Perspective of Rural Landscape Cognition and Research Based on Adaptation [J]. *Chinese Landscape Architecture*, 2020, 36(03): 97–102.

Zheng Wenjun, Zhang Beibei, Wu Zhongjun. The wisdom of creating human settlements in Longji, Guilin [J]. *Chinese Garden*, 2019, 35(09):20–24.

*Advances in Renewable Energy and Sustainable
Development – Liang & Kasmani (Eds)
© 2023 Copyright the Author(s), ISBN: 978-1-032-39407-7*

Permanent basic farmland based on minimum cumulative resistance model

Lu Huang*

Hainan Guoyuan Institute of Land and Mineral Survey Planning & Design, Haikou, Hainan, China

ABSTRACT: Land resources are the most precious natural resources for human beings. As the basic resource for human survival and development, cultivated land plays a key role in ensuring national food security, and permanent basic farmland is the bottom line of cultivated land protection. The situation of cultivated land protection, especially permanent basic farmland protection, is becoming more and more serious. Therefore, a systematic study on land-use conflicts can provide guidance for rational land use, alleviate land-use contradictions and maximize land-use benefits. Based on the minimum cumulative resistance model, this paper rationally divides and evaluates the types and intensity of cultivated land multi-objective use conflicts. Based on extended land suitability evaluation, the types and degrees of land-use conflicts are quantitatively identified and judged, and permanent basic farmland is delimited in combination with the results of agricultural land quality classification. The research has important practical significance for solving the contradictions among cultivated land protection, new construction land layout, and ecological security construction.

1 INTRODUCTION

As the basic resource for human survival and development, cultivated land plays a key role in ensuring national food security. The permanent basic farmland is the bottom line of the cultivated land protection. However, with the accelerated development of urbanization and industrialization, on the one hand, a certain amount of arable land must be retained to ensure national food security. On the other hand, a certain scale of new construction land turnover support is needed (Zhang et al. 2020).

Based on the characteristics of limited area, fixed location and irreplaceable land resources, the contradictions and conflicts in the process of land resource utilization are more extensive. In view of the causes, identification and coordination of land-use conflicts, the existing identification methods mainly include mathematical model methods such as pressure-state-response, BP, CA, ecological risk index method and multi-objective suitability ranking combination method (Zhang 2018). Although the mathematical model method can objectively and quickly reflect the recognition results, the simplification of the model can not correctly reflect the actual situation due to the influence of the research object, the evaluation factors and the rationality of the model construction. The multi-objective suitability ranking and combination method is used to study the combination of regional agricultural land and construction land evaluation. The multi-objective suitability evaluation of cultivated land lacks the consideration of ecological security level and ecological utilization demand (Zhang & Zhang 2020). In a few comprehensive evaluation studies considering ecological land, agricultural land, and construction land, the evaluation of ecological land mainly adopts a

---

*Corresponding Author: 116168547@qq.com

166   DOI 10.1201/9781003349648-24

multi-factor weighted superposition. The index system is single, and the consideration of the ecological expansion level process and the integrity of the ecological security system is lacking (Yuan et al. 2020).

In the fields of basic farmland protection and delimitation, domestic scholars mainly study the quantity determination and spatial layout of basic farmland from the aspects of cultivated land quality evaluation, land suitability evaluation and potential analysis to promote the basic farmland protection from quantity protection to quantity and quality protection (Wei et al. 2019). However, the current basic farmland delimitation method is not very reasonable. Although the existing delimitation results can meet the requirements of quality and quantity, they lack the consideration of the spatial location of basic farmland plots and the judgment and prevention of potential conflict risks of plots. As a result, under the competition of various land-use modes, basic farmland is often distributed in areas with intense competition for land use and high potential conflict risks of land use. As a result, the layout of basic farmland is unstable, and the protection cost is high (The history, methods, systems, subjects and outlooks of rural planning in Japan-Rural Planning in Japan-residents participation, landscape, ecological village 2018).

This paper attempts to make a multi-objective suitability evaluation of cultivated land, using the minimum cumulative resistance model to construct the ecological security pattern of the study area and combining it with the evaluation results of cultivated land suitability and construction suitability. It also uses the multi-suitability permutation and combination method to divide the potential land-use conflict areas. Finally, under the principle of avoiding the potential conflict areas, it combines with the results of agricultural land classification. It provides a new idea for the coordinated development of ecology, cultivation and construction, enriches the method system of permanent basic farmland delimitation, and provides a reference for the rational multi-functional utilization of cultivated land and the realization of permanent basic farmland protection.

2 LAND WEIGHT ANALYSIS OF MINIMUM CUMULATIVE RESISTANCE MODEL

The minimum cumulative resistance model (MCR model) is a derivative application of the cost distance model, which refers to the sum of the minimum cost of species in the process of moving from the "source" through different landscape units to the destination, simulates the horizontal process of factor diffusion, and reflects the accessibility of potential movement and expansion of species, which is an important basis for evaluating the level of ecological security (Yang et al. 2019). The model was first proposed by Knappen in 1992 and is widely used in landscape pattern analysis. The establishment of the model mainly considers the three elements of "source," "distance," and "resistance coefficient of landscape medium" (Yang 2021), which are expressed as follows:

$$MCR = f \min \sum_{j=n}^{i=m} D_{ij} \cdot R_i \tag{1}$$

MCR represents the face value of the minimum cumulative resistance; f represents the positive correlation between the minimum cumulative resistance and the ecological expansion process, and the farther away from the source, the greater the resistance value; D_{ij} refers to the spatial distance from the source j to the landscape unit i; R_i represents the resistance coefficient of the landscape unit i to species movement.

The weight of the resistance factor refers to the relative importance of the resistance factor in the expansion of ecological headwaters. There are many methods to determine the weight, including the qualitative analysis scoring method, principal component analysis, grey correlation method, analytic hierarchy process, etc. This paper uses the analytic hierarchy process to determine the weight of each resistance factor. The steps of the analytic hierarchy process (AHP) include: the first step is to establish a hierarchical structure model based on the factor analysis and the relationship between factors, and arrange the selected factors in the order of target layer (ecological security evaluation), criterion layer (terrain factor, ecological environment factor, distance factor), and index layer (slope, elevation, land-use type, NDVI, distance from rivers and lakes, distance from roads). The second step is to construct a pair of comparative judgment matrixes, in which experts score the evaluation factors according to a scale of 1-9, and then a plurality of judgment matrixes are formed by pairwise comparison. Finally, the consistency test is carried out. First, the maximum eigenvalue λ_{max} of each judgment matrix is calculated, then the consistency index CI is calculated, and finally, the consistency ratio is calculated through $CR = \frac{CI}{RI}$, and whether CR is less than 0.1 is judged. If it is less than 0.1, the consistency test is passed; otherwise, the matrix is adjusted (Rao & Zhang 2021). Based on the above analysis, the resistance factor weight and grading assignment are shown in Table 1.

Table 1. Resistance factor weight and grading assignment table.

First-order factor	Weight	Second-order factor	Weight	Grade	Resistance coefficient
Terrain factor	0.2865	slope	0.5123	225°	1
				15° -25°	3
				6° -15°	5
				2° -6°	7
				<2°	9
		elevation	0.4877	21200	1
				900-1200	3
				600-900	5
				300-600	7
				<300	9
Ecological environmental factor	0.5562	land-use type	0.5532	Forest land, water area, mudflat and marsh	1
				Scenic spots and special land, grassland and garden plot in construction land	3
				Cultivated land	5
				Other agricultural land and nature reserve land	7
				Construction land	9
		NDVI	0.4468	20.9	1
				0.7-0.9	3
				0.5-0.7	5
				0.2-0.5	7
				<0.2	9
Distance factor	0.1573	Distance from rivers and lakes (m)	0.5214	<500	1
				500-1000	3
				1000-2000	5
				2000-5000	7
				25000	9
		From the main road (m)	0.4786	25000	1
				3000-5000	3
				1000-3000	5
				500-1000	7
				<500	9

3 LAND GRADING OF RESISTANCE SURFACE

The resistance surface reflects the comprehensive influence of landscape units on ecological expansion in the horizontal process. The concrete realization process of the establishment of the resistance surface comprises the following steps: firstly, uniformly convert the data of each resistance factor determined above into grid data, and then calculate the ecological expansion resistance surface by using a grid calculator tool in ArcGIS according to the determined weight, classification and corresponding resistance coefficient of the resistance factor and a weighted index sum method. The drag surface is the total drag value for each grid cell (Shu et al. 2020). The calculation formula of the ecological expansion resistance surface is:

$$A_i = \sum w_1 \times w_2 \times f_{ij} \tag{2}$$

In the formula, A_i represents the resistance value of the ith grid unit; w_1 and w_2 represent the primary and secondary weight values of the jth resistance factor; f_{ij} represents the expansion resistance value of the ith grid unit and the jth resistance factor.

The generation of minimum cumulative resistance surface with minimum cumulative resistance model (MCR) is realized by the cost distance tool of Spatial Analyst in ArcGIS. The specific operation steps are as follows: firstly, select cost distance in the Spatial Analyst Tools toolbar, select "ecological source" in the element source data box, and use the "ecological expansion resistance surface" generated in the previous step as the cost grid data. The minimum cumulative resistance surface of ecological land expansion in the study area was obtained by specifying the output path.

Agricultural land classification is a process of comprehensively evaluating the quality of cultivated land according to the natural, social and economic conditions that reflect the relatively stable quality of cultivated land. The calculation process is as follows: the natural quality grade is determined according to the specific land conditions of each grading unit in the index area. The result of the natural quality grade is multiplied by the land-use coefficient to obtain the use grade reflecting the yield level of cultivated land crops. Then, the result is multiplied by the land economic coefficient to obtain the economic grade reflecting the proportional relationship between input and output of cultivated land, and the final economic grade is the cultivated land quality grade. There are 15 grades of cultivated land quality in China, with grade 1 being the best and grade 15 being the worst. Grades 1-4, 5-8, 9-12 and 13-15 are classified as excellent, high, medium and low, respectively. See Table 2 for statistics of cultivated land quality.

Table 2. Classification of cultivated land quality.

Class	8	9	10	11	12
Area /hm^2	1580.81	19935.19	21841.93	16673.00	1227.80
Percent	2.58%	32.54%	35.66%	27.22%	2.00%

Statistics of each region are shown in Figure 1.

Using The overlay analysis tool of Arc GIS, the results of cultivated land potential conflict identification and the results of cultivated land quality grade were overlaid to generate an overlay map and attribute database. According to grade 8, grade 9, grade 10 and grade 11, the cultivated land in the farming advantage area (A), the medium conflict area (M) and the strong conflict area (S3) is selected as the permanent basic farmland. The permanent basic farmland protection index issued by the superior is used as the constraint condition until the delimitation result meets the task index given by the superior.

4 ANALYSIS OF POTENTIAL LAND-USE CONFLICT IDENTIFICATION RESULT

The Arc GIS overlay analysis tool is used to overlay the cultivated land ecological security distribution map, the land cultivation suitability evaluation result map and the land construction suitability evaluation result map so that each evaluation unit can get three classification results. According to

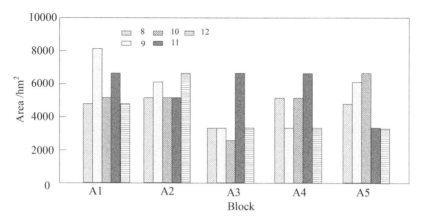

Figure 1. Statistics of the equal area of cultivated land quality.

the permutation combination method, the potential land-use conflict results of the cultivated land are finally obtained. See Table 3 for the specific quantitative relationship statistics.

Table 3. Statistical table of identification results of potential land-use conflict of cultivated land.

	Combination of partition	Area /hm²	Percent
Strong conflict zone (S)	Ecologically dominant area (E)	4085.11	6.67%
	Cultivated area (A)	25873.31	42.24%
	Construction of advantageous area (U)	3195.73	5.22%
	Low conflict zone (W)	645.61	1.05%
	Middle conflict zone (M)	12550.2	20.49%
	Ecology and farming (S1)	833.26	1.36%
	Ecology and Construction (S2)	395.08	0.64%
	Cultivation and construction (S3)	13077.48	21.35%
	Ecology, farming, construction (S4)	602.95	0.98%

In terms of quantity, the ecological advantage area, the farming advantage area, and the construction advantage area account for 6.67%, 42.24% and 5.22% of the total cultivated land area, respectively, which reflects the superior natural resources endowment of cultivated land and the wide-area suitable for farming, which is consistent with the quality of cultivated land. Nearly 46% of the cultivated land has potential land-use conflicts, of which the low conflict area accounts for 1.05%, indicating that only a small part of the cultivated land has low suitability for ecology, cultivation and construction. The area of the middle conflict area is 12550.2 hm², accounting for 20.49% of the cultivated land area, which can be considered a candidate area for permanent basic farmland. The total area of high conflict area is 14908.78 hm², accounting for 24.34%, of which the area of strong conflict between cultivated land and construction is the largest, accounting for 21.35% of the cultivated land area, the area of strong conflict between ecology and cultivated land accounts for 1.36% of the cultivated land area, and the area with highly suitable ecology, cultivation and construction account for about 1%. Therefore, further coupling and coordination should be carried out in this area. From the analysis of the quantitative characteristics of different potential land-use types, it can be seen that the cultivated land has superior farming conditions, and the potential conflict is mainly reflected in the competition between farming and construction land demand.

5 CONCLUSION

The paper considers the contradiction of land use demand under the guidance of different objectives of ecology, construction and cultivation. Then, it considers the problems of unstable layout and the high risk of potential land-use conversion in the process of protection and delimitation of permanent basic farmland. The delimitation of permanent basic farmland based on the identification of potential land-use conflicts is studied. The following conclusions are drawn:

(1) Select the indicators needed for the evaluation of ecological security, cultivation suitability and construction suitability, and determine the threshold and weight of the indicators by collecting and analyzing data.
(2) Use the minimum cumulative resistance model and the weighted index sum method. The ecological security level and the suitable level of cultivated land cultivation and construction were divided, and the potential land-use conflict system was constructed by the permutation combination method.
(3) The evaluation results were analyzed and applied to the delimitation of permanent basic farmland in combination with agricultural land quality classification results.

The construction of the ecological security pattern based on the MCR model still has subjectivity and limitations and is limited by data acquisition. In future research, the impact of policy factors on land use should be considered in the multi-objective suitability evaluation system, and a more perfect and authoritative evaluation index system should be established.

REFERENCES

Rao Haoyu, Zhang Kun. The influencing factors of rural space in China and its evolutionary stage research. *Architecture and Culture*, 2021(6):174–175.

Shu Bo, Xu Jingjing, Chen Yang. (2020) The research progress and prospects of rural planning construction in China-Literature measurement analysis based on the results of the National Natural Science Foundation of China. *Planner*, 36(4):41–49.

The history, methods, systems, subjects and outlooks of rural planning in Japan-Rural Planning in Japan-residents participation, landscape, ecological village. *Small Town Construction*, 2018(4):10–13.

Wei Shuwei, Wang Yang, Chen Kaiyue, etc. (2019) The evolution characteristics and revelation of my country's rural system planning since the reform and opening up. *Planner*, 35(16):56–61.

Yang Jun, Guo Lilan, Li Zheng. (2019) The research progress of rural space planning in my country under the perspective of rural revitalization. *Resource Development and Market*, 35(9):1152–1156.

Yang Jun. (2021) Practical village planning and exploration of practical villages under the land and space planning system-take Shangcun, Jixi County, Anhui Province as an example. *Residential Technology*, 41(4):44–47.

Yuan Yuan, Zhao Xiaofeng, Zhao Yuntai, et al *Land and Space Planning System, the hierarchical planning and vertical conduction study of the planning and establishment of the Village Plan*. 2020(6):43–48.

Zhang Jingxiang, Zhang Shangwu, Duan Dezheng, Chen Qianhu, Ma Xiangming, Shi Huaiyu, Zhao Wei, Zhao Yi, Zhao Yi. (2020) Multi-regulation and one-piece practical village plan. *Urban Planning*, 44(3):74–83.

Zhang Li. The evolution of rural thought in global vision and the construction of rural planning in Japan—also the guide for this period. *Small Town Construction*, 2018(4):5–9, 28.

Zhang Qiuju, Zhang Chaofeng. The theoretical and policy practice of endogenous development in rural areas: Take the direct subsidy system such as the Zhongshan area of Japan as an example. *World Agriculture*, 2020(11):20–28.

Advances in Renewable Energy and Sustainable
Development – Liang & Kasmani (Eds)
© 2023 Copyright the Author(s), ISBN: 978-1-032-39407-7

Causes and countermeasures of poor habitat status of endangered birds

Mingrui Li*
Fenner School, The Australian National University, Canberra, Australia

ABSTRACT: Many species in the world are at risk of extinction, and their habitats are in poor condition, but there are many ways to protect them. Many bird species are also facing extinction, and endangered bird habitats are in poor condition due to various factors. There are many studies on habitat protection of endangered birds, but there are few reviews on the causes and corresponding methods of the poor habitat conditions. This paper discusses the corresponding solutions from the four causes of habitat impact and compares them with each other. The main conclusion showed as follows: (1) Direct nest protection with a massive workload was used to solve the lack of a suitable nest; (2) Invasive species management was used to solve bird habitats invaded by invasive species; (3) The establishment of ecological corridors while long-term effectiveness remains to be explored was used to solve habitat fragmentation of endangered birds; (4) Biodiversity offset policy could be used to offset the loss of the habitat of endangered birds affected by the project and implemented case by case. The advantages and disadvantages of these methods were compared for the reference of the protection decision-makers. In the future, more studies should combine these conservation approaches to explore the long-term impact on mitigating endangered bird habitats.

1   INTRODUCTION

There are more and more species facing the great challenge of being extinct. The reasons leading them to be extinct include habitat loss and fragmentation, over-harvesting, invasive species, altered disturbance regimes, climate change, disease and pollution. The giant panda became endangered because of habitat loss and fragmentation. Climate change will accelerate this process as there will be a less spatial match between bamboos and giant pandas (Zang et al. 2020). The Blue Mountains water skink *(Eulamprus leuraensis)* living in the very restricted swamp were endangered species due to altered fire regimes, which frequency will be affected by climate change (Gorissen et al. 2015) and fire at the wrong time will destroy their habitat irreversibly. Based on the drivers of being endangered, conservation methods can be found to protect endangered species in time. Ecological corridors can be built to connect giant panda habitats as habitat fragmentation can be prevented (Bu et al. 2021). New laws can be made to inhibit people fishing in some areas to protect some endangered fish in their breeding time. In terms of endangered species which almost entirely lose their habitat, ex situ conservation can be applied. Conservation genetics have the ability to imply the genetic structure and find genetic drift of some endangered species (Jiménez-Mena et al. 2015). This could help identify which species are endangered earlier and provide guidance for breeding existing ones. In short, there are many reasons why species become endangered, and many ways to protect them.

Among the numerous endangered animals, the conservation research of endangered birds is also significant. Many factors can influence a bird's life length and fertility rate. Climate change could lead to more droughts or changing spawning seasons, which may also reduce food resources. All these factors will eventually lead to the emergence of endangered birds and continue to affect those

---

*Corresponding Author: u7367713@anu.edu.au

species. The population of Snail Kite showed a decline because of climate change. Longer droughts would lower their food availability, leading to a lower fertility rate with a long life span (Reichert et al. 2012). Many studies also explore ways to protect endangered birds compared to other species. Samuel et al. (2020) researched whether genetic editing can save Hawaiian honeycreepers, which were endangered because of avian malaria. The endangered population slowly becomes resistant to malaria by releasing an individual with a malaria-resistant gene into the population. Habitat loss and fragmentation can also make life difficult for birds. The number of regent honeyeaters living in Australia experienced a dramatic decline as many trees have been cut down (Kvistad et al. 2015). Although the species has scattered populations and habitats, studies have found that their genes are not so different that they can be managed as a population. It would also reduce the cost of protecting this rare bird. There are a number of factors that could work together to have a greater impact, and conservation strategies are under intense study. For birds, a common element that is influenced by a variety of factors is habitat.

Birds have diverse habitats, as do endangered ones. Bird habitat refers to the place where birds breed, and because there are so many birds, there are many kinds of habitats. Some live in cities close to people, some in wild forests, some in swamps, and some by the sea. Bird habitat selection is based on vegetation type, and breeding and predation sites are not necessarily the same (Stokke et al. 2014). Some birds live in marshes, and their pollution of marshes needs to be reduced or banned to protect them, while some other birds breed and survive in forests, and it may be more effective to protect forests or the trees where they build their nests (Manning et al. 2004). Because of the diversity of habitats, different species of endangered birds may have different conservation strategies. However, there is less review literature about the methods of bird habitat protection. Protecting endangered bird habitats is sometimes the most straightforward.

In this article, the causes of the poor state of some habitats are discussed. Useful conservation strategies of bird habitats are summarized, and the effects are discussed. The necessity, as well as the advantages and disadvantages of these strategies, are analyzed. Some suggestions are put forward for future research directions.

2 NEST PROTECTION

The habitat most directly related to birds is the nest. Nests are also the lowest level of habitat, so protecting birds' habitat can start by protecting nests. The superb parrot (*Polytelis swainsonii*) is an endangered species living in Australia. A study of its nests revealed a preference for a type of eucalyptus tree (Blakely's red gum) and dead trees (Manning et al. 2004). Dead trees provide nests because their trunks are hollow. In the latter eucalypts, they die back over time, and the rate of regeneration is slow, resulting in fewer nests available in the long run. The entrance to the cave is not particularly large for protection from predators, but it is too small and not deep enough for breeding, so finding a suitable nest can be difficult. Fortunately, although these holes are a limited resource, Superb Parrot reuses the same hole 55.8% of the time, and they prefer Blakely's Red Gum (Manning et al. 2004). Therefore, it's not so much about protecting these holes, and it's more about protecting this kind of tree. Interestingly, this eucalyptus tree is also endangered. These nesting trees are also adjacent to many agricultural lands, so ways should be found to enhance their regeneration and protect existing nest trees. Birds that nest in grass may be exposed to predators because of changes in the mowing frequency, which can be fatal to them. After reduced mowing frequency, 70% of whinchat (Saxicola rubetra) survived to emerge, as the nest was not exposed due to persecution (Grüebler et al. 2012). Indirectly protecting nests by changing the frequency of mowing allows endangered birds to survive. Moreover, whether directly or indirectly protecting nests, these measures require long-term behavior monitoring, which means time and effort. These conservation measures only improve the nest condition, but do not increase the availability of food and other resources (Buckingham et al. 2004). Artificial nests as an alternative to conventional ones have improved survival but not reproductive rates for some endangered birds. However, each species has different nesting characteristics, and artificial nests are not suitable for all endangered birds (Pichegru 2013). Artificial intervention in nest conditions may not be effective, so these protective measures need to be implemented with caution. More research should focus on the individual endangered bird species themselves and on the effects of long-term nest management.

3 PREVENTING INVASIVE SPECIES FROM INVADING

In an ecosystem, each species has its niche. If multiple species occupy the same space and food resources, their niches will overlap, resulting in interspecific competition. Endangered birds also have their ecological niches and may have adapted to local environments before becoming endangered. However, when invasive species appear, niche overlap occurs, leading to competition for living space. The spotted owl population *(Strix occidentalis)* has declined over the past 200 years due to habitat fragmentation caused by logging. The recent emergence of new invasive species has accelerated the decline (Dugger et al. 2016). This was because new invasive species, called barred owl *(Strix varia)*, occupies a wider range of forest habitats, including those in the spotted owl (Hamer et al. 2007). A bird that lost the competition may lose its ecological niche and become endangered. And it was found that in forest habitats, the colonization rate of spotted owls was negatively correlated with the discovery probability of barred owls, and the extinction probability was positively correlated with the discovery probability of barred owls, as shown in figure 1 and figure 2 (Dugger et al. 2011).

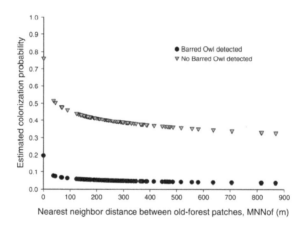

Figure 1. The relationship between estimated colonization probabilities and nearest neighbor distance between old-forest patches when detected or not detected barred owl (Dugger et al. 2011).

In response to this niche overlap, increasing the range of habitats can be adopted. This could ease competition by allowing endangered spotted owls to occupy new ecological niches, but it is doubtful whether this will work. Because barred owls are inherently more adaptable (Wiens et al. 2014), they may be able to occupy new habitats faster than endangered spotted owls, which also has the potential to further expand the scale of competition between species. Another approach is to remove the invasive species directly, either by trapping them or by killing them intensively. The results of experiments showed that the fatal elimination of barred owls could increase the population of spotted owls significantly within a certain range (Diller et al. 2016). This approach simply removes invasive species from the ecosystem and gives endangered birds their niche back. More results are needed to provide the basis for government plans to manage the invasive barred owls over the long term (Diller et al. 2016). Long-term management of invasive species is time-consuming and expensive, and large-scale culling of invasive birds raises ethical debates. However, removing barred owls in lethal ways to protect spotted owls was temporarily justified (Lynn 2018).

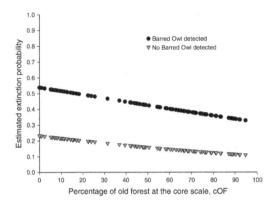

Figure 2. The relationship between mean extinction probabilities and percentage of the old forest at the core scale when detected or not detected barred owl (Dugger et al. 2011).

Therefore, based on the discussion of spotted owls, lethal removal of invasive species like the barred owl is the first choice, and the method of habitat expansion needs to be carefully considered.

4 PREVENTING HABITAT FRAGMENTATION

Habitat loss and fragmentation are one of the most important causes of birds becoming endangered. There are many reasons for habitat fragmentation, such as large-scale deforestation (Tortorec et al. 2013) and the establishment of multiple highways. The disappearance of part of the forest will make the habitat discontinuous. Some birds cannot hide better due to the decrease of trees in the migration process, thus increasing the probability of predators predating. Habitat fragmentation increases the resistance of gene exchange, and when the time is long enough, the sub-populations with incomplete genetic development will be formed in the habitat fragments (Boland & Burwell 2020). With the deepening of fragmentation, the sub-populations of each patch may enter the extinction vortex. One endangered bird, called the black-capped vireo (*Vireo atricapilla*), has experienced a significant decline in genetic diversity due to ongoing habitat fragmentation, along with a significant decline in its population due to genetic bottlenecks (Athrey et al. 2012). To cope with habitat fragmentation, ecological corridors can be established. Potential ecological corridors can be identified by the least cost path method (Guo et al. 2019). For endangered birds living in forests, it seems feasible to achieve habitat continuity through artificial afforestation between habitats, regardless of cost. Existing riparian forests can also be considered ecological corridors for birds, and riparian forests significantly enhance ecological diversity, making them a good stopping point for migratory birds (Mendes 2016). For endangered birds that may live in cities, it is possible to increase the connectivity of various green parks in cities by increasing tree cover on both sides of some streets (Graviola et al. 2021; Liu et al. 2020). Due to the diversity of habitat patch distribution, the selection and establishment of an ecological corridor should be considered according to the actual situation of each patch. And whether endangered birds are at increased risk because of poachers in ecological corridors is unknown. In conclusion, establishing ecological corridors carefully is a feasible method to prevent further habitat fragmentation.

5 BIODIVERSITY OFFSET

To compensate for the biodiversity losses caused by development, biodiversity offset policies can be adopted, i.e., No Net Loss. Offset policies are not preferred when development effects can be avoided or mitigated. An offset can be applied when a development project inevitably affects the habitat of endangered birds, usually by establishing equivalent environmental conditions on another site (Rogers & Burton 2017). But the offset sites have lower biodiversity, and species richness is five to ten times lower than the pre-development plots (Regnery et al. 2013). This also indicates that no net loss is difficult to achieve, and inadequate offset is prevalent (Vanderduys et al. 2016). For birds, the effectiveness of offset may be increased by selecting more numbers and kinds of endangered birds for offset rather than affected common birds (Rogers & Burton 2017). Endangered birds also migrate, so offset sites can be selected at stopover sites along the migration route, but if the site is too far away from the affected site, it is not acceptable (Rogers & Burton 2017). Therefore, a combination of policy requirements and species migration laws is needed to determine offset sites. After the offset is implemented, long-term monitoring management is indispensable to help maintain habitat conditions, increasing the cost of protecting endangered birds. In a word, biodiversity offset is an indirect method of habitat protection for endangered birds, because its protection objects are not affected sites, but to establish the same or better habitat conditions at a different site.

6 FOUR CAUSES AND THEIR COMPARISON

The comparison of the above four solutions is summarized in Table 1. Compared with the prevention of alien species, habitat fragmentation and ecological diversity compensation, the most direct target of endangered birds' protection is the nest. For endangered birds that nest in trees or grass, their nest carriers should be protected, i.e., nesting trees should be protected because they regenerate too slowly. This method requires individual nest surveys, so it's a lot of work, but it probably works best. Follow-up monitoring of nest use can be done using a camera mounted nearby. Future directions could use more artificial nests to help improve poor habitat conditions.

Table 1. A comparison of four approaches to dealing with the poor habitat condition of endangered birds.

Methods	Reasons	Advantages and disadvantages
Nest protection	Lack of nesting carriers, such as trees that regenerate slowly	Direct protection, but a lot of work
Invasive species management	Niche competition causes endangered species to withdraw from their original habitats	Management is segmented according to the level of intrusion, taking public opinion into account
Ecological corridors	Habitat fragmentation	long-term effectiveness remains to be explored
Biodiversity offset	The habitat is already occupied by existing projects	A viable remedy, difficult to achieve, with no net loss. Each country has different policies and needs to discuss them on a case-by-case basis

Invasive birds compete with endangered birds and may occupy dominant ecological niches, leading to niche encroachment, such as gradual removal from suitable habitats. The droppings of invasive birds may even carry seeds from invasive plants (Gosper et al. 2005). Feasibility studies, including public opinion and the possible benefits of invaders, are crucial to implementing any strategy to combat invasive species (Strubbe et al. 2011). In terms of the management of invasive birds, the best is prevention. If there are a small part of the invasion should be strictly controlled in a small area, generally, after the spread of management will become extremely difficult. At the boundaries of different areas, i.e., some people like to keep some birds as pets and take them to travel, which may also become invasive species and need to be strictly checked at the border.

Compared with the nest level, habitat fragmentation of endangered birds is a macro habitat state. This is due to the construction of roads and houses that divide large habitats into smaller patches.

As a result, bird populations are separated, and populations decline into an extinction spiral (Fagan & Holmes 2006). The solution is to build ecological corridors, but their long-term effectiveness remains to be explored.

Biodiversity offset policies can be adopted for habitat loss of endangered birds caused by engineering projects. This method is similar to setting up new habitats to compensate for damage. Still, it is difficult to achieve no net loss and long-term management after compensation is difficult to achieve. Each country has different compensation policies and standards, and many countries do not have biodiversity compensation policies, indicating that birds in these countries lack a certain degree of protection (Bezombes et al. 2018).

7 CONCLUSION

This paper discussed four reasons for endangered birds' poor habitat conditions and compared the corresponding protection strategies. The main conclusions could be summarized as follows: (1) The lack of a suitable nest required direct nest protection with a massive workload; (2) Bird habitats invaded by invasive species required invasive species management with considerations of the level of intrusion and public opinion; (3) Habitat fragmentation of endangered birds required the establishment of ecological corridors while long-term effectiveness remains to be explored; (4) The habitat of endangered birds affected by the project should be compensated by biodiversity offset policy and implemented case by case. The advantages and disadvantages of these methods were compared for the reference of the protection decision-makers. Further research should consider combining these conservation approaches and exploring the long-term impact on endangered bird habitats.

REFERENCES

Athrey, G., Barr, K.R., Lance, R.F., and Leberg, P.L., Birds in space and time: genetic changes accompanying anthropogenic habitat fragmentation in the endangered black-capped vireo (Vireo atricapilla). *Evolutionary Applications*, vol. 5, no. 6, pp. 540–552, 2012.

Bezombes, L., Gaucherand, S., Spiegelberger, T., Gouraud, V., and Kerbiriou, C., Development of a standardized framework for the evaluation of biodiversity in the context of biodiversity offsets. *Ecological Indicators*, vol. 93, pp. 1244–1252, 2018.

Boland, C.R.J. and Burwell, B.O., Habitat modeling reveals extreme habitat fragmentation in the endangered and declining Asir Magpie, Pica asirensis, Saudi Arabia's only endemic bird (Aves: Passeriformes). *Zoology in the Middle East*, vol. 66, no. 4, pp. 283–294, 2020.

Bu, H. et al., Not all forests are alike: the role of commercial forest in the conservation of landscape connectivity for the giant panda. *Landscape Ecology*, vol. 36, no. 9, pp. 2549–2564, 2021.

Buckingham, D.L., Atkinson, P.W., and Rook, A.J., Testing solutions in grass-dominated landscapes: a review of current research. *Ibis* (London, England), vol. 146, no. s2, pp. 163–170, 2004.

Diller, L.V. et al., Demographic response of northern spotted owls to barred owl removal. *The Journal of Wildlife Management*, vol. 80, no. 4, pp. 691–707, 2016.

Dugger, K.M. et al., The effects of habitat, climate, and Barred Owls on long-term demography of Northern Spotted Owls/Efectos del habitat, del clima y de Strix varia sobre la demografia a largo plazo de Strix occidentalis caurina. *The Condor* (Los Angeles, Calif.), vol. 118, no. 1, p. 57, 2016.

Dugger, K.M., Anthony, R.G., and Andrews, L.S., Transient dynamics of invasive competition: Barred Owls, Spotted Owls, habitat, and the demons of competition present. *Ecological Applications*, vol. 21, no. 7, pp. 2459–2468, 2011.

Fagan, W.F. and Holmes, E.E., Quantifying the extinction vortex. *Ecology Letters*, vol. 9, no. 1, pp. 51–60, 2006.

Gorissen, S., Mallinson, J., Greenlees, M., and Shine, R., The impact of fire regimes on populations of an endangered lizard in montane south-eastern Australia. *Austral Ecology*, vol. 40, no. 2, pp. 170–177, 2015.

Gosper, C.R., Stansbury, C.D., and Vivian-Smith, G., Seed dispersal of fleshy-fruited invasive plants by birds: contributing factors and management options. *Diversity & Distributions*, vol. 11, no. 6, pp. 549–558, 2005.

Graviola, G.R., Ribeiro, M.C., and Pena, J.C., Reconciling humans and birds when designing ecological corridors and parks within urban landscapes. *Ambio*, vol. 51, no. 1, pp. 253–268, 2021.

Grüebler, M.U., Schuler, H., Horch, P., and Spaar, R., The effectiveness of conservation measures to enhance nest survival in a meadow bird suffering from anthropogenic nest loss. *Biological Conservation*, vol. 146, no. 1, pp. 197–203, 2012.

Guo, R., Wu, T., Liu, M., Huang, M., Stendardo, L., and Zhang, Y., The Construction and Optimization of Ecological Security Pattern in the Harbin-Changchun Urban Agglomeration, China. *International Journal of Environmental Research and Public Health*, vol. 16, no. 7, p. 1190, 2019.

Hamer, T.E., Forsman, E.D., and Glenn, E.M., Home range attributes and habitat selection of Barred Owls and Spotted Owls in an area of sympatry/Atributos del ambito de hogar y seleccion de habitat de Strix occidentalis y Strix varia en un area de simpatria. *The Condor* (Los Angeles, Calif.), vol. 109, no. 4, p. 750, 2007.

Jiménez-Mena, B., Hospital, F., and Bataillon, T., Heterogeneity in effective population size and its implications in conservation genetics and animal breeding. *Conservation Genetics Resources*, vol. 8, no. 1, pp. 35–41, 2015.

Kvistad, L., Ingwersen, D., Pavlova, A., Bull, J.K., and Sunnucks, P., Very Low Population Structure in a Highly Mobile and Wide-Ranging Endangered Bird Species. PloS one, vol. 10, no. 12, pp. e0143746-e0143746, 2015.

Liu, Z., Huang, Q., and Tang, G., Identification of urban flight corridors for migratory birds in the coastal regions of Shenzhen city based on three-dimensional landscapes. *Landscape* ecology, vol. 36, no. 7, pp. 2043–2057, 2020.

Lynn, W.S., Bringing Ethics to Wild Lives: Shaping Public Policy for Barred and Northern Spotted Owls. *Society & animals*, vol. 26; 2018, no. 2, pp. 217–238, 2018.

Manning, A.D., Lindenmayer, D.B., and Barry, S.C., The conservation implications of bird reproduction in the agricultural "matrix": a case study of the vulnerable superb parrot of south-eastern Australia. *Biological Conservation*, vol. 120, no. 3, pp. 363–374, 2004.

Mendes, A.I.d.S., *"The use of riparian forests as ecological corridors by passerine birds in the south of Portugal,"* ProQuest Dissertations Publishing, Dissertation/Thesis, 2016.

Pichegru, L., *Increasing breeding success of an Endangered penguin: artificial nests or culling predatory gulls*? Bird conservation international, vol. 23, no. 3, pp. 296–308, 2013.

Regnery, B., Couvet, D., and Kerbiriou, C., Offsets and Conservation of the Species of the E.U. Habitats and Birds Directives. *Conservation biology*, vol. 27, no. 6, pp. 1335–1343, 2013.

Reichert, B.E., Cattau, C.E., Fletcher, R.J., Kendall, W.L., and Kitchens, W.M., Extreme weather and experience influence reproduction in an endangered bird. *Ecology* (Durham), vol. 93, no. 12, pp. 2580–2589, 2012.

Rogers, A.A. and Burton, M.P., Social preferences for the design of biodiversity offsets for shorebirds in Australia. *Conservation biology*, vol. 31, no. 4, pp. 828–836, 2017.

Samuel, M.D., Liao, W., Atkinson, C.T., and LaPointe, D.A., Facilitated adaptation for conservation – Can gene editing save Hawaii's endangered birds from climate-driven avian malaria? *Biological Conservation*, vol. 241, p. 108390, 2020.

Stokke, S., Motsumi, S.S., Sejoe, T.B., and Swenson, J.E., *"Cascading Effects on Smaller Mammals and Gallinaceous Birds of Elephant Impacts on Vegetation Structure,"* in Elephants and Savanna Woodland Ecosystems, C. Skarpe, J. T. du Toit, and S. R. Moe, Eds. Chichester, UK: John Wiley & Sons, Ltd, 2014, pp. 229–250.

Strubbe, D., Shwartz, A., and Chiron, F., Concerns regarding the scientific evidence informing impact risk assessment and management recommendations for invasive birds. *Biological Conservation*, vol. 144, no. 8, pp. 2112–2118, 2011.

Tortorec, E. et al., Habitat fragmentation and reproductive success: a structural equation modeling approach. *The Journal of animal ecology*, vol. 82, no. 5, pp. 1087–1097, 2013.

Vanderduys, E.P., Reside, A.E., Grice, A., and Rechetelo, J., Addressing Potential Cumulative Impacts of Development on Threatened Species: The Case of the Endangered Black-Throated Finch. *PloS one*, vol. 11, no. 3, pp. e0148485-e0148485, 2016.

Wiens, J.D., Anthony, R.G., and Forsman, E.D., Competitive interactions and resource partitioning between northern spotted owls and barred owls in western Oregon. *Wildlife Monographs*, vol. 185, no. 1, pp. 1–50, 2014.

Zang, Z. et al., Climate-induced spatial mismatch may intensify giant panda habitat loss and fragmentation. *Biological Conservation*, vol. 241, p. 108392, 2020.

Advances in Renewable Energy and Sustainable Development – Liang & Kasmani (Eds)
© 2023 Copyright the Author(s), ISBN: 978-1-032-39407-7

Evaluation of construction waste recycling schemes under the life cycle: A proposed weight-TOPSIS approach improved by GRA

Runfei Chen*, Jiawei Xu*, Yudong Xie* & Shaokun Zhang*
Fuzhou University, College of Civil Engineering, Fuzhou, China

ABSTRACT: The promotion of the construction waste (CW) recycling industry in China is facing enormous challenges. There are few comprehensive evaluation studies on the existing CW recycling system, which are difficult to be used as an optimization tool in engineering design. To construct a comprehensive evaluation index system for the resource utilization of CW, a combination model of TOPSIS and entropy weight, which is further improved by grey relation analysis (GRA), is established, and a study is made on the comprehensive benefits of natural aggregate concrete (NAC) and recycled aggregate concrete (RAC) with different substitution rates in the two recycling methods. A lifecycle-based evaluation system for the recycling of CW in a building conservation project in the Mawei District is constructed, and the calculation method of the environmental and economic indicators is specified. The proposed method can better support the selection of different resource utilization schemes by fading out the drawbacks of subjective weight in the traditional TOPSIS and reflecting the change in internal factors.

1 INTRODUCTION

The unmanageable large amount of difficult-to-treat CW has become a problem facing the world with the growing demand for infrastructure. In 2020, China's CW reached 235.12 million tons (Ministry of Housing and Urban-Rural Development of the People's Republic of China 2020), 15.46% higher than the figure for 2016 (Ministry of Housing and Urban-Rural Development of the People's Republic of China 2016). In the process of realizing the goal of "carbon emissions peak" and "carbon neutrality" in China, the traditional informal CW disposal methods such as simple landfills and random stacking are no longer suitable, which will cause serious problems such as occupying a large amount of land and affecting the environment. To pursue sustainable development in civil engineering, the consensus has been reached that recycling construction waste could become a sensible way to tackle the dilemma. It is immediately necessary to establish a comprehensive benefit evaluation technology and adopt a more effective and economical way to select a resource utilization scheme to ensure smooth implementation.

At present, the research on resource utilization of CW mostly focuses on the performance of recycled products and the technology of resource utilization. To study the benefits of resource utilization, some scholars have assessed CW recycling from various aspects to study resource utilization benefits. Kurda (2019) (2019) and Andrea Di Maria (2018) (2020) conducted the environmental life cycle cost theory to systematically measure the costs. In terms of environmental impacts, the assessment methods are mostly pollutant leaching experiment evaluation (Hjelmar 2001), AHP method (Ma et al. 2018)and life cycle assessment (LCA) (Lyu et al. 2021). However, most of the existing evaluation models consist of separable models evaluating discrete indexes, and there is relatively little research on the comprehensive benefits of resource treatment and utilization. A few scholars, such as Zhang Fan (2021)(2021), put forward a fuzzy heterogeneous multi-criteria decision-making approach and, combined with expert scoring, constructed a comprehensive index

---

*Corresponding Authors: 051901502@fzu.edu.cn; 051901601@fzu.edu.cn; 051901513@fzu.edu.cn and 051901111@fzu.edu.cn

DOI 10.1201/9781003349648-26

system for CW utilization schemes from four dimensions: society, environment, capital, and technology. Limitations could be obviously found in current comprehensive evaluation methods when it comes to selecting diversified indicators, and factors are commonly static values, without considering the potential effects of each period from crushing to resource product output. Meanwhile, the subjectivity of traditional expert scoring greatly affects the determination of the weight of each index.

Taking a protection project in the Mawei District as an example, based on LCA, this paper calculates the environmental and economic indicators of the whole life cycle of recycled concrete with different replacement rates produced by two different recycling methods: outbound crushing and in-situ crushing. The entropy weight method of the weight (Shannon 1951) is applied to determine the weight of the index, thus avoiding the error caused by human subjective. In addition, the GRA (Deng 1993) can better reflect the change of the internal factors even when the sample size is small, which can make up for the difficulty of TOPSIS (Hwang & Yoon 1981) in determining enough size of samples to characterize the influence degree of the index. Therefore, this improved method is introduced in this paper to quantitatively analyze the comprehensive environmental and economic benefits of the recycling of CW, which would have good guiding significance for the promotion of CW recycling.

2 ESTABLISHMENT OF A COMPREHENSIVE EVALUATION INDEX SYSTEM FOR RESOURCE UTILIZATION OF CW

2.1 *Comprehensive evaluation index system*

RAC utilization is characterized by smaller primary mineral or natural resources than NAC. Studying the actual effect of waste concrete resource utilization in environmental aspects has important reference value for its implementation. This study mainly discusses the indicator impacts of greenhouse gas emissions, energy consumption and related pollutant emissions. Economic benefits are also an important indicator for promoting the recycling of CW. Few studies combine economic and environmental benefits, and it is worth thinking about whether the benefits of RAC are overvalued. The comprehensive index system of CW resource utilization constructed in this study is illustrated in Figure 1.

Whether the selection of impact indicators is reasonable or not directly affects the accuracy of the evaluation results. In addition to satisfying the hierarchy, scientific rationality, comprehensiveness and comparability, the dynamism should also be satisfied. Therefore, based on the LCA methodology, this study analyzes the index values in the entire life cycle of concrete from design to scrap to obtain more systematic results. This paper focuses on the recycling process of waste, that is, recycling the waste after demolition, obtaining recycled aggregate through magnetic separation, screening, and then preparing recycled concrete. The calculation boundary is shown in Figure 2. To make each concrete have a unified input and output function measurement in the life cycle, concrete with the same strength and workability is selected as the functional unit.

2.2 *Index calculation method*

2.2.1 *Environmental indicators*
(1) Emissions of greenhouse gas equivalent

The greenhouse gas equivalent is the sum of the emissions of CO_2, CH_4 and N_2O multiplied by the corresponding global warming potential (GWP) value. The life cycle calculation process refers to the model studied by Chen Runfei et al. (2022). Based on the field survey, the calculation parameters of greenhouse gas equivalent in each stage of the concrete life cycle are described in detail in Table 1.

Apart from that, during the raw material transportation, waste treatment and recycling stages, the transport vehicles are all 9.6-meter heavy-duty trucks mainly powered by diesel, and the fuel

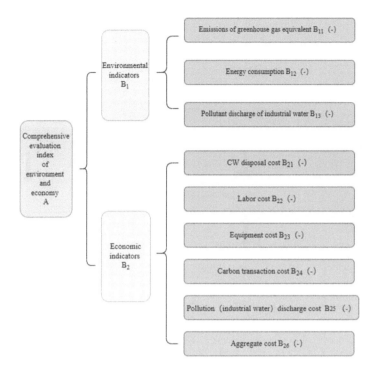

Figure 1. The comprehensive evaluation system.

Table 1. Calculation parameters of greenhouse gas equivalent of RAC.

Material/t	Power Consumption/ KWh	Raw coal consumption/ kg	Diesel consumption/ L	Crude oil consumption/ kg	Gas Consumption/ kg
Cement	148.07	286.46	–	–	–
Natural aggregate	2.67	–	1.523	–	–
Concrete	–	0.519	–	2.67E-03	4.31E-3
Construction waste	9.11	–	–	–	–
Recycled aggregates	1.8	–	–	–	–

consumption unit is 1.85L/(100km*t). Concerning the concrete transportation stage, the ready-mixed concrete is transported by a six-cubic mixer truck using diesel, with average fl consumption of 0.6L/(6t* km). The greenhouse gas emission factors and GWP of each energy source are obtained from a database developed by the China Statistics Bureau (National Bureau of Statistics 2012) and IPCC guidelines for national greenhouse gas inventories (Paustian, Ravindranath, Amstel, 200).
(2) Energy consumption
E_X is the energy consumption generated by the Xth stage in the life cycle, and the calculation formula is shown in (1):

$$E_X = \sum_{i=1}^{n} E_i \times M_i \quad (1)$$

E_i is the calorific value of the ith energy used in the Xth stage of the concrete life cycle (National Bureau of Statistics 2012) and M_i is the consumption of the ith energy. In this study, only standard

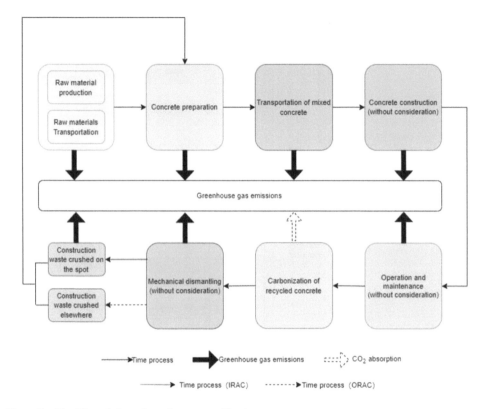

Figure 2. The life cycle boundary of resource utilization.

coal, raw coal, crude oil, natural gas, diesel and electric energy with large consumption in the project are considered.

(3) Calculation of pollutant discharge

D_X is the discharge amount generated in the Xth stage of the concrete life cycle, obtained from (2) below:

$$D_X = \sum_{j=1}^{n} D_j \times M_j \tag{2}$$

D_j is the discharge factor of the jth energy used for the discharge amount of the ith pollutant used by concrete in stage X of its life cycle, and M_j is the consumption of the jth energy. The pollutant studied in this paper is industrial wastewater, and the pollution factors of common energy sources are shown in Table 2.

Table 2. Pollutant emission factors of common energy SOURCES (Xiao et al. 2020; General Office of the State Council of the People's Republic of China 2008).

Energy consumption	waste water/g	Energy consumption	waste water/g
Electricity/KWh	14.6815	Crude oil/kg	87.7
Diesel/kg	476.98	Gas/m^3	0
The raw coal/kg	480	–	–

2.2.2 *Economic indicators*

(1) CW disposal cost

The CW disposal cost is composed of land requisition compensation cost, site pre-treatment cost, and various insurance costs. According to the "Notice of Fuzhou Municipal People's Government on Adjusting the Compensation Standards for Land Requisition in Four Urban Areas" (Fuzhou.gov.cn, 2022), a landfill 1 ton of CW will cost ¥9.6/t for land acquisition compensation, and the figure for the pre-treatment cost is 13.42. Simultaneously, an insurance premium of ¥0.25/t and a tax of ¥4.51/t are also required (Zhou 2018).

(2) Labor cost

The data for monthly employment costs were collected from the cost of construction information website (Cecn.org.cn, 2022). NAC production requires three natural aggregate miners (¥4000/person), three concrete production workers (¥4000/person), four drivers (¥3000/person), one manager (5000 /person), and one technical director (¥5000/person). The number of recycled aggregate recyclers (¥5000/person) increases with the rise of the replacement rate. A ton of concrete is calculated based on a day's work.

(3) Equipment cost

The equipment cost in this study will be replaced by the rental cost (0.6% of the equipment cost). Since the price of the mobile crushing plant applied in IRAC production fluctuates in a large range, the most common price in the market is chosen for simplified calculation. See table 3 below for the equipment costs required in each stage of the life cycle.

Table 3. Equipment and price required for each life cycle stage.

Equipment	Cost/ ¥	Equipment	Cost/¥
Natural aggregate production	9365.5	Landfill equipment	6800
Cement production	25030	Recycled aggregate production equipment (in situ crushing)	15449.9
Concrete mixing station	4180	Recycled aggregate production equipment (outbound crushing)	10000

(4) Carbon transaction cost

Carbon trading is the trading of greenhouse gas emission rights. Generally speaking, it is a brand new environmental and economic policy tool that uses carbon dioxide equivalent per ton as a calculation unit. Under the premise of total emission control, the emission rights of greenhouse gases, including carbon dioxide, have become a scarce resource and thus have the commodity attribute. To stimulate enterprises to develop low-carbon technologies and products, China has made corresponding decisions and pilot arrangements for the establishment of a carbon emission trading system. In this paper, the average carbon market transaction price on Carbon Exchange (tanjiaowang.net, 2022) is ¥51.66/ton.

(5) Pollution discharge cost

According to the standard management method issued by the Ministry of Ecology and Environment of the People's Republic of China (Ministry of Ecology and Environment of the People's Republic of China, 2003), pollutant discharge fees should be charged according to the type and quantity of pollutants and calculated by the number of pollution equivalents, and the pollution equivalent factor is the ratio of the pollutant discharge (kg) to the pollution equivalent value (kg). Industrial wastewater is the most important pollutant in the concrete production process, and its equivalent pollution value is 4kg, while the pollution equivalent price is ¥0.7.

(6) Aggregate cost

According to the data from the sand and aggregate information website(cssglw.com 2022)and the cost of construction information website, the buying rate of natural aggregate is ¥104/ton, and the selling price of recycled aggregate is ¥29/ton. The profit of recycled aggregate is the difference between the sales income of recycled aggregate after resource recovery and the purchase cost of raw material recycled aggregate.

3 COMPREHENSIVE BENEFIT EVALUATION MODEL OF RESOURCE UTILIZATION OF CW UNDER LIFE CYCLE

This paper uses the entropy weight-TOPSIS model (Fu et al. 2018) to quantitatively evaluate the comprehensive benefits of the above CW recycling schemes. This method can sort a limited number of evaluation objects according to their closeness to the idealized target. On this basis, we calculate the comprehensive level of energy saving, emission reduction and economic benefits in the life cycle of resource utilization of CW. The calculation steps are as follows:

3.1 *Calculation of indicator weights*

Firstly, an n × m original matrix X is constructed, with n (number of schemes) = 9 and m (number of indicators) = 10. Then the matrix should be regularized and standardized; secondly, the entropy weight of each indicator can be obtained from the normalized information effect value after calculating the information entropy according to the theory that the larger the information utility value is, the more important the indicator is. The entropy weight calculation method is shown in (3) - (5):

$$e_j = -\frac{1}{\ln(n)} \sum_{i=1}^{n} p_{ij} \ln(p_{ij}), \quad (j = 1, 2, \ldots, m) \tag{3}$$

$$d_j = 1 - e_j \tag{4}$$

$$W_j = \frac{d_j}{\sum_{j=1}^{m} dj} \tag{5}$$

Where e_j is the information entropy; d_j is the information utility value; W_j is the entropy weight; p_{ij} is the original value of the jth index of the ith scheme.

3.2 *Fusion TOPSIS method with GRA*

3.2.1 *Determination of the Ideal Solution*

The weight of W of each index determined by the entropy weight method is used to modify the original matrix, as shown in (6):

$$Y = \begin{bmatrix} y_{11} & y_{12} & \cdots & y_{1m} \\ y_{21} & y_{22} & \cdots & y_{2m} \\ \vdots & & \cdots & \cdots \\ y_{n1} & y_{n2} & \cdots & y_{nm} \end{bmatrix}$$

$$= \begin{bmatrix} p_{11 \times w_1} & p_{12 \times w_2} & \cdots & p_{1m \times w_2} \\ p_{21 \times w_1} & p_{22 \times w_2} & \cdots & p_{2m \times w_2} \\ \vdots & & \cdots & \cdots \\ p_{n1 \times w_1} & p_{n2 \times w_2} & \cdots & p_{nm \times w_2} \end{bmatrix} \tag{6}$$

The maximum and minimum values of each column of the revised matrix constitute the optimal and worst vectors, respectively, as in (7) and (8):

$$Y^+ = (\max\{y_{11}, y_{21}, \ldots, y_{n1}\}, \ldots, \max\{y_{1m}, y_{2m}, \ldots y_{nm}\}) \tag{7}$$

$$Y^- = (\min\{y_{11}, y_{21}, \ldots, y_{n1}\}, \ldots, \min\{y_{1m}, y_{2m}, \ldots y_{nm}\}) \tag{8}$$

Where Y^+ represents the ideal positive solution while Y^- represents the negative one.

3.2.2 Calculate the Euclidean distance

The distance between each scheme and positive or negative ideal solutions can be calculated with:

$$d_i^+ = \left[\sum_{j=1}^{m}(Y^+ - y_{ij})^2\right]^{1/2} \tag{9}$$

$$d_i^- = \left[\sum_{j=1}^{m}(Y^- - y_{ij})^2\right]^{1/2} \tag{10}$$

3.2.3 Determine the grey correlation coefficient between each scheme and positive and negative ideal solutions

The correlation coefficient matrix Y^+ of each scheme and positive ideal solution $R^+ = (r_{ij}^+)_{n*m}$ can be obtained by (11):

$$r_{ij}^+ = \frac{|Y_j^+ - Y_{ij}|_{min} + \rho|Y_j^+ - Y_{ij}|_{max}}{|Y_j^+ - Y_{ij}| + \rho|Y_j^+ - Y_{ij}|_{max}} \tag{11}$$

Similarly, the correlation coefficient matrix Y^- of each scheme and the negative ideal solution $R^- = (r_{ij}^-)_{n*m}$ is:

$$r_{ij}^- = \frac{|Y_j^- - Y_{ij}|_{min} + \rho|Y_j^- - Y_{ij}|_{max}}{|Y_j^- - Y_{ij}| + \rho|Y_j^- - Y_{ij}|_{max}} \tag{12}$$

ρ is the resolution coefficient, usually 0.5.

3.2.4 Determine the correlation degree between each scheme and the positive and negative ideal solutions

According to (13) and (14), the correlation degree between each scheme and positive and negative ideal solutions can be obtained as follows:

$$s_i^+ = \frac{1}{m}\sum_{j=1}^{m}R_{ij}^+, i \in n \tag{13}$$

$$s_i^- = \frac{1}{m}\sum_{j=1}^{m}R_{ij}^-, i \in n \tag{14}$$

3.2.5 Carry out dimensionless treatment on Euclidean distance and correlation degree

Euclidean distance and correlation degree are dimensionless according to (15)-(18):

$$D_i^+ = \frac{d_i^+}{max_i d_i^+} \tag{15}$$

$$D_i^- = \frac{d_i^-}{max_i d_i^-} \tag{16}$$

$$S_i^+ = \frac{s_i^+}{max_i s_i^+} \tag{17}$$

$$S_i^+ = \frac{s_i^+}{max_i s_i^+} \tag{18}$$

3.2.6 *Combine the dimensionless Euclidean distance with the grey correlation degree*
According to (19) and (20), the correlation degree E_i can be obtained:

$$E_i^+ = \alpha D_i^- + \beta S_i^+ \tag{19}$$

$$E_i^- = \alpha D_i^+ + \beta S_i^- \tag{20}$$

Where $\alpha = \beta = 0.5$. The concept of correlation degree can be led by further fusing the positive and negative correlation degree γ_i, as shown in (21). Its size can reflect the trend degree of the scheme to the positive or negative ideal solutions, and the larger its value is, the more ideal the scheme is.

$$\gamma_i = \frac{E_i^+}{E_i^+ + E_i^-} \tag{21}$$

4 A CASE STUDY OF A RESOURCE UTILIZATION PROJECT

4.1 *Project Overview*

This paper takes a building conversation and refurbishment project in Mawei District, Fujian province, China, as an example. The comparison of concrete with a strength grade of C30 with different recycled aggregate replacement rates and by different resource utilization methods (in-situ crushing and recycling and outbound crushing recycling). Compared with outbound crushing, in-site crushing can crush and sieve waste concrete at the construction site without transporting it to the crushing station, which greatly shortens the production process of RAC. To quantify the specific differences between the two methods, it is particularly important to analyze the case studies on resource utilization. Nine research schemes are taken into consideration: RAC is divided into two sorts according to the crushing site. Each category is divided into four types according to different recycled aggregate replacement rates (30%, 50%, 70%, 100%), and NAC is also set as a control group.

4.2 *Calculation by integrating the TOPSIS and GRA*

According to equations (3)-(5), the weights of each indicator are calculated as shown in Table 4 below.

Table 4. Weights of each index.

Indicators	The weights	Indicators	The weights	Indicators	The weights
B_{11}	8.62%	B_{21}	6.30%	B_{24}	8.10%
B_{12}	6.59%	B_{22}	11.71%	B_{25}	5.57%
B_{13}	8.63%	B_{23}	36.65%	B_{26}	7.83%

The final relative similarity degree γ_i results by applying equations (6) - (21) are listed in Table 5:

Table 5. The relative closeness of concrete.

Concrete type	γ_i	Concrete type	γ_i	Concrete type	γ_i
NAC	0.5714	ORAC-70	0.3940	IRAC-50	0.4974
ORAC-30	0.3847	ORAC-100	0.4017	IRAC-70	0.5056
ORAC-50	0.3889	IRAC-30	0.4897	IRAC-100	0.5176

4.3 *Analysis of Evaluation Results*

4.3.1 *Analysis of the evaluation results of secondary indicators*
The improved entropy weight-Topsis model with the grey correlation degree can also analyze the secondary indicators, and the specific results are demonstrated in Figure 3.

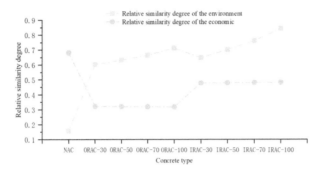

Figure 3. Evaluation results of secondary indicators.

(1) The unilateral evaluation results of environmental indicators show that the environmental benefits of RAC are better than that of NAC, among which IRAC is particularly outstanding. It would be inferred from the data in table 6 that the difference in raw material transportation and carbonization absorption is the main reason for the environmental protection potential of RAC. Under the same resource recovery method, an average increase of 1% in recycled aggregate replacement rate can reduce emissions by about 0.15kg greenhouse gas equivalent, energy consumption by about 4865.59KJ, and industrial wastewater discharge by the discharge of approximately 136.01g. Comparing the two resource utilization schemes, the reduction in the transportation distance of crushed concrete makes the discharge of various pollutants in the recycling and resource utilization stage of in-situ crushing largely less than that of outbound crushing.

(2) With regard to unilateral evaluation results of economic indicators, it can be seen that the life cycle boundary of RAC is wider than that of NAC, and the equipment used is more abundant. Aggregate has a higher economic cost due to the increase in manpower and material resources during recycled aggregate processing. Although the economic benefits of IRAC are higher than those of ORAC, the equipment cost fluctuates widely, and the perceived impact of the mobile crushing plant cost in different projects has to be determined.

4.3.2 *Analysis of Comprehensive Evaluation Results*
It can be found from the ranking of the relative similarity degree γ_i in Table 5 that the comprehensive benefit of recycled concrete increases with the growth of the replacement rate under the same resource recovery method. However, although RAC has a certain potential for energy saving and emission reduction, the burden of its economic cost makes its comprehensive benefits inferior to NAC. At the same time, reviewing the value in table 4, it is obvious that the economic indicators have the dominant weight, thus playing a leading role in the comprehensive evaluation of CW recycling. Therefore, the control of economic costs is particularly critical for selecting high comprehensive benefit resource recycling projects.

Most notably, there are many types of mobile crushing stations on the market, and the price difference is large (¥50,000~1,000,000), making the equipment cost index of IRAC fluctuate widely, and its comprehensive benefit is difficult to be accurately measured. Compared with outbound crushing, on-site crushing reduces the links of waste concrete transportation and has better environmental protection potential. However, from the perspective of comprehensive benefits, if the equipment price exceeds a certain range, the in-site crushing method will decrease in relative similarity degree to the ideal solution. In order to ensure the comprehensive benefit of in-site resource utilization, when the replacement rate is 30%, 50%, 70%, and 100%, the corresponding equipment price should not exceed¥92425.4, ¥94025.4, ¥95225.4, ¥96075.4; simultaneously, if the corresponding price is lower than¥68425.5, ¥68725.5, ¥69025.5, and ¥69245.5, its comprehensive benefit will be better than that of NAC. As a result, the reasonable control of equipment costs will have long-term significance for improving the production efficiency of recycled concrete, the development and reform of the traditional concrete industry and the prosperity of the sustainable building circular economy.

5 CONCLUSION

Although resource recycling has considerable potential for energy conservation and emission reduction, its expensive recycling cost prohibits enterprises, and the evaluation of its comprehensive benefits will be of great significance to its orderly and universal implementation. In this paper, the entropy weight method is used to determine the index weight, which avoids the subjectivity of experts in traditional evaluation. This paper integrates the TOPSIS and GRA to establish a comprehensive evaluation model for the utilization of CW resources. This method not only solves the shortcomings of the traditional TOPSIS method that it is difficult to determine the degree of influence of indicators but also reflects the changing trend of internal factors. Calculating and ranking the benefits of different schemes is more feasible in this comprehensive evaluation model. Through the analysis of the results of examples, this study shows that the comprehensive evaluation model of entropy weight TOPSIS improved based on the grey relational degree method is reasonable and effective and can provide method support for the evaluation of construction and waste utilization indicators in various places.

REFERENCES

A. Di Maria, J. Eyckmans, and K. Van Acker, "Downcycling versus recycling of construction and demolition waste: Combining LCA and LCC to support sustainable policy making," *Waste Manage.* (Oxford), vol. 75, pp. 3–21, 2018.

C. E. Shannon, "Prediction and Entropy of Printed English," *Bell Syst. Tech. J.*, vol. 30, no. 1, 1951.

C. L. Hwang, K. Yoon, *Methods for multiple attribute decision making, Multiple attribute decision making*, 1981, 58–191.

C. Xiao, M. Chang, P. Guo, et al., "Analysis of air quality characteristics of Beijing–Tianjin–Hebei and its surrounding air pollution transport channel cities in China," *J. Environ. Sci.*, vol. 87, pp. 213–227, 2020.

Cecn.org.cn. 2022. *Cost Information Network.* [online] Available at:<http://www.cecn.org.cn/>. (Chinese version)

cssglw.com. 2022. *Sand and gravel aggregate mesh.* [online] Available at:<https://cssglw.com/> [Accessed 1 February 2022]. (Chinese version)

F. Zhang, Y. Ju, EDRS. Gonzalez, et al., "Evaluation of construction and demolition waste utilization schemes under uncertain environment: A fuzzy heterogeneous multi-criteria decision-making approach," *J.Cleaner Prod.*, vol. 313, pp. 127907, 2021.

Fuzhou.gov.cn. 2022. *government of Fuzhou website.* [online] Available at: <http://www.fuzhou.gov.cn/> [Accessed 8 April 2017]. (Chinese version)

General Office of the State Council of the People's Republic of China, *Manual of the Coefficient of Production and Discharge of Industrial Pollution Sources for the first National Survey of Pollution Sources*, Beijing: Chinese Research Academy of Environmental Sciences, 2008.

H. Ma, S. Li, and C. S. Chan, "Analytic Hierarchy Process (AHP)-based assessment of the value of non-World Heritage Tulou: A case study of Pinghe County, Fujian Province," *Tourism Management Perspectives*, vol. 26, pp. 67–77, 2018.

J. L. Deng, "Introduction to grey system theory," *The electricity Journal of Inner Mongolia*, vol. 03, pp. 51–52, 1993. (Chinese version)

K. Paustian, N. H. Ravindranath, and A. V. Amstel, 2006 IPCC Guidelines for National Greenhouse Gas Inventories, *International Panel on Climate Change*, 2006.

M. C. Zhou, *Research on cost accounting model and application of construction waste resource Recovery*, Master's Thesis, Chongqing University, 2018. (Chinese version)

Ministry of Ecology and Environment of the People's Republic of China, Standard management methods for collection of pollutant discharge fees, Beijing: China Statistical Publishing House, 2003.

Ministry of Housing and Urban-Rural Development of the People's Republic of China, 2020 China Urban Construction Statistical Yearbook, Beijing: China Planning Press; ISBN, 2021.

Ministry of Housing and Urban-Rural Development of the People's Republic of China, 2016 China Urban Construction Statistical Yearbook, Beijing: China Planning Press; ISBN, 2017.

National Bureau of Statistics, China Statistical Yearbook, Beijing: China Statistical Publishing House, 2012.

O. Hjelmar, *"Development of Acceptance Criteria for Landfilling of Waste: An Approach Based on Impact Modeling and Scenario Calculations,"* Proceedings of the Eighth International Waste Management and Landfill Symposium, Sardinia, 2001, pp. 711–721.

R. F. Chen, Y. D. Xie, D. H. Wang, et al., "Life cycle energy consumption and carbon emission evaluation of construction waste resource utilization," *Fujian Architecture & Construction*, vol. 02, pp. 22–27, 2022. (Chinese version)

R. Kurda, J. de Brito, and JD. Silvestre, "CONCRETop method: Optimization of concrete with various incorporation ratios of fly ash and recycled aggregates in terms of quality performance and life-cycle cost and environmental impacts," *J.Cleaner Prod.*, vol. 226, no. 20, pp. 642–657, 2019.

tanjiaowang.net. 2022. *Carbon trading.* [online] Available at:<https://www.tanjiaowang.net/> [Accessed 23 January 2022]. (Chinese version)

Y. Lyu, Y. Gao, H. Ye, et al., "Quantifying the life cycle environmental impacts of water pollution control in a typical chemical industrial park in China," *J. Ind. Ecol.*, vol. 25, no. 6, pp. 1673–1687, 2021.

Z. G. Fu, B. H. Liu, L. Liu, et al., "The comprehensive evaluation method of thermal power unit is combined with entropy weight TOPSIS method and grey relational degree method," *Journal of North China Electric Power University*, vol. 45, no. 6, pp. 68–75, 2018. (Chinese version)

Advances in Renewable Energy and Sustainable
Development – Liang & Kasmani (Eds)
© 2023 Copyright the Author(s), ISBN: 978-1-032-39407-7

Research on noise reduction of driving vehicles in Sponge City

Zhou-Fu Liang* & Xin-yu Li
School of Environment and Life Sciences, Nanning Normal University, Nanning, China

Jie Pan
Fujian Hongsheng Materials Technology Co., LTD. R. D., Fuzhou, China

Zhi-ling Yang
Taiwan Affairs Office of the People's Government of Guangxi Zhuang Autonomous Region, Nanning, China

Si-yu Tang
Guangxi Zhuang Autonomous Region Transportation Comprehensive Administrative Law Enforcement Bureau 6th detachment, Yulin Branch, China

Xiu-qin Yang
Guangxi Transportation Investment Group Yulin Expressway operation Co., Ltd. Yulin Branch, China

Zhan-hui Chen
School of Environment and Life Sciences, Nanning Normal University, Nanning, China

ABSTRACT: This paper mainly discusses the noise reduction effect of the road surface commonly used by motor vehicles in sponge cities. The noise reduction effect of various road surfaces is compared by noise experiments inside the vehicles. Under the same speed of driving vehicles, the noise reduction effect of Sponge City pavement is at least 1 dB compared with cement pavement, which shows that the pavement after sponge transformation has a noise reduction effect.

1 INTRODUCTION

With the development of human civilization, noise pollution has been a kind of energy pollution. Generally speaking, noise is not fatal, but it can cause a certain degree of impact on people's work and learning and even endanger people's physical and mental health. If the surrounding environment has noise interference for a long time, people will have memory loss, insomnia, deafness, irritability and other symptoms. If it is serious, it will even cause permanent damage to people.

2 ANALYSIS OF ROAD STRUCTURE PERFORMANCE

2.1 *Structural analysis of roads in Sponge City*

Differences between permeable roads and drainage roads:

A) Permeable pavement: The water passes through each layer from the road surface to the subgrade.
   Drainage pavement: some layers can't allow water to pass through.
B) Permeable pavement: The water flow is in the vertical direction.
   Drainage pavement: The flow has a horizontal movement.

---

*Corresponding Author: 3101054413@qq.com

C) Permeable pavement: The water flow will be discharged to both sides of the road before reaching the subgrade, and rainwater will be discharged through an additional artificial drainage structure.

Drainage pavement: They often have greater intensity than permeable roads.

The upper layers of drainage asphalt pavement and permeable asphalt pavement all belong to the multi-porosity structure. The particle size of the multi-porosity asphalt mixture is related to its acoustical absorption coefficient.

According to relevant research and test results, the asphalt surface structure can absorb the traffic noise of 40~90dB most effectively when its void ratio is 20%~25%. The coarse aggregate used in the surface layer is usually small single-size grading. That is because the main factor determining the permeability and mechanical strength is the grain series of aggregate. The nominal standard of the mixture in China is 16mm or lower.

2.2 *Analysis of the sound absorption mechanism of the road*

When the sound wave enters the road, part of the sound energy is reflected and flowed into the gap of the pavement, where the air vibrates and turns into other energy. And finally, those energy is absorbed and consumed under the combination of viscous force and friction. The sound absorption effects of the sound-absorbing material are shown in Figure 1.

2.3 *Sound absorption performance of permeable asphalt pavement*

When the void ratio is 24%, the sound absorption effect is the best; with the increase of thickness, the low-frequency sound absorption performance of the material is improved, and the peak value of its sound absorption coefficient moves to low frequency. The permeable asphalt concrete material prepared with small particle sizes has better sound absorption performance.

3 ANALYSIS AND CONTROL OF THE NOISE GENERATED BY DRIVING VEHICLES

The noise produced by the interaction of the car tires and the road surface is the most important part of the noise produced by the driving vehicle. The higher the speed, the greater the noise produced. The air pump effect and vibration caused by the contact between the tire and the road are the main causes of the noise.

4 EXPERIMENT AND DATA ANALYSIS

This paper mainly studies the noise reduction effects of cement concrete pavement before and after sponge transformation in Nanning.

4.1 *Selection of the experimental field*

In order to ensure the accuracy of the test results, the selection of test sites is crucial. Nanning Jinqiao Road and Jianxing Road were selected as the test sections to study the noise reduction effect of the road of sponge city construction in Nanning City. The information on these roads is shown in Table1.

4.2 *Main test equipment*

The test vehicles are in good condition and meet the standards of Class M1 ordinary passenger cars. The microphone used for the microphone test is AWA 5636 multi-function sound level meter. Its

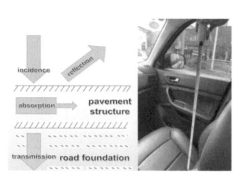

Figure 1. Sound absorption mechanism.

Figure 2. Microphone position point.

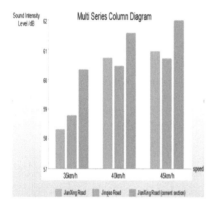

Figure 3. Front vehicle noise.

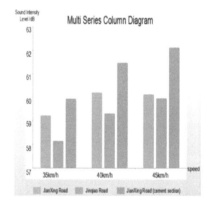

Figure 4. Rear vehicle noise.

main performance meets the requirements of the IEC6172 standard for type II sound level meter, with high reliability and a wide range of applications.

The vertical coordinates of the multi-function acoustic intensity meter at the seat should be at 0.7m±0.05m above the intersection line of the (unmanned) seat surface and the backrest surface. And the horizontal coordinates should be on the central surface (or symmetrical surface) of the seat. For manned seats, the horizontal abscissa to the right to the center surface of the seat is 0.20m±0.02m. The position of the microphone during the test is shown in Figure 2.

4.3 *Analysis of the experimental result*

The background noise in the experimental fields is shown in Table 2.

An appropriate speed should be selected during the test to minimize the impact of the engine and various noises, and excessive speed will also lead to aerodynamic noise. Therefore, we choose 35km/h, 40km/h and 45km/h to test the road noise generated on different types of road structures at different speeds. The test results are shown in Table 3 and Table 4.

Table 1. Conditions of test sites.

	Temperature (°C)	Relative Humidity (%)	Wind Speed (m/s)	Traffic Flow (vehicle/min)
Jianxing Road (Spongy)	31	78	2	20
Jinqiao Road	25	95	2	14
Jianxing Road (Cement Section)	23	100	2	16

Table 2. Background noise in the experimental fields.

Payment Type	Jianxing Road (Sponge Section)	Jinqiao Road	Jianxing Road (Cement Section)
Front Row Background Noise/dB(A)	39.4	39.7	40.0
Rear Background Noise/dB(A)	37.4	38.1	39.2

Table 3. Front-row experimental data.

	35km/h	40km/h	45km/h
Jianxing Road (Sponge) Average	58.32	60.72	60.94
Jinqiao Road Average	58.79	60.45	60.70
Jianxing Road (Cement) Average	60.33	61.56	61.98

Table 4. Back-row experimental data.

	35km/h	40km/h	45km/h
Jianxing Road (Sponge) Average	59.60	60.21	60.14
Jinqiao Road Average	58.15	59.32	59.97
Jianxing Road (Cement) Average	59.95	61.48	62.13

5 CONCLUSION

This paper analyzes the structure and function of drainage and permeable pavement, which are often used in sponge urban roads. The type of traffic noise caused by modern vehicles and its impact on human health are analyzed. The study also found out the factors affecting the noise produced by the action of vehicles and roads.

According to the analysis of Table 2,3,4 and Figure 3,4, we draw the following conclusions:

(a) At the same speed, the noise reduction effect of the sponge urban pavement is at least about 1dB compared to the cement pavement, indicating that the road surface after the sponge transformation indeed has the noise reduction effect.
(b) With the increase in the vehicle speed, it can be seen that the road noise reduction effect is more obvious after the transformation of sponge city.
(c) The front seat noise is higher than the rear seat noise because the front seats are affected by the engine.

REFERENCES

Chen Ye (2019). Analysis of Automotive Noise Source and Noise Reduction Method, *Internal combustion engine and accessories*, pp. 163–165.

Ding Qingjun, Shen Fan, Liu Xinquan, Liu Motherland and Hu Shuguang (2010). Noise reduction properties of permeable asphalt pavement material, *Proceedings of Chang'an University* (Natural Science Edition), pp. 24–28.

Lee Hui (2012). *Analysis* [D]. Hunan Agricultural University.

Liu Shengnan, Cai Jun, Wang Yaqin, Yu Xiaojuan and Liu Ling (2015). A field test method for the sound absorption coefficient of the road surface, *Acoustic Technology*, pp. 535–539.

Sun Hua (2019). *Study on the Difference between Asphalt and Drainage Pavement*, Shanxi Technology, pp. 36–40.

Tao Zhiwen (2018). *Research on urban roads under the concept of "sponge city,"* Xi'an University of Technology.

Yu Wentao (2016). *Study on the Noise Reduction Effect of low-noise asphalt pavement reconstruction*, Chongqing Jiaotong University.

Zheng Xin, Lei Xuekun, Zhang Jianlong, and Zhong Chunyao (2007). Overview of low-noise asphalt pavement research at home and abroad, *Highway & Transportation*, pp. 67–69.

Advances in Renewable Energy and Sustainable
Development – Liang & Kasmani (Eds)
© 2023 Copyright the Author(s), ISBN: 978-1-032-39407-7

Construction and reflection of a new electricity market mechanism

Yuhui Xing, Maolin Zhang, Qinggui Chen, Ling Chen, Meihan Jin & Xuan Yang
Kunming Power Exchange Center Company Limited, Kunming, Yunnan Province, China

Yiguang Zhou* & Haoyue Wu
Beijing Tsintergy Technology Company Limited, Beijing, China

ABSTRACT: The goal of "carbon peaking and carbon neutral" will promote the green transformation of energy in all aspects of the source, network, load, and storage, and new energy will become the main power source. This paper first summarizes the experience of foreign new energy market mechanism design to provide experience for domestic low-carbon energy transformation, then analyzes the main changes of new electricity market under the goal of "double carbon" from the aspects of market subjects, market mode, and trading mode, and analyzes the key mechanisms of new electricity market with the characteristics of the new electricity system. Finally, the key mechanisms of the new electricity market are discussed in light of the characteristics of the new power system. It is pointed out that the new electricity market needs to guide the priority consumption of new energy through market price signals, adjust the generation capacity guarantee mechanism in the light of the development of China's electricity market, and use market competition to pull the demand-side resource regulation potential, aiming to provide support and reference for the construction of the new domestic electricity market mechanism.

1 INTRODUCTION

Accelerating the construction of new power systems with new energy sources as the mainstay and achieving clean energy substitution is an essential way to achieve the goal of "carbon peaking and carbon neutrality" (Zheng et al. 2021). Considering the randomness and volatility of new energy sources, new energy sources as the main power source will bring more challenges to the safe and stable operation of the power system. How to promote the consumption of new energy and ensure the adequacy of system power generation is an issue that must be considered in the design of the new power market mechanism (Zhang et al. 2021). At the same time, with the rapid development of distributed power sources, flexible loads, and energy storage, the market subjects are more diversified. Enriching market trading varieties and improving the market trading system are also urgent requirements for the construction of a new electricity market (Chen et al. 2018; Feng et al. 2020; Xuan et al. 2021).

A considerable amount of research has been done on how new energy, distributed energy, flexible loads, and energy storage participate in the market. Chen et al. and Zhang et al. presented the trading methods and corresponding price formation mechanisms of new energy sources in domestic pilot provinces, and gives the path design and related construction suggestions for new energy sources to participate in the market in China (Chen et al. 2020; Zhang 2021). Chen clarified the role of each type of power source in the new power system based on the characteristics of different types of power sources. The corresponding market trading schemes are proposed accordingly, reflecting the value attributes of different types of power sources (Chen 2020; 2021). Many scholars described market

---

*Corresponding Author: auanke@outlook.com

DOI 10.1201/9781003349648-28

incentives abroad in the face of the low-carbon energy transition, providing a reference for domestic market mechanism innovation (Chen & Liao 2021; Federal Energy Regulatory Commission 2019; Wang & Li 2016; Zhao 2019). Liu et al. considered the construction of an electricity market with Chinese characteristics and clarifies the principles for the construction of a new type of electricity market (Liu et al. 2021). A number of scholars also focused on analyzing how user-side resources such as distributed resources and flexible loads can participate in the market, and proposed a market integration mechanism with multiparty interaction between source, network, load, and storage (Luo & Shen 2021; Wang et al. 2020; Zhang et al. 2021).

On the basis of existing research, this paper presents an introduction to the way new energy sources participate in market transactions in typical foreign markets. The incentives to promote the consumption of new energy in foreign countries are summarised to provide a reference for the construction of new electricity markets in China. Subsequently, this paper analyses and summarises the main changes in the new electricity market under the background of "double carbon". Finally, considering the characteristics of the new electricity market, the key mechanisms for constructing the new electricity market are discussed, and suggestions for market construction are given.

2 MECHANISMS FOR FOREIGN NEW ENERGY PARTICIPATION IN THE MARKET

2.1 *United States of America*

The US has proposed a series of policies to promote the development of new energy sources, mainly including production taxes, investment tax credit policies, renewable energy quota systems, and green certificates (Wang & Li 2016). This section presents a detailed description of the renewable energy quota system and the green certificate system.

A renewable energy quota system is an electricity-based incentive. The government imposes a mandatory and legally binding quantity requirement on energy production, requiring power generators to provide renewable energy electricity at no less than a certain percentage of the electricity supply by a set date. Renewable energy companies land subsidies through the sale of certificates, and the government adjusts the level of subsidies to renewable energy companies by adjusting the ratio factor between electricity production and certificates and the buyout price of the quota. This enables renewable energy companies the ability to gain both from the electricity market and the certificate market. The framework of the renewable energy quota system and green certificates system is shown in Figure 1.

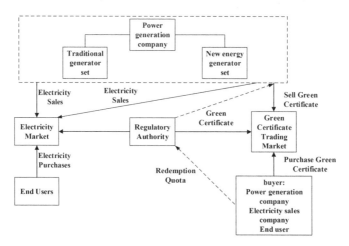

Figure 1. The framework of the renewable energy quota system and green certificates system.

2.2 *United Kingdom*

The main new energy incentives in the UK include the Renewable Energy Obligation, the Feed-in Tariff subsidy system, and the Contract for Difference mechanism.

1) The Renewables Obligation is similar to the quota system in that electricity suppliers are required to purchase Obligation Certificates from renewable energy generators or markets. Obligation certificates are issued by the Office of Gas and Electricity Markets (OFGEM) to eligible renewable energy generators based on the amount of electricity generated. Generators can sell Renewable Energy Obligation Certificates to electricity suppliers or trading organizations to receive a subsidy in addition to the wholesale price of electricity.
2) The feed-in tariff subsidy system was established to promote the development of small-scale renewable energy and low-carbon power generation technologies, which requires electricity suppliers to provide a fixed subsidy for the amount of electricity generated and fed into the grid for eligible renewable energy sources. This is relatively similar to China's subsidy policy for distributed photovoltaic power generation.
3) The core of the Contract for Difference mechanism is to allow new energy sources to participate directly in the electricity market and to protect their revenues by entering into medium and long-term contracts for difference. The new energy sources will be financially responsible for the balancing costs arising from deviations in forecasts before, during, and after the day. The Contract for Difference mechanism uses bidding to determine the contract price, which ensures a reasonable return for new energy companies while avoiding excessive incentives for new energy.

2.3 *Germany*

1) Fixed tariff system (FIT), also called the compulsory power purchase method, means that the government sets the feed-in tariff for all types of new energy generators and the power is purchased by the grid operator. New energy is not required to participate directly in market transactions, but is purchased and sold by the grid operator at a fixed tariff, and is directly marginalized in the spot market at a lower price.
2) Premium Subsidy (FIP) is a promotion of a market-based approach to the consumption of new energy based on a fixed tariff. New energy sources can choose to participate in the market through either the FIT or FIP system, and sell their electricity through direct participation in the spot market. New energy sources that choose the premium subsidy system receive a "fixed premium subsidy" in addition to the market price through competitive bidding, with generators receiving a fixed subsidy per unit of electricity in addition to the proceeds from market sales. In order to better respond to price signals in the electricity market, the German premium subsidy is gradually being adjusted from a "fixed premium subsidy" to a "floating premium subsidy". The implementation price of the project is determined through competitive bidding and the subsidy is the gap between the implementation price and the market price, beyond which there is no subsidy.

2.4 *Summary of experience and inspiration*

In overseas practice, the mechanisms and policies to support the development of new energy are mainly divided into electricity-based and price-based incentives. The main forms of electricity-based incentives are renewable energy quota systems and renewable energy certificates, while the main forms of price-based incentives are fixed feed-in tariffs, premium subsidies, and contracts for difference. China stands at the early stage of electricity market construction. In order to promote low-carbon energy transition, based on foreign experience, innovation of market mechanism should be strengthened and the market trading system enriched, taking into full consideration the characteristics of China's energy structure and the market mechanism.

Firstly, the construction of a spot market with new energy as the mainstay should be accelerated, the price formation mechanism of new energy should be improved, the full consumption of new energy should be promoted, and the cost recovery of new energy capacity should be guaranteed. Secondly, take the weight of responsibility for renewable energy consumption as a policy constraint, use green power trading as the main means of fulfilling responsibility, and build a market mechanism system that reflects green values. Finally, co-ordinate the development of a new electricity market, carbon trading market, and green certificate trading market. The research will be conducted in terms of the trading behavior and price formation mechanisms of various market players to build a synergistic mechanism for the electricity and carbon markets to promote the development of renewable energy and achieve low-carbon energy transformation.

3 MAJOR CHANGES IN THE NEW ELECTRICITY MARKET

According to the requirements of the new electricity system, the construction of the new electricity market will change from "safety, economy, and green co-ordination" to "a combination of low carbon and safety, while taking into account the economy". This section analyses the main changes in the new electricity market under the "double carbon" objective. Finally, complete content and organizational editing before formatting. Please take note of the following items when proofreading spelling and grammar:

The position of the main market players has undergone an important change. New energy generation as the main power source has become the main power supply, coal power, hydropower, and other conventional power sources gradually play a supporting and regulating function, and natural gas power generation and pumped storage power plants will become important flexible resources. A large number of load-side distributed producers and consumers will emerge, and network-load interaction and demand-side responsiveness will continue to improve.

The value of electricity commodities is gradually multi-dimensional. Under the new power system, the value of power commodities is more subdivided than life. Gradually, the value of electricity is changing from mainly electrical energy value to a multi-dimensional value system such as electrical energy value, reliability value, flexibility value, and green environment value.

Refinement of the market organization mode. Under the new power system, the market operation mode needs to adapt to the characteristics of the new energy generation. The market organization should be transformed into a refined one to ensure the integration of market and operation, medium and long-term and spot.

Electricity market space polarization. The rapid development of new energy sources shows a trend of centralized development of energy bases and the construction of load centers on a branch basis going hand in hand. In line with this, the power market space will show the characteristics of overall expansion and local decentralization. It is necessary to rely on the large power grid mutual aid ability to achieve centralized new energy range optimized allocation, at the same time with the micro-grid flexible regulation ability to achieve distributed new energy local consumption, to enhance the power grid new energy consumption capacity.

Demand-side resource aggregation. With the new power system to speed up the construction, distributed power supply, diversified load, energy storage, and other emerging market players continue to emerge, the power system from the original line "power following the load" to "power, grid, load, energy storage synergy interaction" change. The continuous development of demand-side resource regulation potential and increase in load elasticity can promote participation in market interaction and grid regulation, to achieve coordinated development of the source, network, load, and storage.

4 EXPLORATION OF KEY MECHANISMS FOR THE CONSTRUCTION OF NEW ELECTRICITY MARKETS

The new electricity market takes new energy as the main supply body, thermal power, energy storage, load, and other flexible resources as the main auxiliary services, associated with the carbon trading market and green certificate market, fully reflecting the market value of each constituent body. A panoramic view of the construction of the new electricity market is shown in Figure 2.

The new electricity market is being built from four perspectives: reliability, flexibility, stability, and environmental value, progressing in corresponding chronological order. This section discusses and analyses the key mechanisms in the new electricity market construction panorama.

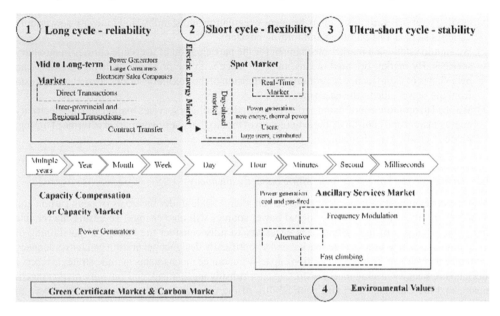

Figure 2. New electricity market construction panorama.

4.1 *Building a new energy-based electricity market mechanism*

As the scale of new energy access continues to expand, the existing policy system and market mechanism cannot satisfy the demand for consumption. It is necessary to further enhance the mechanism system of new energy participation in the market, to better adapt to the characteristics of new energy randomness and volatility, and to guide the priority of new energy consumption through market price signals.

1) It is important to establish a more refined mid- and long-term trading mechanism, so as to promote the extension of mid to long-term trading to shorter periods, change finer time periods, increase the frequency of trading, shorten the trading cycle, and improve the formation mechanism and flexible adjustment mechanism of new energy mid to long-term trading curves.
2) It is needed to improve the new energy spot market price competition mechanism and encourage new energy quotations to participate in the spot market, with the advantage of low variable costs, to achieve priority dispatch.
3) It is attempted to enrich the organization of new energy trading, improve the new energy bundle trading mechanism, carry out the new energy alternative conventional energy generation rights trading, and continuously expand the scope and scale of green power trading.
4) Commitment is made to strengthen the construction of a real-time balancing market mechanism, with an attempt to refine the design of the balancing market, establish and improve the balancing mechanism, better meet the fluctuating characteristics of new energy to the higher requirements of real-time balancing of power supply, establish a clear balance of responsibility between the two sides of the generation and use, and fully mobilize the system to flexibly adjust resources to respond to the demand for new energy consumption.

4.2 Building a synergistic auxiliary service system with multiple resources

W should optimize the design of auxiliary service varieties in response to the characteristics of new energy sources, with a multi-pronged approach to thermal power, energy storage, and customer-side resources, working together to satisfy system flexibility requirements.

1) We ought to improve the market mechanism for auxiliary services, optimize the design of auxiliary service varieties such as frequency regulation and standby, and innovate new trading varieties such as fast climbing and turning constant.
2) We should establish a market mechanism for the participation of user-side entities in auxiliary services. By market-oriented means, in accordance with the principle of "who benefits, who bears", it is of significance to promote the extension of the auxiliary service market mechanism from the power side to the user side.
3) We need to promote coordinated action between the auxiliary services market and the electricity energy market and coordinate the articulation of the auxiliary services market in terms of timing, process, clearing mechanism, and the price mechanism.

4.3 Guarantee mechanism for generating capacity adequacy

With the development of new energy sources on a large scale under the goals of peak carbon and carbon neutrality, the role of conventional power sources will also change. To ensure the reliable and stable operation of the system, it is necessary to fully consider the functional positioning of conventional power sources in the market and scientifically design a generation capacity adequacy guarantee mechanism. Typical generation capacity guarantee mechanisms include strategic reserve mechanism, scarcity pricing mechanism, capacity subsidy mechanism, and capacity market mechanism, and the characteristics and applicability of the corresponding mechanisms are shown in Table 1.

According to the characteristics of the four typical generation capacity guarantee mechanisms in Table 1, it can be illustrated that peak tariffs are suitable for regions with more developed market mechanisms and a higher tolerance for electricity price fluctuations. Capacity compensation mechanisms are relatively easy to implement and have a low risk of reforming wind, but market-based instruments are not sufficiently applied. The strategic standby mechanism can mitigate power supply risks, but has certain limitations. The capacity market mechanism requires high market infrastructure conditions and is suitable for regions with relatively well-developed market systems.

Taking into account the development of China's electricity market, it is recommended that a capacity cost compensation mechanism be adopted at the early stage of development. The scope of compensation will be reasonably determined, and capacity compensation standards and prices will be set, considering generation costs, system reliability, and other requirements. Capacity costs are incorporated into the public service costs of market operation and apportioned to the customer side. At a later stage of development, when the market system is mature, a capacity market mechanism will be established and a capacity demand curve formation mechanism will be studied to effectively release price signals and guide power generation enterprises in their investment and operation. The capacity and regulation capacity prices will be flexibly adjusted through the market formation to ensure capacity adequacy and regulation capacity adequacy. Explore ways for load-side market players to provide capacity resources, and develop methods for assessing the effective capacity of various types of resources.

4.4 Demand-side resources participate in market mechanisms

The rapid development of distributed energy, energy storage, flexible loads, and other emerging market players will be instrumental in enhancing system flexibility and tapping into demand-side regulation capabilities. Incentive-compatible market mechanisms need to be designed to guide

Table 1. Typical generation capacity guarantee mechanism.

Mechanisms	Advantages	Disadvantages	Applicability
Strategic Spare Capacity Mechanism	1. The Bid-winning unit is under the full control of dispatch, which is conducive to ensuring a reliable power supply.	1. Excessive market intervention, which may create artificial shortages of generation capacity.	1. Areas with an urgent need for flexible power.
	2. We should avoid the scrapping of some old or inefficient units.	2. Difficulties in determining a reasonable price for payment, resulting in unreasonable or unfair phenomena. 3. The freedom of dispatch to mobilize strategic reserve capacity, with implications for different market players.	2. Areas with low future load growth.
Scarcity Pricing Mechanism	Fully motivated user-side demand response.	1. Uncertainty about the number and timing of peak tariffs, making investment incentives less reliable.	1. Suitable for systems where market mechanisms are well established and tariff-distorting policies or mechanisms are rare.
		2. High demands on the regulatory system.	2. Areas where demand-side resources are sensitive to electricity prices and can adjust their electricity consumption behavior in a timely manner.
Capacity Subsidy Mechanism	Easy to achieve and low risk of modification.	1. The inability to provide targeted incentives in the event of capacity shortages 2. The need to monitor the operation of each unit in order to prevent units from misrepresenting capacity availability, which makes it more costly. 3. The process of determining the fixed capacity of a generating unit is complex.	Suitable for situations where the electricity market is in its early stages of development and the cost of stranding is severe.
Capacity Market Mechanisms	It is up to the system supplier to determine future capacity requirements and to ensure the reliability of long-term capacity supply.	1. Difficulties for system operators to forecast future capacity demand 3-4 years in advance. 2. Calculating demand prices based on historical unit revenues, resulting in inaccurate capacity prices.	Areas with more mature electrical energy markets.

demand-side flexible regulation resources to participate in the market and promote the coordinated development of source, network, load, and storage.

1) We should explore ways to organize distributed new energy trading and promote the participation of distributed new energy resources in market transactions in various ways, including aggregated transactions, direct transactions, agency transactions, etc.
2) It is significant to explore the participation of demand-side resources in the market trading mechanism, promote the participation of various demand-side resources in the electricity market as well as in the auxiliary services market such as frequency regulation and standby, fully mobilize the regulation potential of demand-side resources through bilateral transactions, and use market competition to promote demand-side "peak-shifting and valley-filling".
3) It is required to expand the scale of demand-side resource participation in the market, improve the market access rules, trading varieties, price mechanism and deviation assessment mechanism, and further expand the range of demand-side resource market participation trading varieties.

4.5 *Inter-provincial and intra-provincial market synergy mechanism*

We should continuously enhance the coupling degree of inter-provincial and intra-provincial markets, achieve rapid response and organic interaction between inter-provincial and intra-provincial markets, make full use of the advantages of inter-provincial markets in allocating resources on a large scale, and promote market integration.

1) It is needed to strengthen inter-provincial and intra-provincial market coupling and improve the coordinated operation mode of the inter-provincial and intra-provincial markets of "two levels of reporting and two levels of clearing". Form a standardized interface in terms of trading mechanism, settlement mechanism, and agency mechanism, we should gradually promote "unified reporting and unified clearing" in provinces that are qualified for this purpose.
2) Efforts should be made to implement market optimization and clearing based on inter-provincial channel ATC, take the available transmission capacity (ATC) of the channel as a constraint, and focus on optimizing the clearance of inter-provincial power, thereby satisfying the market operation needs of each province and enhance the effect of resource allocation of the power grid.
3) To promote the establishment of an inter-provincial standby sharing mechanism and establish a balanced market with provinces as the main body, it is attempted to explore the establishment of an inter-provincial standby sharing mechanism and promote inter-provincial auxiliary service resources to a large extent.

4.6 *Green power trading mechanism*

Green power trading is a significant mechanism innovation, which will effectively utilize the market, reflect the environmental value of the new power system, and facilitate the further promotion of the "double carbon" goal.

In the electricity market, policies such as the renewable energy quota system, green power trading, and green certificates reflect the environmental value attributes of renewable energy. In the upper reaches of the market, electricity generating entities are guaranteed income, and investment is guided, while in the lower reaches, price signals are transmitted to end-users to guide them to change their electricity consumption behaviors.

We are committed to strengthening the integration of the electricity market and the carbon market. With thermal power as a common market entity, connect through price and jointly promote the development of renewable energy, and strengthen coordination in terms of market space, the price mechanism, and green certification.

5 CONCLUSION

Building a new type of electricity market is an important way to optimize the allocation of resources and realize "carbon peaking and carbon neutral". Promoting the innovation of market mechanisms and enabling all kinds of power resources to realize their economic value in the market transaction are important driving forces to promote low-carbon power transition.

This paper first summarizes and analyzes the trading methods and incentives for new energy participation in foreign typical electricity markets, and provides a reference for the construction of new electricity market mechanisms in China. Then it focuses on the main features of the market mechanism in the context of double carbon, discusses the construction of the new electricity market mechanism from the aspects of the electric energy market, auxiliary service market, generation adequacy guarantee mechanism, and demand-side distributed resources, and gives relevant suggestions. The new electricity market is encouraged to link up with the carbon trading market and the green certificate market, actively respond to the demand for new energy consumption, and improve the auxiliary service system from the thermal power, energy storage, and customer side, so as to enhance the flexibility of the power system. This paper is expected to provide an effective reference for the construction of a new power system market mechanism with new energy as the mainstay, and to help achieve the goal of "carbon peaking and carbon neutrality".

REFERENCES

B. Luo, X. Shen. "Bidding model and trading mechanism of shared energy storage joint market based on blockchain," *Electrical Measurement & Instrumentation*: pp. 1–10, 2021.

C. Wang, Q. Li. "Foreign New Energy Consumption Market Mechanism and Its Enlightenment to my country," *Energy of China*, vol. 38, no. 08, pp. 33–37, 2016.

Federal Energy Regulatory Commission. "State of the markets report 2018," *Federal Energy Regulatory Commission, Management*, vol. 22, no. 06, pp. 69–74, 2019.

G. Chen, Z. Liang, Y. Dong. "Analysis and Reflection on the Marketization Construction of Electric Power With Chinese Characteristics Based on Energy Transformation," *Proceedings of the CSEE*, vol. 40, no. 02, pp. 369–379, 2020.

G. Zhang, S. Xue, M. Fan, H. Zang. "Design of Demand-Response Market Mechanism in Accordance with China Power Market," *Electric Power Construction*, vol. 42, no. 04, pp. 132–140, 2021.

H. Chen, Y. Liao. "The Enlightenment of Foreign New Energy's Participation in Power Market Innovation Mechanism," *China Power Enterprise Management*, no. 16, pp. 28–31, 2021.

H. Chen. "Power system and power market transformation under the goal of carbon neutrality," *China Power Enterprise Management*, no. 28, pp. 19–23, 2020.

H. Chen. "Electricity Value Analysis and Market Mechanism Design Under Carbon-Neutral Goal," *Power Generation Technology*, vol. 42, no. 02, pp. 141–150, 2021.

J. Xuan, X. Liu, et al. "Distributed Energy Transaction Based on Multi-chain Collaborative BlockChain," *Electric Power Construction*, vol. 42, no. 11, pp. 34–43, 2021.

L. Wang, Y. Jiang, J. Wang. "Optimization of Grid-Side Energy Storage Considering Multi-Market Transaction," *Power System and Clean Energy*, vol. 36, no. 11, pp. 30–38, 2020.

Q. Chen, K. Wang, et al. "Transactive Energy System for Distributed Agents: Architecture, Mechanism Design and Key Technologies," *Automation of Electric Power Systems* vol. 42, no. 03, pp. 1–7+31, 2018.

S. Zhang. "Thinking of New Energy's Participation in Power Market Mechanism in New Power System," *China Power Enterprise Management*, no. 16, pp. 23–27, 2021.

T. Zhang, Y. Hu, J. Zhang, L. Han, P. Wang, Y. Chen. "Analysis of Power Market Capacity Remuneration Mechanisms Adapted to High Penetration of Renewable Energy Development," *Electric Power Construction*, vol. 24, no. 03, pp. 117–125, 2021.

Y. Feng, X. Li, J. Fan, Y. Jiang. "Practice of the source-grid-load-storage interactive transaction considering clean energy consumption," *Power Demand Side Management*, vol. 22, no. 06, pp. 69–74, 2020.

Y. Liu, L. Chen, X. Han. "The Key Problem Analysis on the Alternative New Energy under the Energy Transition," *Proceedings of the CSEE*, pp. 1–10, 2021.

Y. Zhao. "Discussion on International Experience of New Energy Electricity Marketization and China's Implementation Path," *Price: Theory & Practice*, no. 10, pp. 9–13+166, 2019.

Y. Zheng, Z. Yang, et al. "Analysis of the Key Issues of the National Unified Power Market Under the Target Scenario of Carbon Peak," *Power System Technology*, pp. 1–19, 2021.

Advances in Renewable Energy and Sustainable Development – Liang & Kasmani (Eds)
© 2023 Copyright the Author(s), ISBN: 978-1-032-39407-7

A differential electricity tariff for high-energy-consuming industry based on comprehensive energy consumption indicators

Zhixun Wang* & Rengcun Fang
State Grid Hubei Electric Power Company Limited Economic Research Institute, Wuhan, China

Xianguo Fan
State Grid Hubei Electric Power Company Limited, Wuhan, China

Tingting Hou
State Grid Hubei Electric Power Company Limited Economic Research Institute, Wuhan, China

Wei Liao
State Grid Hubei Electric Power Company Limited, Wuhan, China

Lanfei He
State Grid Hubei Electric Power Company Limited Economic Research Institute, Wuhan, China

Chao Luo
State Grid Hubei Electric Power Company Limited, Wuhan, China

ABSTRACT: The goal of "carbon peaking and carbon neutrality" accelerates the development of renewable energy, and also places higher demands on industries to improve energy efficiency. On the other hand, the high proportion of renewable energy raises the operating costs of the power system, and expands the cross-subsidy to guarantee stable tariffs for residents and agriculture. To reasonably evacuate the increasing costs, a differential electricity tariff for the high-energy-consuming industry is designed in this paper. The total tariff cross-subsidy of residential and agricultural users is first calculated based on the average feed-in tariff and transmission and distribution tariffs by voltage levels. Then, a comprehensive energy consumption index (CECI) is designed to evaluate the energy consumption efficiency of each industry. Finally, the tariff increases for different high-energy-consuming industries are calculated according to the CECI of each industry, forming a differentiated apportionment of emission reduction costs for the power system.

1 INTRODUCTION

With the increasingly stringent requirements of carbon emission assessment, China proposed to build a new power system to achieve the goal of "carbon peak and carbon neutrality" in 2021(Ymwab, Kcab, Jnkab, et al, 2022; Xu, Dong, Xu, et al, 2022), which clearly defines the development direction of future energy and power systems. The power grid is the hub platform for energy conversion, utilization, and transmission. As the construction of the new power system progresses, the overall investment and operation costs of the power system increase, and it is urgent to establish a scientific and reasonable cost-evacuation mechanism.

In the electricity tariff link, to speed up the electricity tariff reform, the state is to promote all commercial and industrial users to join the electricity market. The commercial and industrial

---

*Corresponding Author: seeewzx@foxmail.com

DOI 10.1201/9781003349648-29

catalog tariff has been canceled while maintaining stable prices for residential and agricultural users.

In addition, the development of renewable energy also brings about an increase in construction and operation costs of the power system (Li, Ho, 2022; Cole, Gates, Mai, 2021). Since the price of residential agriculture needs to be stabilized to shield the lowest-income households from increasing energy costs (Pacudan, Hamdan, 2019), a corresponding differential tariff mechanism needs to be established to allow commercial and industrial users to reasonably bear the cross-subsidy brought by guaranteeing a stable electricity price for residential and agriculture users.

From the perspective of promoting users to reduce emissions, raising the sales tariff of high-energy-consuming industry users is one of the ways to evacuate power system costs. Ref. (Yupeng, Junwen, 2018) has verified that differential tariff policy has a significant incentive effect on reducing the intensity of electricity consumption. Casper and Niels (2022) investigated Danish time-series data for 99 industries and found that industries will save electricity and other energy when electricity prices rise. The latest policy also advocates the establishment of a unified differential tariff system for high energy-consuming industries, with a multistep tariff based on different energy efficiencies, and no tariff increase for enterprises whose energy efficiency reaches the benchmark level.

However, the current energy consumption differentiation management measures are few, and the tariff increase for high energy-consuming industries is usually artificially formulated and lacks a parameter basis.

This paper first calculates the total tariff cross-subsidy based on the average feed-in tariff, transmission and distribution tariffs by voltage levels, and electricity consumption by residential and agricultural customers. Then, a differential electricity tariff model based on comprehensive energy consumption indicators (CECI) is designed. The comprehensive energy consumption of each industry is evaluated in two dimensions, namely, total electricity consumption (EC) and electricity consumption intensity (ECI). Finally, the tariff increases for different high-energy-consuming industries are calculated according to the CECI of each industry, forming a differentiated apportionment of power system costs.

2 DIFFERENTIAL ELECTRICITY TARIFF MODEL BASED ON CECI

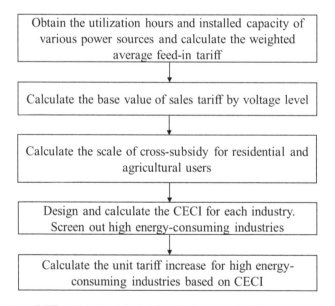

Figure 1. Flow chart of differential electricity tariff model based on CECI.

The process of the differential electricity tariff model is shown in Figure 1, which includes the following steps:

Step 1: Obtain the utilization hours and installed capacity of each type of generation unit. Based on the actual feed-in tariff of each power source, the weighted average feed-in tariff is calculated.

$$\bar{C}_{ge} = \frac{\sum_{i \in \Psi} C_{ge,i} P_{ge,i} H_{ge,i}}{\sum_{i \in \Psi} P_{ge,i} H_{ge,i}} \tag{1}$$

Where $C_{ge,i}$ is the feed-in tariff of type i generation unit, $P_{ge,i}$ is the rated power of type i generation unit, $H_{ge,i}$ is the annual power utilization hours of type i generation unit, and Ψ is the collection of power generation types.

Step 2: To obtain the total revenue for the transmission and distribution grid of each voltage level, the unit transmission and distribution tariff based on the transmission power is calculated. The transmission and distribution tariff for a certain voltage level is calculated as:

$$C_{j,sp} = \frac{S_j}{E_{j,tol}} \tag{2}$$

Where $C_{j,sp}$ is the tariff of voltage level j, S_j is the total revenue of voltage level j, and $E_{j,tol}$ is the transmission power of voltage level j

Based on the distribution tariff of each voltage level, the base value of the sales tariff by voltage level is calculated.

$$C_{j,ref} = \bar{C}_{ge} + C_{j,sp} + C_{fj} \tag{3}$$

Where $C_{j,ref}$ is the base value of the sales tariff for voltage class j, $C_{j,sp}$ is the transmission and distribution tariff of voltage class j, C_{fj} is the governmental funds and surcharges.

Step 3: Obtain the total electricity consumption and actual catalog tariff of residential and agricultural users, and calculate the scale of cross-subsidy according to the benchmark value of sales tariff and its actual catalog tariff.

$$C_{btsum} = E_{re} \left(C_{r,ref} - C_{re,cat} \right) + E_{ag} \left(C_{a,ref} - C_{ag,cat} \right) \tag{4}$$

Where C_{btsum} is the scale of cross-subsidies received for residential and agricultural customers. $C_{r,ref}$ and $C_{a,ref}$ are the base value of sales tariff for residential and agricultural voltage levels, respectively. E_{re} and E_{ag} are the electricity consumption for residential and agriculture, respectively.

Step 4: Design and calculate the CECI for each industry.

The CECI is divided into two parts, which are EC and ECI. Considering the difference in characteristics of the two indicators, a flexible evaluation index should be adopted for the total electricity consumption, i.e. a convex function is used to express the impact of the total amount on the comprehensive energy consumption evaluation index. A log function is used to normalize the data by taking the upper and lower limits of total electricity consumption.

$$m_{i,tol} = \frac{\log_{E_{tol\,min}} \left(E_{i,tol} \right)}{\log_{E_{tol\,min}} \left(E_{tol\,max} \right)} \tag{5}$$

Where $E_{tol\,max}$ and $E_{tol\,min}$ are the upper and lower limits of total electricity consumption, $E_{i,tol}$ is the annual average energy consumption of i industry. $m_{i,tol}$ is the evaluation indicators, which range is between 0 and 1. The above parameters are taken as the annual average of the distribution tariff regulation cycle.

Taking the average electric energy consumption intensity of non-energy-consuming industries as the benchmark value, the ECI evaluation index is calculated as:

$$x_{i,\text{its}} = \frac{E_{i,\text{tol}}}{C_{i,\text{GDP}}} \tag{6}$$

$$m_{i,\text{its}} = \frac{x_{i,\text{its}} - \bar{x}_{\text{low}}}{x_{\text{its max}} - \bar{x}_{\text{low}}} \tag{7}$$

Where $x_{i,\text{its}}$ is the ECI of i industry. $E_{i,\text{tol}}$ is the total annual electricity consumption of i industry, $C_{i,\text{GDP}}$ is the annual GDP of i industry. $\bar{x}_{\text{low}}$ is the average ECI of non-energy-consuming industries. $x_{\text{its max}}$ is the maximum value of the ECI, $m_{i,\text{its}}$ is the ECI evaluation index of i industry.

The CECI of i industry M_i is calculated as

$$M_i = m_{i,\text{tol}} m_{i,\text{its}} \tag{8}$$

The commercial and industrial users are ranked, and the industries whose CECI is higher than the average value are identified as high energy-consuming industries, while the industries below the average are uniformly classified as non-high energy-consuming industries.

Step 5: Based on the CECI, the differential tariff to be levied in high energy-consuming industries can be calculated.

First, the total amount of subsidies levied among each high-energy-consuming industry should be equal to the sum of the total subsidies received by residents and agriculture:

$$\sum_{i=1}^{n} R_i E_{i,\text{tol}} = C_{\text{btsum}} \tag{9}$$

where R_i is the unit tariff increase of i industry.

The amount of subsidy for a certain industry should be proportional to its CECI.

$$\frac{R_i}{M_i} = \frac{R_q}{M_q} \quad \forall i, q \in \Omega \tag{10}$$

where Ω is the collection of high-energy-consuming industries.

The unit differential tariff for each high-energy-consuming industry can be calculated by Equations (8)-(10).

3 CASE STUDY

A regional grid in China is used as a case study to measure the cross-subsidies shared by residential and agricultural users in agriculture and the implementation of differential tariffs for industries with different CECI.

3.1 Calculation of cross-subsidy

It is assumed that thermal power break-even utilization hours of 4,000 hours and its power generation benchmark price is ¥0.418/kWh. Considering that with the increase of renewable energy, thermal power generation hours will decrease in the future, the unit feed-in tariff of thermal power will increase to remain its annual operating income at a constant level. Assuming that the total load demand of the regional grid is 305 billion kWh, by obtaining the annual utilization hours, the utilization hours and its average feed-in tariff can be calculated in Table 1.

It can be seen that due to the increase in renewable energy penetration, thermal power units have to reduce the operating hours to meet the system peaking demand. On the other hand, to protect

Table 1. Average feed-in tariff of different generations.

Generation type	Installed capacity (MW)	Annual utilization hours	Average feed-in tariff (¥/kWh)
Wind	10,000	1960	0.416
PV	27,000	1103	0.416
Hydro	19,130	3460	0.305
Nuclear	13,000	3300	0.390
Thermal	39,250	3719	0.450
Weighted average feed-in tariff			**0.404**

its reasonable income needs its feed-in tariff is increased accordingly, which leads to the rise in the system's indirect cost brought by renewable energy.

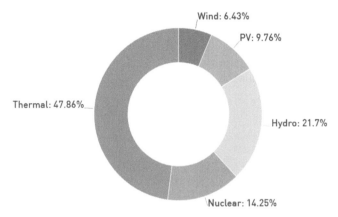

Figure 2. The electricity proportion of different generation types.

The electricity proportion of different generation types is shown in Figure 2. The weighted average feed-in tariff for each type of power source is calculated as ¥0.409/kWh.

Assuming that the voltage level of residential and agricultural users is 1 kV, the base values of the sales tariff for residential and agricultural users are shown in Table 2.

Table 2. The base value of the sales tariff.

Type	Tariff (¥/kWh)
Average feed-in tariff	0.404
Distribution tariff of 1kV	0.285
governmental funds and surcharges	0.03
the base value of the sales tariff	0.719

The scale of cross-subsidy for the residential agricultural users can be calculated based on the residential agricultural electricity consumption, catalog tariff, and the base value of the sales tariff. Assuming that the catalog tariff of residential and agricultural users is ¥0.619/kWh and ¥0.419/kWh respectively, the scale of cross-subsidy is shown in Table 3.

As can be seen in Table 3, the residential and agricultural catalog electricity tariffs are significantly lower than the true cost of electricity supply. This is due to the cross-subsidy policy implemented in China to promote economic development in rural areas, and to reduce the burden

Table 3. Cross-subsidy of residential and agricultural users.

Parameter	Residential	Agricultural
Annual electricity consumption (billion kWh)	15	5
catalog tariff (¥/kWh)	0.619	0.419
the base value of the sales tariff (¥/kWh)	0.719	0.719
cross-subsidy (billion ¥)	1.50	1.50

of electricity consumption in residential agriculture, i.e., the cross-subsidy structure of electricity consumption in industry and commerce subsidizing electricity consumption by residents and agriculture.

3.2 Calculation of CECI and unit tariff increase

The industrial and commercial industries are divided into 15 industries in this paper, and the EC and ECI of each industry are shown in Figure 3.

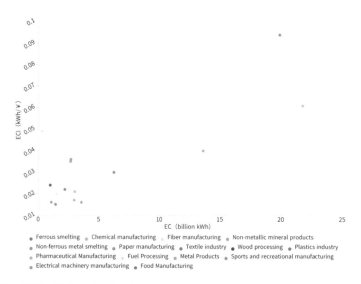

Figure 3. EC and ECI of different industries.

It is found in Figure 3 that ferrous smelting, chemical manufacturing, and non-metallic mineral products are typical high energy-consuming industries, whose EI and ECI are significantly higher than the average level. The average ECI of the industries is 0.033 kWh/¥, and according to equation (8)-(10), the high energy-consuming industries can be screened out. Then the CECI for each industry and the tariff increases for different high-energy-consuming industries are calculated, which is shown in Table 4.

From Table 4, industries with high EC and ECI need to bear a higher standard of unit tariff increase, which helps to support enterprises to improve energy efficiency and reduce carbon emissions. In addition, the funds from the tariff increase can also be used to guarantee a stable price level for residential agriculture users.

Table 4. CECI for each industry.

Industry Classification	Industries	ECI (kWh/¥)	EC (billion kWh)	CECI	Unit tariff increase (¥/kWh)
high energy-consuming industries	Ferrous smelting	0.094	19.89	0.983	0.083
	Chemical manufacturing	0.061	21.77	0.553	0.046
	Fiber manufacturing	0.049	0.33	0.086	0.007
	Non-metallic mineral products	0.040	13.63	0.246	0.021
	Non-ferrous metal smelting	0.036	2.73	0.132	0.011
	Paper manufacturing	0.035	2.70	0.123	0.010
non-high-energy-consuming industries		0.020	23.25	0	0

4 CONCLUSION

To achieve a reasonable sharing of additional costs brought by cleaner energy transition while ensuring stable tariffs for residential and agricultural users, this paper proposes a differential electricity tariff model for power systems based on a comprehensive energy consumption index. the conclusions are obtained as below:

(1) The real cost of electricity supply for residential agricultural customers is higher than the catalog tariff, resulting in additional costs borne by the other commercial and industrial users.
(2) The method achieves a reasonable cost-sharing for all high-energy-consuming industries, and industries with CECI below the benchmark level do not need to pay the additional tariff. Under the mechanism, enterprises will be motivated to improve their energy efficiency and reduce energy consumption.
(3) On the basis of differential tariffs, further research can be conducted on the evaluation of the effect of differential tariffs and dynamic adjustment mechanisms of energy consumption benchmark to improve the incentive effect and refinement of differential tariffs.

ACKNOWLEDGMENTS

This work was financially supported by the 2022 Research Project of State Grid Hubei Electric Power Company Limited (521538220005).

REFERENCES

Casper B, Niels F. M. (2022) The influence of electricity prices on saving electricity in production. Automated multivariate time-series analyses for 99 Danish trades and industries [J]. *Energy Economics*, 107, pp. 105444.

Cole W, Gates N, Mai T. (2021) Exploring the cost implications of increased renewable energy for the U.S. power system [J]. *The Electricity Journal*, 34, pp. 106–113.

Li J, Ho M S. (2022) Indirect cost of renewable energy. Insights from dispatching[J]. *Energy Economics*, 105, pp. 105778.

Pacudan R. and M. Hamdan. (2019) Electricity tariff reforms, welfare impacts, and energy poverty implications." *Energy Policy* 132, pp. 332–343.

Xu G, Dong H, Xu Z, et al. (2022) China can reach carbon neutrality before 2050 by improving economic development quality [J]. *Energy*, 243, pp. 123–130.

Ymwab C, Kcab C, Jnkab C, et al. (2022) Policy and management of carbon peaking and carbon neutrality. A literature review [J]. *Engineering* (Available online).

Yupeng Z, Junwen C. (2018) Influence of China's Differential Price Policy on Electricity Consumption Intensity Based on Difference-in-Difference Method [J]. *Power & Energy* 39(6), pp. 725–730.

Advances in Renewable Energy and Sustainable Development – Liang & Kasmani (Eds)
© 2023 Copyright the Author(s), ISBN: 978-1-032-39407-7

Analysis of requirements and key problems of national carbon market construction

Xiaoxuan Zhang*, Zheng Zhao, Yu Wang & Su Yang
State Grid Energy Research Institute Co., Ltd, Beijing, China

ABSTRACT: The goals of "carbon peak" and "carbon neutralization" require to speed up the construction of the national carbon market, and also put forward new requirements for the construction of the national carbon market. Effectively implementing the national carbon emission reduction target, promoting the green transformation of industrial development mode, accelerating the low-carbon transformation of energy and power, and strengthening the integration and link with the international market have become the new requirements for the construction of the national carbon market. To improve the carbon emission trading mechanism, we should focus on solving six key problems, such as solving the problems of the reasonable setting of total amount target, fair and efficient distribution of quota, orderly expansion of coverage, effective formation and regulation of carbon price, improvement of CCER offset mechanism and link between domestic and international markets.

1 INTRODUCTION

In September 2011, the National Development and Reform Commission issued the notice on the pilot work of carbon emission trading, which officially approved the pilot of carbon emission trading in seven provinces and cities: Beijing, Tianjin, Shanghai, Chongqing, Shenzhen, Guangdong, and Hubei (Lu et al. 2020). The pilot markets have started trading since 2013. In 2014, under the organization and guidance of the National Development and Reform Commission, we started the top-level design and construction of the national carbon market system by learning from the experience of pilot carbon market construction. On December 19, 2017, the department in charge of carbon trading under the State Council and its main supporting institutions was transferred from the National Development and Reform Commission to the Ministry of Ecological Environment. The government tried to steadily promote the construction of the national carbon market from the aspects of promoting carbon trading legislation, with a view to establishing and perfecting the system, accelerating infrastructure construction and strengthening basic capacity construction. In 2020, after Xi Jinping put forward the "3060" target, the Ministry of Ecology and Environment issued the relevant documents densely, and the construction of the national carbon market continued to accelerate. On January 1, 2021, the first performance cycle of the national carbon market was officially launched.

2 REQUIREMENTS FOR THE CONSTRUCTION OF A NATIONAL CARBON MARKET

The goals of "carbon peak" and "carbon neutralization" require accelerating the construction of the national carbon market, strengthening the total carbon emission control, promoting the low-carbon

---

*Corresponding Author: zhangxiaoxuan@sgeri.sgcc.com.cn

transformation of energy and power, forcing the optimization and upgrading of industrial structure, promoting green and low-carbon technological innovation, and effectively implementing the national carbon emission reduction goal. The 2020 central economic work conference proposed to accelerate the adjustment and optimization of the industrial structure and energy structure, promote the peak of coal consumption as soon as possible, vigorously develop new energy, accelerate the construction of a national energy use rights and carbon emission trading market, and improve the dual control system of energy consumption. The 14th five-year plan for national economic and social development and the outline of long-term goals for 2035 proposes to formulate an action plan for peaking carbon emissions by 2030; we should implement a system dominated by carbon intensity control and supplemented by total carbon emission control, and support qualified localities, key industries, and key enterprises to take the lead in reaching the peak of carbon emissions; we should strengthen the control of greenhouse gases such as methane and hydrofluorocarbons. We are determined to achieve carbon neutrality by 2060, so we will adopt more powerful policies and measures. Overall, the construction of the national carbon market under the vision of carbon neutralization mainly faces the following requirements:

First, we should effectively implement the national carbon emission reduction target. We should speed up the construction of the national carbon market, improve the carbon pricing mechanism, promote the shift from intensity-based total target to absolute total target control, and effectively decompose and implement the task of carbon reduction. The carbon market helps to reduce carbon dioxide emissions per unit of GDP by 18% in the 14th five-year plan, reach the peak carbon by 2030 and neutralize carbon by 2060.

Second, it is attempted to promote the green transformation of industrial development mode. The Fifth Plenary Session of the 19th CPC Central Committee proposed to promote green and low-carbon development, build a modern economic system and vigorously develop strategic emerging industries. Through the construction of a carbon market, we will force the low-carbon development of industrial structure, strictly control high energy consumption and high emission projects, and guide funds from high-carbon industries to low-carbon industries.

Third, it is important to promote the acceleration of the low-carbon transformation of energy and power. China's coal consumption accounts for 57.6% of primary energy consumption, higher than the global average of 27%. The carbon emission intensity of energy consumption is 30% higher than the world average. Curbing coal power, accelerating the development of renewable energy, vigorously developing smart grid and energy storage, and accelerating the commercial use of CCS and CCUs are the main ways to decarbonize the power industry.

Fourth, it is needed to strengthen the integration and link with the international market. We should promote the development of international carbon emission trading mechanisms such as CORSIA, explore links with international markets such as the EU carbon trading system, and adapt to international carbon pricing competition. China has accepted the Kigali amendment to the Montreal Protocol, trying to increase the control of non-carbon dioxide greenhouse gases such as methane and hydrofluorocarbons on the basis of carbon dioxide control.

3   KEY ISSUES IN THE CONSTRUCTION OF A NATIONAL CARBON MARKET

To deepen the design of the national carbon emission market mechanism, we should focus on solving six key problems, such as the reasonable setting of the total amount target, the fair and efficient distribution of quotas, the orderly expansion of coverage, the effective formation and regulation of carbon price, the improvement of CCER offset mechanism, and the link between domestic and international markets.

3.1   *Total target setting*

The total amount of the national carbon market is set based on coverage, greenhouse gas emission control objectives, economic growth expectations, industrial development layout, etc. (Engle et al. 2020). The total quota is determined by combining top-down and bottom-up methods. The implementation of the total amount is realized through quota allocation.

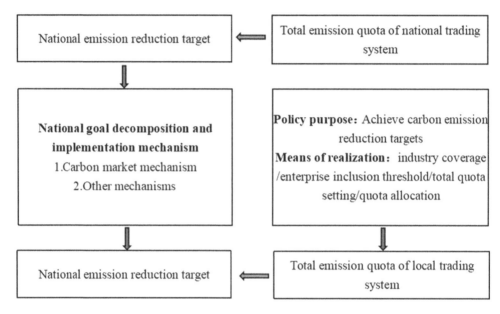

Figure 1. Connection diagram of national, local, and industrial total amount control objectives.

The total amount of the national carbon market is a flexible total considering both industrial development and national carbon emission control objectives. The total amount target of the national carbon market should be connected with the national, local, and industrial total amount control targets. After the national total carbon emission target is determined, it is necessary to distinguish the total national carbon market and the noncarbon market. For the total local carbon emission target, it is necessary to distinguish the total part covered by the national carbon market, the part covered by the local carbon trading pilot, and other mechanisms. For the total amount control target of the industry, it is necessary to distinguish the industrial enterprises included in the national carbon market from other enterprises.

3.2 *National carbon market coverage*

The national carbon market only includes CO_2 in the initial stage and may cover N_2O, PFC, and other greenhouse gases in the future. Considering that the power industry is the main emission source, with large carbon emissions and a good database, and the foreign carbon market usually starts from the power industry, the only industry covered by the carbon market is the power generation industry. (Yang et al. 2020). At present, 2162 power generation enterprises are included, with a total annual quota of about 4.5 billion tons, accounting for 40% of China's carbon emissions. After all the eight industries are included, the total quota is expected to reach 7-8 billion tons, accounting for more than 70% of the total national emissions. It is expected that all the eight industries in the 14th five-year plan will be included in the national carbon market. Considering the emission reduction potential, database, and industrial policy requirements, building materials (cement), steel, and nonferrous metals (electrolytic aluminum) will be included first, and petrochemical, chemical, papermaking, and aviation will be included if they are mature.

3.3 Quota allocation method

A carbon emission quota is the total amount limit of carbon dioxide emissions that can be legally caused by the emission sources included in the carbon emission trading system in a specific period of time. It is the initial carbon dioxide emission right issued to relevant enterprises by the competent government department in response to climate change. Quota allocation methods include benchmark method, enterprise historical intensity reduction method, and auction method. The benchmark method is a method to calculate the quota of legal person enterprises according to the two elements of physical output and the industry benchmark of legal person enterprises. The enterprise historical intensity reduction method is a method to calculate the quota of legal person units according to the three elements of physical output, historical intensity value, and emission reduction coefficient of legal person enterprises. Different emission trading systems choose appropriate quota allocation schemes according to their economic development, industries covered, and emission reduction targets.

3.4 Carbon valence formation mechanism

The current carbon price formation mechanism mainly includes three ways: first, the price formation mechanism based on market pricing; Second, the price formation mechanism based on government pricing, such as setting prices according to pricing authority; the third is the mixed pricing mechanism, that is, the price formation mechanism combined with the market, government, intermediaries and other forces (Zhu et al. 2019). From the pilot experience, there are some problems, such as inactive carbon trading, concentrated trading during the performance period, a high proportion of offline transactions, opaque market price information, and so on, which affect the formation of an effective carbon price. Therefore, it is necessary to establish and improve the carbon price formation mechanism. The carbon price is mainly affected by the total quota, economic development, distribution methods, etc. The carbon price fluctuates greatly, so it is necessary to establish a monitoring and regulation mechanism. In the future, carbon trading systems such as spot trading, forward trading, futures trading, and options trading should be gradually established. We should promote the formation of long-term carbon price signals to guide investment, expand the strategic choice space of market participants, and meet their diversified needs such as locking in investment risks.

3.5 Improve CCER mechanism

In 2012, the general office of the National Development and Reform Commission issued "the interim measures for the administration of voluntary greenhouse gas emission reduction transactions" and "the guidelines for the approval and certification of voluntary greenhouse gas emission reduction projects". These two documents basically established the work processes of declaration, approval, filing, certification, and issuance of China's voluntary emission reduction projects, which means that the domestic voluntary emission reduction market has taken a substantive step. In 2015, the trading of voluntary emission reduction projects was officially launched. However, in March 2017, the National Development and Reform Commission issued an announcement to suspend the filing application for CCER projects and emission reductions. At present, the Ministry of Ecology and Environment has yet to clarify the final reform plan of voluntary emission reduction trading and restart the approval of CCER projects and emission reductions.

The entry of CCER into the market enriches the trading varieties of the carbon market, reduces the performance cost of key emission units, improves the activity and operation efficiency of the carbon market, and also provides a broader space for carbon market participants such as emission control enterprises and investment institutions. Therefore, the CCER mechanism should be improved. establish and improve the CCER offset mechanism. According to the national carbon market rules, emission control enterprises are allowed to use certified voluntary emission reductions (CCER) to achieve the carbon trading performance goal.

3.6 International market connection

The expansion of the global carbon trading system and the implementation of the EU carbon border regulation tax require accelerating the connection between the national carbon market and other carbon markets such as the EU (Ilhan et al. 2021).

At present, the most successful connection in the world is the carbon market in California and Quebec. The market design, policy framework, and operation mechanism of both sides are similar, and the integration is completed according to the implementation plan. Both sides are participating members of the Western climate initiative. Links include mutual recognition and use of compliance tools (including emission quotas and offset credits), a unified registration system and auction mechanism, and information sharing among market participants. The link between the two sides is realized in the form of regional legislation. Relevant exploration has been carried out in the domestic carbon pilot market. On March 18, 2019, Guangzhou carbon emissions exchange and European energy exchange (EEX) launched business promotion simultaneously to promote cross-border carbon market cooperation and strengthen the connection between China and European carbon markets. On November 12, 2019, the Guangzhou carbon emission rights exchange issued the business guidelines for the swap transaction between the Guangdong carbon emission quota and the EU carbon emission quota. If China's carbon market is linked with the international carbon market in the future, it will help to reduce the performance cost of both sides and achieve win-win results.

4 CONCLUSION

The goals of "carbon peak" and "carbon neutralization" put forward new requirements for the construction of ecological civilization, and curbing the greenhouse effect and reducing carbon dioxide emissions are also important measures for China to practice green development. Under the new development background, new requirements are put forward for the hypothesis of the national carbon market. Solving the key problems in the construction of the national carbon market can accelerate the pace of carbon emission reduction and make the carbon emission trading mechanism a powerful tool to achieve the goals of "carbon peak" and "carbon neutralization". The construction of the carbon market and the goals of "carbon peak" and "carbon neutralization" are interactive. On the one hand, the construction of a carbon market is an important tool to achieve these goals. On the other hand, the introduction of the goals of "carbon peak" and "carbon neutralization" has accelerated the construction and transformation of the carbon market.

ACKNOWLEDGMENTS

This work was financially supported by the State Grid Corporation of China's science and technology project "Research on the key mechanism, operation simulation, and evaluation technology of power grid carbon reduction contribution of national carbon market construction under the goal of" 3060.

REFERENCES

Engle R., Giglio S., Lee H., et al. (2020) Hedging climate change news. *J.Ssci. Review of financial studies.*, 33: 1184-1216.

Ilhan E., Sautner Z., Vilkovi G., et al.(2021) Carbontail risk. *J.Ssci. Review of financial studies.*, 34: 1540-1571.

Lu H.F., Ma X., Huang K., et al. (2020) Carbon trading volume and price forecasting in China using multiple machine learning models. *J. Sci. Journal of cleaner production.*, 249:1–46.

Yang W., Pan Y.C., Ma J.H., et al.(2020) Effects of allowance allocation rules on green technology investment and product pricing under the cap-and-trade mechanism.*J.Sci. Energy policy.*, 139: 1–12.

Zhu B.Z., Ye S.X., Han D., et al.(2019) A multiscale analysis for carbon price drivers.*J.Ssci. Energy economics.*, 78: 202–216.

Advances in Renewable Energy and Sustainable
Development – Liang & Kasmani (Eds)
© 2023 Copyright the Author(s), ISBN: 978-1-032-39407-7

Design of a market trading mechanism for energy storage in the context of a new power system

Yuhui Xing, Zhelin Yang, Maolin Zhang, Qinggui Chen, Ling Chen, Meihan Jin & Xuan Yang
Kunming Power Exchange Center Company Limited, Kunming, Yunnan Province, China

Yiguang Zhou*
Beijing Tsintergy Technology Company Limited, Beijing, China

ABSTRACT: In the context of the double carbon target and the construction of a new type of power system, the power system has been developing rapidly in the direction of new energy sources as the mainstay. Energy storage has a significant role to play in the reliable grid connection and efficient consumption of a high proportion of new energy sources, and can also fundamentally solve the problems of insufficient grid regulation and difficult frequency stability that it brings. Therefore, it is important for the development of energy storage to participate in electricity market transactions and to improve the market transaction mechanism. This paper introduces the domestic and foreign energy storage market trading mechanisms from four aspects: foreign energy storage platform, key issues of energy storage participation in the market, energy storage participation in the electricity market trading mechanism, and energy storage participation in the auxiliary services market trading mechanism, and focuses on the design and example analysis of energy storage participation in the electricity market trading mechanism. The results show that it is not easy for energy storage to participate in the electricity energy trading market, but the revenue in the frequency regulation auxiliary service market is considerable and higher than that in the electricity energy market.

1 INTRODUCTION

As global climate change brings more and more adverse effects to human society, many countries are responding positively to the development goal of "carbon neutrality". In 2020, China set the strategic goal of "striving to reach peak carbon by 2030 and carbon neutrality by 2060". Since then, the development of a new power system with new energy sources as the mainstay has become a key factor in achieving the "carbon neutrality" goal. However, the intermittent and fluctuating nature of renewable energy sources such as wind and solar energy poses a challenge to the stability and flexibility of power systems. Energy storage systems have the advantages of islandable operation, fast regulation rate, and flexible configuration, which can help to improve the peak regulation, frequency regulation, and voltage regulation of power grids and ensure the safe and stable operation of power systems (Li 2021). Therefore, the birth of energy storage technology can solve the various challenges brought by new energy sources, and has become the focus of research by scholars and experts.

With the advancement of electricity market reforms in major developed countries, the development and operation of energy storage in the future will mainly be realized in the context of marketization (Chen et al. 2021). At present, scholars have conducted many studies on the participation of energy storage in the market. In terms of energy storage business models, as China's electricity system reform is still in its infancy, the existing business operation models are highly

---

*Corresponding Author: auanke@outlook.com

DOI 10.1201/9781003349648-31

consolidated, while many developed countries abroad have relatively well-developed energy storage business models. In order to discover a potentially feasible energy storage commercial operation model for China, scholars in China have summarised and analyzed the energy storage business models of developed countries such as the UK (Zhu et al. 2022), the US and Germany (Li et al. 2022; Wu et al. 2019). Literature (Qiu et al. 2021) designed a comprehensive evaluation index system based on the shared energy storage business model and verified the superiority of the index system through the analysis of actual data calculation examples. The paper (Hu et al. 2019) makes recommendations on the business model and policies for grid-side energy storage in China, taking into account factors such as investment costs, electricity prices, and main bodies. In terms of the participation of energy storage in the auxiliary service market, the current participation of energy storage in power market transactions is the most important application area of energy storage technology due to its superior performance in frequency regulation and peak regulation (Li et al. 2019; Wei & Jiang 2020) and peak regulation (Nan et al. 2020; Lin et al. 2019). The literature (Liu et al. 2022) puts forward corresponding suggestions for the future development of energy storage participation in the auxiliary service market in four dimensions: policy reform, battery research, collaborative control, and centralized control of distributed energy storage in the context of double carbon. The literature (Zhang & Wang 2021) analyzed foreign cases of energy storage participation in the FM auxiliary service market and summarised the inspiration of foreign experience for China. With the development of integrated energy systems, energy storage systems that operate jointly with wind power (Wang et al. 2019; Xing 2020), photovoltaic (Xing 2020), hydrogen (Kong et al. 2022) and other energy sources have also been studied extensively.

This paper first classifies foreign energy storage business models and explains the key issues of energy storage participation in the market. Based on this, the mechanism of energy storage participation in the spot energy and ancillary services market is designed, and then the benefits of energy storage participation in the spot energy and ancillary services market are measured and analyzed based on the operating environment of the settlement of the spot trading market in a certain month in a province.

2 FOREIGN ENERGY STORAGE BUSINESS MODELS

In addition to participating in system operation and control, energy storage can also generate corresponding economic benefits. Compared to the domestic market, the development of foreign energy storage markets is relatively mature, so it is necessary to classify and explore foreign energy storage business models. Based on market size, revenue model, commercial operation, target areas, and markets, cost structure, and other aspects, the current business models of foreign distributed energy storage projects are grouped into the following categories: "rent to sell" model, shared revenue model, virtual power plant model, community energy storage model, hybrid or integrated model.

2.1 *"Rent to Sell" model*

The "rent to sell" is currently the most widely used investment and operating model in the field of energy storage abroad. This model means that the developer of the energy storage project leases the energy storage system to the customer, who receives the electricity savings and pays a fixed monthly rent according to a certain percentage, including equipment usage, operation and maintenance costs, software costs, installation costs, taxes, and fees. In addition, the lease term can be set flexibly depending on the objectives of the user or the product.

2.2 *Shared Revenue model*

This model refers to a strategy of sharing energy storage revenues between the energy storage project developer and the user, and its main application is for commercial and industrial users. Similar to the leasing model, the user pays a regular fee to the energy storage developer, but unlike the leasing model where a fixed amount of rent is paid to the developer, the rent paid to the developer in the shared revenue model is a variable amount and is split in proportion to the revenue generated.

Furthermore, in terms of the actual market development for both the leasing and shared revenue models, the latter, when applied on its own, generally has a longer contract period of 10 years and above, so the model is usually run in combination with other business models.

2.3 Virtual power plant model

In this model, the utility or a third-party company integrates the customer with the smart grid through a central control room and participates in grid services by unifying control and optimizing system operation for application benefits. The participation of energy storage allows the model to be valuable in a number of ways, including participation in the ancillary services market, demand-side response for revenue after unified dispatch and management, active support on the customer side, and dynamic expansion of transformers.

The virtual power plant model is a unified and optimized control of the energy storage system through a central control room, which can find optimization points when the utility or third party company is responding to demand and help customers to reduce expenditure. Figure 1 shows the virtual power plant model developed by Fenecon, BYD's German distributor, in cooperation with Ampard of Switzerland.

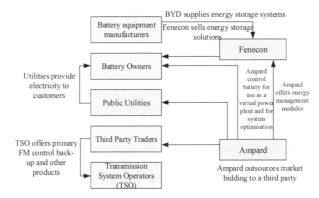

Figure 1. The virtual power plant model of Fenecon/Ampard.

2.4 Community energy storage model

The community storage model is currently mainly used by distributed PV users, who pay a fixed rental fee to use the energy storage and store the PV power in battery storage, which is then used for self-consumption or traded between community users. The advantage of this model is that the user pays a lower lease fee than the cost of purchasing electricity from the grid. The main regions where this model is currently used are Germany, the USA, and Australia. Figure 2 shows MVV Engergie, a German district energy provider, applying the community energy storage model to the MVV Strombank project it has developed.

2.5 Hybrid or integrated model

For the hybrid model, current foreign practice shows that Stem in the US, for example, aggregates the leasing model it carries out to build virtual power plants that participate in spot market transactions and respond to dispatch. In addition, the combination of shared revenue and virtual power plants is also a more common hybrid model.

For the integrated model, the South Australian government has contracted with an energy storage builder, for example, for a third party to operate the plant. Based on the principle of "who pays, who

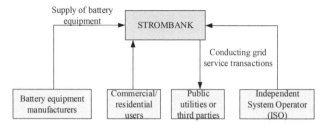

Figure 2. Community energy storage model.

pays", the storage system earns revenue through frequency control, ancillary services, low storage, and high discharge, and can also provide a variety of grid services for a fee based on results.

3 KEY ISSUES FOR ENERGY STORAGE PARTICIPATION IN THE MARKET

The following issues have been explored by academia and industry: the identity of energy storage when participating in market transactions, the market framework when energy storage is a market player, the revision of the energy market clearing model when energy storage participates in the wholesale market, the price mechanism and capacity value accounting, and the trading mechanism of distributed energy storage when it is combined with distributed energy storage.

3.1 *The question of the identity of energy storage in the market*

However, in the existing market, transmission assets are regulated entities. Therefore, it is a new issue to be explored as to what kind of entity energy storage will participate in the market.

1) As a single market player, although it is possible to obtain revenue from the market, this does not reflect the value of energy storage as a transmission asset, making it difficult to invest in energy storage projects that have an advantage in transmission.
2) When acting as a single regulated subject, it is necessary to verify its advantages in transmission and specify the appropriate scale of investment, and after verification of feasibility, it also needs to be controlled by the International Organisation for Standardisation (ISO) in order to avoid blind investment behavior.
3) When it is a market player and a regulated player at the same time, in this case, energy storage is able to demonstrate its full potential value. However, there are also a number of issues and challenges that need to be addressed: for example, energy storage needs to be prioritized among the functions of the market and those of the regulated entity in a given situation; the amount of revenue it can generate in both capacities needs to be clarified to prevent duplication of revenue capture, etc.

3.2 *Framework design issues for energy storage participation in the market*

When energy storage is involved in the market, a suitable framework and convergence model are needed for resource allocation. To address this issue, existing research on convergence models falls into 2 main categories.

1) The first is the unified operation model (Xiao et al. 2022), where a physical model of energy storage resources is established in the existing market system and suitable market rules are designed. The market will allocate power generation, electricity consumption, and energy storage resources at the same time.

2) The second type is the independent operation model (Wang et al. 2018), i.e. the purchaser is the ISO, the customer or the power producer, which improves its own balancing capacity, output curve or power consumption curve by purchasing the right to use energy storage.

In terms of transmission capacity selection, the Unified Operating Model uses an implicit auction, while the Independent Operating Model is an explicit auction. This allows the objective function of the former clearing model to be globally optimal, but therefore requires the simultaneous optimization of the charging and discharging schedules, generation output, and consumption curves of the storage.

3.3 *Energy market model correction issues*

In the market resource allocation system, the largest share of transactions is in the wholesale capacity market. From practical experience abroad, in the US, its energy market requires the direct generation of executable dispatch and price plans, resulting in energy storage's load state constraints and the cost of device aging being difficult to reflect in the payout model. In the UK, on the other hand, the market's dispatch plan is determined by each entity, so the participation of energy storage does not pose significant problems.

1) The different technical routes have their own advantages and disadvantages in terms of market model amendments. In the case of the bid model, for example, the advantage of the bid model is that it can accurately assist energy storage in reflecting the cost of one charge/discharge cycle, but it poses difficulties in modelling the objective function.
2) Multiple market participation models can cater to the diversity of energy storage types. The former can facilitate the ability of storage to "shift peaks and fill valleys", which is more applicable to pumped storage with larger individual units, while the latter can allow storage to test the peak market price, but this is more applicable to small-scale, costly electrochemical storage. It is important to clarify that these two models do not force a particular type of storage to correspond to a particular market participation model, maintaining the neutrality and fairness of ISO.

3.4 *Market price mechanism design*

In the energy market under the marginal tariff settlement model, there is the problem of difficulties in matching the amount of revenue obtained from energy storage with the amount it contributes. Therefore it is important to design a suitable market price mechanism. From the current research, scholars have already made studies on the settlement mechanism and the optimization of its incentive compatibility. Among them, the Vickery-Clark-Gloves (VCG) mechanism, which controls the allocation of resources on a demand basis to achieve a rational use of resources, has received a lot of attention from scholars. The advantage of this mechanism is that it can effectively suppress strategic behavior while preserving the offer rights of market players. Under this mechanism, the revenue generated by energy storage is controlled by the social value it creates, which ultimately aligns individual benefits with collective benefits, but requires a suitable capital regulation scheme to address the problem of under-achievement of the mechanism for energy storage. In addition, the use of VCG mechanisms for energy storage alone may be questionable in the face of many other market players.

3.5 *Accounting for the value of energy storage capacity*

The problem of accounting for the capacity value of energy storage has always been a difficult issue. There are currently three ways of accounting for the capacity value of energy storage as follows.

1) Firstly, set the continuous discharge time requirement for energy storage, and then directly discount the discharge power of energy storage. The advantage of this approach is that the standard is clear and simple to implement, but in practice, energy storage will reduce or increase the

discharging power during the corresponding period with the fluctuation of system load, resulting in unreasonable continuous charging and discharging time requirements, and eventually the power after direct discounting of energy storage will be lower than its capacity value.

2) The effective load carrying capability (ELCC) is determined by market simulation, which is already being used in the UK. The advantage of this approach is that the capacity value obtained through simulation is more accurate, but the disadvantage is that it is more difficult to embed in the simulation system. In addition, the simulation accuracy may not be accurate enough as it only relies on input parameters to get the simulation results.

3) Set the corresponding capacity factor for energy storage with a different energy to power ratio. This approach has the advantages and disadvantages of the first two approaches and is a neutralization of the first two approaches.

3.6 *Model issues for distributed energy storage participation in the market*

Distributed is one of the most promising development directions of the power system, but the huge amount of distributed energy storage, and the unknown willingness to participate in the market offer, thus, it is difficult to optimize it through the market model, and the exploration of distributed energy storage participation market model is necessary. At this stage, there are two main models for distributed energy storage participation in the market.

1) Aggregator model (Hupez et al. 2021). This model uses aggregators as a communication medium between distributed energy storage and the wholesale market, but the adoption of this model requires consideration of 2 relationships, namely the relationship between aggregators and the wholesale market and distributed energy. In addition, in order to make the aggregators participating in the market competitive, they need to be nurtured.

2) Direct trading model for distribution networks (Chen et al. 2018). Compared to the first model, it has the advantage of facilitating the local balancing of energy on the distribution grid side, as well as reducing transmission tide pressure and market clearing calculation pressure. However, this model is more complex and difficult to apply to small-scale markets.

4 ENERGY STORAGE PARTICIPATION IN THE ELECTRICITY ENERGY MARKET TRADING MECHANISM

The electricity market is a market with electricity as the trading commodity, which generally includes the energy market and auxiliary service market, and assumes that these markets are competitive. As the electricity market has the characteristics of competition and openness, through reasonable allocation and optimization of resources, it can meet the maximization of economic returns between the market subjects.

4.1 *Design of trading mechanism for energy storage participation in the electricity energy market*

From overseas experience, the Pennsylvania, New Jersey, and Maryland (PJM) market, for example, consists of a day-ahead energy market and a real-time energy market. The day-ahead market and the real-time energy market complement each other. After the day-ahead market is cleared to obtain the next day's clearing price and generation capacity, the real-time energy market optimizes the allocation of the next day's actual grid operation and the difference between the day-ahead generation plan to meet the day's trading demand. Most domestic electricity market models use centralized bidding and, in addition to the two trading cycles mentioned above, the domestic electricity energy market also includes medium and long-term trading, based on which the following electricity energy trading mechanisms are designed.

A spot market with medium to long-term financial contracts for difference and full power reporting and centralized optimized clearing, provides a good risk hedge for power producers, and is settled in accordance with the principle of "day-ahead basis, real-time difference, contract difference". During the operation period of the spot market settlement, the declaration and clearing method adopt a unilateral daily 10-segment declaration on the generation side, and each unit's energy revenue is settled at the nodal tariff on the generation side. During the operation period, the settlement mode adopts a zero-sum model on the generation side, and the market share refund costs include energy share refund funds, cost compensation costs, and market-based auxiliary service costs, which are shared by the participating units in accordance with the proportion of their respective government-authorized contract tariffs.

Four simulation scenarios have been designed for energy storage participation in the spot electricity energy market.

1) Scenario 1: Charging and discharging based on peak and trough periods, two charges and two discharges per day, charging during the low hours with a settlement price of low valley electricity, and discharging during the peak hours with a settlement price of coal-fired benchmark electricity.
2) Option 2: Two charging and two discharging according to the spot market price, charging during the lowest average electricity price on the electricity consumption side during the early morning hours, discharging during the high nodal electricity price on the generation side during the morning peak, secondary charging during the lowest average electricity price on the electricity consumption side during the midday load drop, and discharging during the high nodal electricity price on the generation side during the afternoon peak.
3) Option 3: Energy storage participates in the spot market with self-scheduled units as the recipient of the price. Based on the average load-side price and generation-side price during the 48 daily periods of spot operation, the self-planned operating hours of the storage plant are set as follows: morning and afternoon periods: 8:45-11:00 and 14:15-16:45 to declare the generation curve, early morning period and noon period: 2:30-5:00 and 11:00-13:30 for the charging curve.
4) Option 4: Charging and discharging is quoted separately to participate in the market, and the quotation is based on a daily uniform quotation. The best daily offer is measured under the condition that the total kWh difference is greater than the sum of the kWh transmission and distribution cost and the kWh loss cost.

The spot market energy settlement method is Eq. (1)-Eq. (4), and the contracted electricity is 0 when energy storage participates in the market simulation.

$$F_{CSM} = F_{DA}^1 + F_{RT}^1 + F_B^1 \tag{1}$$

$$F_{DA}^1 = \sum \left(Q_{DA,t}^1, P_{DA,t}^1 \right) \tag{2}$$

$$F_{RT}^1 = \sum \left(Q_{RT,t}^1 - P_{DA,t}^1 \right) P_{RT,t}^1 \tag{3}$$

$$F_B^1 = \sum Q_{B,t}^1 \left(P_{B,t}^1 - P_{DA,t}^1 \right) \tag{4}$$

where F_{CSM} is the centralized spot market electricity energy settlement cost; F_{DA}^1 is the unit's day-ahead full electricity settlement cost; F_{RT}^1 is the unit's real-time deviation settlement cost; F_B^1 is the unit's medium and long-term contracted electricity price difference cost; $Q_{DA,t}^1$ is the day-ahead market unit's T-time winning power; $P_{DA,t}^1$ is the day-ahead market unit's T-time nodal electricity price; $Q_{RT,t}^1$ is the unit's real-time market T-time feed-in tariff; $P_{RT,t}^1$ is the unit's real-time market T-time settlement price; $Q_{B,t}^1$ is the unit's T-time medium and long-term time-sharing power; $P_{B,t}^1$ is the unit's T-time contracted integrated price.

4.2 Analysis of the benefits of energy storage participation in the electricity energy market trading

Now take a 10kv energy storage power station as an example, the energy storage battery is a lead carbon battery. Design rated charge and discharge power 9MW, energy storage power for 24MWh, depth charge and discharge cycle limit 2200 times, the life of about 5-10 years, charge and discharge efficiency 88%, the final decay after the residual 80%, the average decay 0.91%/100 times cycle, project investment of about 80 million yuan. Table 1 shows the peak and valley level segmentation table to be implemented in a province from January 1, 2022. Using the above electricity energy trading mechanism as a basis, the revenue of the energy storage power plant is simulated and measured based on the relevant data of a month of spot market settlement operation in a province, and the revenue of energy storage participation in the electricity energy market is analyzed.

Table 1. Peak and valley weekday segments.

Timeslot name	Time period
Peak periods	11:00-12:00
	14:00-21:00
Flat periods	7:00-11:00
	12:00-14:00
	21:00-23:00
Low periods	23:00-7:00

Based on the above, and based on data such as the amount of electricity and tariffs for a province's spot market settlement operation, the full month's revenue of energy storage plants participating in the spot electricity energy market under each scenario is shown in Table 2. Scenario 4 has the highest return under consideration of energy storage conversion efficiency, capacity decay, and transmission and distribution costs. Scenario 1 uses the scheme model, where energy storage valley charging is settled at the grid sales tariff, and is subject to cross-subsidy, which makes its charging costs smaller and only largely maintains returns. Although Scenarios 2 and 3 select the right price hours and load hours for charging and discharging, the overall costs are higher than the overall benefits due to the smaller overall market spread space, the larger proportion of transmission and distribution costs, and the increasing power loss costs as generation increases. Option 4 gives full play to the flexibility of energy storage regulation capacity by actively responding to market demand and reasonably declaring generation tariffs and load tariffs, and still has room for revenue after deducting transmission and distribution costs.

Table 2. Simulation results for a full month of energy storage under each scenario.

Programme		1	2	3	4
Monthly expenditure/RMB 10,000	Electrical energy costs	50.78	22.56	22.87	1.07
	Transmission and distribution costs	50.78	32.07	32.07	13.06
Monthly charging volume/MWh		1440.83	1440.83	1440.83	572.44
Monthly income / RMB 10,000		52.11	36.56	36.1	16.56
Monthly electricity generation/MWh		1262.53	1262.53	1262.53	497.78
Monthly Revenue/RMB 10,000		1.31	−17.34	−18.45	2.56

5 ENERGY STORAGE PARTICIPATION IN THE AUXILIARY SERVICE MARKET TRADING MECHANISM

Common auxiliary services are frequency regulation, peak regulation, standby, voltage, and black start. Among them, frequency regulation is a typical application scenario of energy storage in the auxiliary service market, and the revenue of this part occupies a larger part of the revenue of energy storage in the auxiliary service market. Therefore, the trading mechanism is designed for the common frequency regulation auxiliary service function of energy storage.

5.1 *Design of the trading mechanism for energy storage participation in the fm auxiliary services market*

According to the impact of FM performance indicators on FM costs, mainly for the capacity price and mileage price in the clearing whether the unit FM performance indicators are considered, the spot FM auxiliary service market trading mechanism is now divided into three categories. The characteristics of each trading mechanism and the representative markets are shown in Table 3.

Table 3. Classification of trading mechanisms in the spot FM ancillary services market.

Code number	Features	Market represented
1	Both capacity and mileage prices are cleared taking into account the unit's FM performance indicators and are calculated in a more comprehensive manner.	PJM Market
2	Neither the capacity price nor the mileage price is calculated without taking into account the FM performance indicator.	Shandong and Guangdong Power Markets in China
3	Both the capacity price and mileage price clearances take into account FM performance indicators, but the FM performance indicators are calculated in a simpler way.	New England, California, Central & New York State Markets

The Guangdong and PJM power markets are more comprehensive and rigorous in their consideration of the calculation of FM performance indicators, and both use multiple regulation indicators to participate in the calculation. For example, regulation rate, regulation timing, and regulation accuracy are weighted and calculated as shown in equation (5).

$$S = k_1 M^{rate} + k_2 M^{del} + k_3 M^{pre} \tag{5}$$

where S is the FM performance index; M^{rate} is the regulation rate; M^{del} is the regulation time; M^{pre} is the regulation accuracy; PJM corresponds to the weighting factor $k_1 = k_2 = k_3 = 1/3$, generally take the weighting factor $k_1 = 0.5$, $k_2 = 0.25$, $k_3 = 0.25$.

The FM performance indicators during the operation of the spot market are measured, mainly considering the regulation rate mark, regulation response delay time indicator, and regulation accuracy indicator. When the FM performance indicators of the unit are not lower than the threshold value of the performance indicators, the actual FM performance indicators of each settlement period of the FM unit are normalized and used in the FM settlement calculation. Based on the above, the design of energy storage participation in the FM auxiliary services market trading mechanism.

During the electricity market settlement operation, the day-ahead market carries out the declaration of FM auxiliary services, including the declaration of FM capacity, the declaration of FM capacity price, and the declaration of FM mileage price; the real-time market carries out the joint clearing of FM auxiliary services and electric energy. The system adjusts the quotation according to the historical FM performance normalization index of each unit to obtain the FM combination ranking price, ranks the FM units from lowest to highest according to the FM combination ranking price, and carries out marginal clearing according to the declared capacity of the FM units and the FM capacity demand of the system to determine the winning FM units, as shown in Equations (6) and (7).

$$S'_{ZJ,i} = S_{ZJ,i} / S_{ZJ,max} \tag{6}$$

$$P'_0 = (P_{C0} + \gamma P_{m0} + C_{OC}) / S'_{ZJ,i} \tag{7}$$

where $S'_{ZJ,i}$ is the unit's historical FM performance normalized index, P'_0 is the FM portfolio ranking price; P_{C0} is the FM unit FM capacity offer; P_{m0} is the FM unit FM mileage offer; γ is the system

FM mileage capacity ratio, taken as the average of the ratio of last month's system hourly FM mileage to last month's system hourly FM capacity; C_{OC} is the FM unit estimated opportunity cost.

The actual clearing of the individual prices in the winning bid process is calculated by equations (8)-(11). If the FM unit's FM capacity and FM mileage declarations are both zero, its actual FM opportunity cost is zero by default and the FM unit settles as the price recipient. If the FM unit's FM declaration is not zero, the actual FM opportunity cost is calculated using the real-time market power price. The impact of the actual FM performance normalization metric will be factored into the FM capacity revenue and FM mileage revenue in the real-time market settlement as shown in equations (12) and (13).

$$P_0'' = \left(P_{c0} + \gamma P_{m0} + C_{OC}'\right) / S_{ZJ,i}' \tag{8}$$

$$P_{Reg} = \max\left(P_{0,j}''\right) \tag{9}$$

$$P_m = \max\left(P_{m0,j}/S_{ZJ,i}'\right) \tag{10}$$

$$P_c = P_{Reg} - \gamma P_m \tag{11}$$

$$E_{c-ZJ} = P_c C_c S_{ZJ}^R \tag{12}$$

$$E_{m-ZJ} = M_R P_m S_{ZJ}^R \tag{13}$$

where C_{OC}' is the actual opportunity cost of the winning unit; P_0'' is the FM pricing ranking price; P_{Reg} is the FM clearing price, P_m is the FM mileage clearing price, P_c is the FM capacity clearing price; $P_{0,j}''$ is the FM pricing ranking price of the winning FM unit, $P_{m0,j}$ is the FM mileage declared price of the winning FM unit, S_{ZJ}^R is the FM performance normalized index, E_{c-ZJ} is the FM capacity revenue; E_{m-ZJ} is the FM mileage revenue.

It is assumed that energy storage and other types of FM auxiliary service units, such as gas-fired units and coal-fired units, participate in the market with the same capacity offer and mileage offer, without considering the impact of the electricity energy market (opportunity cost is 0), and all the resources declared in the market are won, and only the impact of other parameters on the participation of energy storage in the FM market is judged according to different rule situations.

5.2 *Example analysis of the benefits of energy storage participation in the fm ancillary services market*

Again, the benefits of energy storage participation in the FM ancillary services market are analyzed using the example in 3.2 as a basis and in conjunction with Table 4, which simulates the benefits of energy storage plants. Table 5 shows the FM pricing in the ancillary services market. As can be seen from Table 4 and Table 5, the FM capacity offer and mileage offer of each FM unit is consistent, but the probability that energy storage can obtain priority clearance in the ancillary service market increases due to its superior FM performance indicators. Combining equations (8)-(13), the measured market FM clearing price under the rules is RMB 77.88/MWh and the FM mileage price is RMB 7.08/MW, and the clearing cost of each FM unit is shown in Table 6.

Table 4. Basic parameters of energy storage participation in the FM market.

FM units	Gas-fired unit 1	Gas-fired unit 2	U0	Energy storage
FM performance indicators	0.44	0.12	0.47	1.01
FM capacity quotation (RMB/MWh)	2	2	2	2
FM capacity (MW)	1	1	1	1
FM mileage quotation (RMB/MW)	0.79	0.79	0.79	0.79
Mileage to capacity ratio (1/h)	10	10	10	10

According to Table 5, it can be seen that with the introduction of the FM performance indicator, the FM portfolio ranking price of energy storage has a significant advantage and can be prioritized

Table 5. FM pricing in the ancillary services market.

FM units	Gas-fired unit 1	Gas-fired unit 2	U0	Energy storage
FM capacity offer (RMB/MWh)	4.86	17.72	4.65	2.02
FM mileage offer (RMB/MW)	1.95	7.08	1.88	0.79
FM mileage offer * mileage capacity ratio (MWh)	19.49	70.8	18.8	7.9
FM opportunity cost (RMB/MWh)	0	0	0	0
FM portfolio sequencing price (RMB/MWh)	21.44	77.88	20.68	8.69
Clearance order	3	4	2	1

Table 6. Power market FM unit settlement costs.

FM units	Capacity charge(RMB)	Mileage charge(RMB)
Gas-fired unit 1	2.93	29.34
Gas-fired unit 2	0.80	7.98
U0	3.05	30.49
Energy storage	7.08	70.80
Total	13.86	138.31

to get out of the market, while it will also be prioritized to be called when the FM call is made, which increases the FM mileage gain of energy storage. Therefore, it is assumed that the storage station, as the best-regulated unit in the FM market, can be given priority in the FM market to win the bid to get the call. According to the spot market FM capacity and FM mileage price, considering the energy storage conversion efficiency and capacity decay, the simulation of energy storage participation in the spot FM market is carried out, and the benefits are shown in Table 7. In the simulation measurement, the monthly revenue of energy storage participating in the FM auxiliary service market can reach 1.675 million yuan, which is much higher than the revenue of energy storage participating in the electric energy market.

Table 7. Participation of energy storage plants in the FM auxiliary service market revenue.

Monthly earnings (RMB)	Annual earnings (RMB)	Life (years)	Annual Return on Investment
1675000	2010.2	5	25.13%

6 CONCLUSION

Based on the analysis of foreign energy storage business models and the key issues of energy storage participation in the market, this paper has conducted a series of explorations on the trading mechanism of energy storage participation in the spot market, which can help guide energy storage to participate in spot market trading and achieve better benefits. The benefits of energy storage participation in the spot electricity energy market and participation in the clearing of the FM auxiliary services market have been measured and analyzed respectively, and the main conclusions reached are as follows.

1) Energy storage has an elastic decoupling function in electricity production and consumption in time, and when it is involved in electricity energy market transactions, it needs to seek market price difference space in order to obtain profit, and under the influence of energy storage conversion efficiency, capacity loss and transmission and distribution and additional costs, it

is more difficult for energy storage to participate in the electricity energy market to make a profit.

2) Although the market mechanism for FM auxiliary services varies across regional markets, most focus on compensation for the actual FM effect. After the benefit analysis, it was found that with the introduction of FM performance parameters into the market clearing and price settlement of these two aspects, for energy storage to participate in the FM auxiliary service market, the probability of winning the bid has been greatly improved.

3) The benefits of energy storage in the FM ancillary services market are much higher than those in the spot electric energy market, so energy storage should be used as a flexible dispatch resource in the ancillary services market, with the FM market as the main focus and the electric energy market as a supplement when participating in electricity market transactions.

REFERENCES

D. Li, Y. Shi, Y. Li. "Research on the business model of the energy storage system in energy Internet," *Power Demand Side Management*, vol. 22, no. 02, pp. 77–82, 2022.

G. Nan, L. Zhang, Z. Guo, et al. "Design of Trading Mode for Grid-side Energy Storage Participating in Peak-shaving Assistant Service Market," *Journal of Electrical Engineering*, vol. 15, no. 03, pp. 88–96, 2020.

H. Zhang, Y. Wang. "Mechanism experience of foreign grid-side storage participating in frequency regulation auxiliary service market and its enlightenment to China," *Energy Storage Science and Technology*, vol. 10, no. 02, pp. 766–773, 2021.

H. Zhu, J. Xu, G. Liu, et al. "UK policy mechanisms and business models for energy storage and their applications to China," *Energy Storage Science and Technology*, vol. 11, no. 01, pp. 370–378, 2022.

Hupez M, Toubeau J F, De Greve Z, et al. "A new cooperative framework for a fair and cost-optimal allocation of resources within a low voltage electricity community," *IEEE Transactions on Smart Grid*, vol. 12, no. 03, pp. 2201–2211, 2021.

J. Hu, Q. Li, B. Huang, et al. "Business Model Research of Energy Storage on Grid Side Adapted to Application Scenarios and Policy Environment in China," *Journal of Global Energy Interconnection*, vol. 2, no. 04, pp. 367–375, 2019.

J. Li. "Key technologies and applications for energy storage systems under the double carbon target," *Electric Power Engineering Technology*, vol. 4, no. 03, pp. 1, 2021.

J. Wang, R. Fang, B. Wen, et al. "Collaborative Optimization in Energy and Frequency Regulation Markets for Wind Farm with Energy Storage," *Electrical & Energy Management Technology*, no. 20, pp. 51–57+63, 2019.

J. Wei, Y. Jiang. "Research on Compensation Mechanism of Energy Storage Participating in Auxiliary Services and Operation of Multi-Business Model," *Electrical & Energy Management Technology*, no. 05, pp. 78–85, 2020.

K. Wang, Q. Chen, H. Guo, et al. "Mechanism Design and Trading Strategy for Capacity Contract of Energy Storage Towards Transactive Energy," *Automation of Electric Power Systems*, vol. 42, no. 14, pp. 54–60+90, 2018.

L. Lin, Z. Mi, Y. Jia, et al. "Distributed energy storage aggregation for power grid peak shaving in a power market," *Energy Storage Science and Technology*, vol. 8, no. 02, pp. 276–283, 2019.

L. Wu, F. Yue, A. Song, et al. "Business models for distributed energy storage," *Energy Storage Science and Technology*, vol. 8, no. 05, pp. 960–966, 2019.

M. Li, F. Jiao, C. Ren, et al. "China's Power Auxiliary Service Market Mechanism and the Economics of Energy Storage Systems Participating in Auxiliary Services," *Southern Energy Construction*, vol. 6, no. 03, pp. 132–138, 2019.

P. Kong, Z. Jiang, L. Yang, et al. "Mechanism and Risk Quantification Model Design of Hydrogen Storage System Participation in Power Market," *Distribution & Utilization*, vol. 39, no. 01, pp. 31–39, 2022.

Q. Chen, C. Fang, H. Guo, et al. "Participation Mechanism of Energy Storage in Electricity Market: Status Quo and Prospect," *Automation of Electric Power Systems*, vol. 46, no. 16, pp. 14–28, 2021.

Q. Chen, K. Wang, S. Chen, et al. "Transactive Energy System for Distributed Agents: Architecture, Mechanism Design and Key Technologies," *Automation of Electric Power Systems*, vol. 42, no. 03, pp. 1–7+31, 2018.

T. Xing. "Research on Optimal Operation Strategy of Combined Photovoltaic and Storage System in Real Time Energy and Frequency Regulation Market," *Nanjing University of Science and Technology*, 2020.

T. Xing. "The Research on the Transaction Mechanism and Optimization Mode for Large Scale Wind Power Participating in the Power Market," *North China Electric Power University*, 2020.

W. Qiu, M. Wang, Z. Lin, et al. "Comprehensive evaluation of shared energy storage towards new energy accommodation scenario under targets of carbon emission peak and carbon neutrality," *Electric Power Automation Equipment*, vol. 41, no. 10, pp. 244–255, 2021.

Y. Xiao, L. Zhang, X. Zhang, et al. "The Coordinated Market Clearing Mechanism for Spot Electric Energy and Regulating Ancillary Service Incorporating Independent Energy Storage Resources," *Proceedings of the CSEE*, vol. 40, no. S1, pp. 167–180, 2022.

Z. Liu, D. Peng, H. Zhao, et al. "Development prospects of energy storage participating in auxiliary services of power systems under the targets of the dual-carbon goal," *Energy Storage Science and Technology*, vol. 11, no. 02, pp. 704–716, 2022.

Advances in Renewable Energy and Sustainable
Development – Liang & Kasmani (Eds)
© 2023 Copyright the Author(s), ISBN: 978-1-032-39407-7

The carbon market research hotspot and trend in China – Based on citespace visualization analysis

Yujie Wu*, Yue Cao & Yinying Duan

School of Business and Travel, Sichuan Agricultural University, Ningbo, China

ABSTRACT: In the context of the new low-carbon era, in order to understand the research status of the domestic carbon market, this paper takes 1029 documents on China's carbon market included in the CNKI database as the research object. This paper uses the CiteSpace 5.8 visualization tool to analyze China's carbon market from the following aspects: the number of published papers, publishing institutions, author distribution, research hotspots, and development trends. Finally, this paper puts forward suggestions for the future in-depth research and follow-up development of China's carbon market.

1 INTRODUCTION

At the United Nations General Assembly in September 2020, President Xi Jinping solemnly proposed China's "30●60" carbon peaking and carbon neutrality goals. This indicates that low-carbon development is urgent. The carbon market, as an important policy tool to regulate carbon emissions, has been heatedly discussed, especially after China launched the carbon market. From a macro perspective, the research on the carbon market mainly focuses on discussing the operation mechanism of the carbon market and the impact of the carbon market on the environment, economy, and society. From a micro perspective, it mainly discusses the relationship between the carbon market and the energy market, as well as the impact of the carbon market on enterprises. However, as China's carbon market is still in the initial stage of development, it is difficult to obtain such data. At present, China's research on the carbon market is not mature, and more in-depth research and exploration are needed.

In recent years, scholars have sorted out and summarized the research results of China's carbon market from different perspectives. But mainly used qualitative methods and subjective analysis to discuss the current research status, existing problems, and future prospects of carbon market research. This paper uses CiteSpace software to carry out a quantitative analysis of the literature on the carbon market in China. The purpose is to find its development context, research status, and research popularity, and to explore its future development trends, so as to provide certain references for future academic research and policy formulation.

2 DATA SOURCES AND PROCESSING METHODS

The literature in this paper is selected from the CNKI database. In order to ensure the quality of the analyzed literature, the search subject is "carbon market", the search condition is "precise", and the source categories are SCI source journals, EI source journals, Peking University Core, CSSCI, and CSCD. A total of 1029 records were obtained, and the retrieval date was April 4, 2022. Export all bibliographic data in Refworks format. Based on the 1029 literature data on the carbon

---

*Corresponding Author: 876995086@qq.com

market, CiteSpace software is used to sort out the research context of scholars on the carbon market from the aspects of research institutions, authors, research hotspots, and trends, so as to provide experience and reference for subsequent scholars' research.

3 TEST RESULTS AND DISCUSSIONS

3.1 *Temporal distribution of carbon market research literature*

Figure 1 shows the overall trend of the number of publications on the carbon market from 2004 to the present. The number of publications on the theme of the carbon market is divided into three stages: the first stage was from 2004 to 2008, during which the number of publications was relatively small. The second stage is from 2009 to 2020. The number of published articles fluctuates and rises, but it has not exceeded 100 articles every year. The third stage is from 2021 to the present, and the number of publications in 2021 will surge and reach its peak. The carbon market has seen a surge in attention since 2009. The reason should be that China announced in 2009 that by 2020, carbon dioxide emissions per unit of GDP will be reduced by 40% to 45% compared with 2005, and it will be incorporated into the national economy as a binding indicator. The year 2021 can be called the "Double Carbon Target Action Plan Formulation Year". The central government has clarified the "1+N" policy system to achieve carbon peaking and carbon neutrality goals and research on the carbon market has become more and more enthusiastic.

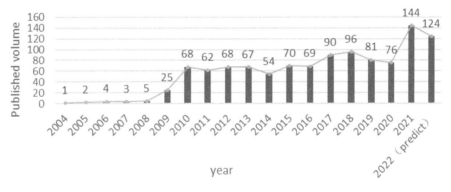

Figure 1. Construction and geometrical dimensions of specimens.

3.2 *Cooperative Analysis of Carbon Accounting Research Institutions*

Run CiteSpace, set the node type to institutions, and get the knowledge graph of institutions co-occurrence in Figure 2 on the next page, where the size of the circle indicates the size of the node. The larger the node, the more the volume of publications; the thicker the connection between nodes, the higher the degree of mutual communication between institutions. On the whole, there are many types of issuing institutions, but they are mainly concentrated in economic and management colleges and research institutes. From Figure 2, it can be seen that institutions such as Tsinghua University (55 articles), Jilin University (27 articles), Wuhan University (27 articles), and Renmin University of China (26 articles) have published a large number of articles. It provides important support for the development of research theory on the carbon market. In areas with rapid carbon market development, schools and research institutes in the area have also published a considerable number of papers, which are inseparable from their carbon trading practices, such as Beijing, Shanghai, Hubei, and other regions. In addition, it can be seen from the figure that the number of connections between each node is relatively small, indicating that there are few connections between various institutions. China has not formed extensive institutional cooperative research in carbon market research, and cooperative research needs to be strengthened.

Figure 2. Research institution cooperation map.

3.3 *Carbon Market Research Author Collaborative Network Analysis*

Figure 3. Research institution cooperation map.

It is needed to run CiteSpace, set the node type to the author, and get the author co-occurrence knowledge graph in Figure 3. As can be seen from Figure 3, generally speaking, there are very few authors with a high number of publications, they show the characteristics of decentralization. In addition, there are four relatively obvious sub-network structures formed in the graph. The team led by Zhenqing Sun and the team led by Jiao He, Xiangzhao Feng, and Kai Yuan have no obvious relationship with other network structures, or they are related but the volume of posts is less. Finally, through the size of the nodes, we can know that the ones with the largest number of articles are: Jiao He, Zhenqing Sun, Caixia Yan, and Xiangzhao Feng. Among them, Xiangzhao Feng published the article earlier. The earliest one was in 2009. Jiao He, Zhenqing Sun, and Caixia Yan posted their articles after 2014, which is relatively late. Zhenqing Sun et al. first started to study the allocation mechanism of carbon emission rights in China (Sun, Zhang, Jia, Liu, 2014), analyzed the advantages, disadvantages, and potential risks of carbon allowance allocation methods adopted in pilot areas, and put forward countermeasures. Then, topics such as the economic and environmental impact of the carbon market on the pilot cities (Liu et al. 2016) and the disclosure of carbon market information (Sun et al. 2018) were discussed. In the past two years, the impact of carbon trading on green total factor productivity (Sun et al. 2022) has also been discussed. Xiangzhao Feng mainly focuses on climate change (Teng et al. 2011) and carbon dioxide emissions (Feng & Zou 2008), while Jiao He and Caixia Yan mainly made contributions in the fields of carbon quota planning (Ye et al. 2018) and system optimization (Yao et al. 2022).

3.4 *Analysis of hotspots in carbon market research*

Firstly, it is required to run the software to obtain the keyword co-occurrence network, then perform cluster analysis on the co-occurrence network, and extract keywords according to the correlation to obtain the cluster map. Selecting the LLR algorithm to name the clusters, the resulting label names will be more in line with the actual situation, and the possibility of repetition is reduced (Chen et al. 2012), and finally the clustering map of the carbon market research literature based on the LLR algorithm is obtained in Figure 4. Q value = 0.4875, S value=0.8194, so the cluster structure of the cluster map is very significant and the result is convincing. For carbon market research, there are 11 clusters, namely: carbon market, carbon trading, low-carbon economy, carbon neutrality, CDM, low-carbon financing, carbon emission rights, climate change, carbon price, application areas, price volatility, and low carbon.

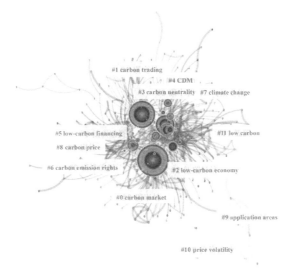

Figure 4. Keyword clustering knowledge graph.

On the basis of the keyword clustering knowledge graph, the keyword co-occurrence clustering table is obtained in the "cluster browsing", see Table 1 for details. By analyzing the keywords in each cluster, it can be found that the research content of each cluster is overlapping. The research on China's carbon market can be summarized into four research themes: the development process of the carbon market, the trading mechanism of the carbon market, the impact of the carbon market, and the development results and prospects of the carbon market.

3.5 *Evolution path of carbon market research themes*

Keywords are used to generate a time series view to analyze the evolutionary path of carbon market research (see Figure 5). By studying the content of the literature and combining the knowledge map of the evolution path, the research on China's carbon market can be divided into three stages:

Table 1. Keyword co-occurrence cluster table.

Cluster number	Cluster name	Cluster size	Keywords (extract the top five)
0	carbon market	65	carbon market; electricity market; mechanism design; market equilibrium; strategic behavior
1	carbon trading	51	carbon trading; carbon tax; green total factor; mechanism test; the double difference
2	low-carbon economy	46	low carbon economy; carbon finance; low carbon city; low carbon finance
3	carbon neutrality	41	carbon neutrality; carbon peaking; green finance; pledged loans for carbon emission rights; clean development
4	CDM	36	CDM; emission reduction; carbon emission reduction; mandatory emission reduction; cer
5	low-carbon financing	31	Low-carbon financing; supply chain; optimized allocation; unified credit line; carbon market
6	carbon emission rights	29	carbon emission rights; fossil energy; energy economy; regional differences; var model
7	climate change	24	climate change; carbon price; power sector; forecasting; stock cap and trade
8	carbon price	21	carbon price; electricity price; allowance allocation; spot market; copula model
9	application areas	6	Application field; co2; market; prospect; carbon market
10	price volatility	5	price volatility; correlation; price forecast; volatility clustering; EU carbon market
11	low carbon	5	low carbon; innovation; keywords; international climate cooperation; carbon bonds

2004-2012, the initial stage of China's carbon market research. This stage is the introduction of the concept and the preliminary exploration of the carbon market mechanism. At this stage, a large number of probabilities about the carbon market emerged, which laid a theoretical foundation for future research. From 2012 to 2017, the rapid development stage of China's carbon market research. The rapid development of the carbon market is due to the country's emphasis on energy conservation and emission reduction. During this period, China has intensively issued a series of documents on energy conservation and emission reduction, such as the "New Environmental Protection Law" promulgated in 2014, the "Implementation Opinions on Accelerating the Promotion of Green Lifestyles" in 2015, and the "General Plan for Ecological Civilization Reform System". From 2017 to the present, it is the precipitation stage of China's carbon market research. At this stage, there are no new keywords with a high degree of co-occurrence, and most of the research on the carbon market is a continuation of the previous research.

3.6 Carbon market research trend analysis

Emerging words are keywords that are cited suddenly in a certain period of time, and can be used to reflect the research trend in a certain period of time. In view of this, running CiteSpace to obtain the keyword emergence graph in Figure 6, the analysis shows that: (1) From 2000 to 2009, the most strongly emerging keywords basically appeared in this interval. It is mainly the construction of the theoretical system. At this time, there is no mature carbon market for empirical research. Most people are beginning to explore a mechanism to solve environmental problems and carry out low-carbon development. (2) From 2009 to 2014, with the empirical research on the carbon market, everyone favored the Clean Mechanism (CDM) and recognized the European Union as a representative of the development of the carbon market. They carried out a lot of analysis and research on it and summarized its practical experience. (3) From 2014 to 2022, my country opened

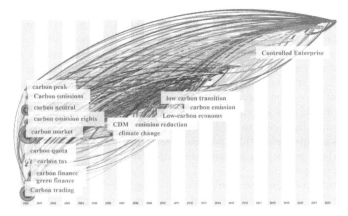

Figure 5. Time series map of frontier keywords in carbon market research.

the carbon market. During this period, the emergence of such emerging words as emission control enterprises and carbon prices shows that there are more studies on the carbon market from a micro perspective.

Top 25 Keywords with the Strongest Citation Bursts

Keywords	Year	Strength	Begin	End	2000 - 2022
Technical emission reduction	2000	4.89	2000	2009	
Runs test	2000	4.72	2000	2008	
echanism inspection	2000	4.72	2000	2008	
climate finance	2000	4.72	2000	2008	
Methodology	2000	4.72	2000	2008	
Clean Development	2000	4.72	2000	2008	
mechanism design	2000	4.38	2000	2008	
spot market	2000	4.38	2000	2008	
evolutionary game	2000	4.21	2000	2008	
electricity price	2000	4.21	2000	2008	
market equilibrium	2000	4.07	2000	2009	
carbon capture	2000	4.05	2000	2007	
two-carbon target	2000	4	2000	2008	
auction	2000	3.66	2000	2009	
emission reduction	2000	4.18	2008	2013	
CDM	2000	7.64	2009	2012	
climate change	2000	5.57	2009	2010	
EU	2000	4.19	2009	2014	
Low-carbon economy	2000	20.22	2010	2013	
carbon finance	2000	11.57	2010	2013	
Energy conservation	2000	3.5	2010	2011	
greenhouse gas	2000	3.48	2011	2013	
carbon emission	2000	4.59	2014	2020	
carbon price	2000	3.62	2017	2020	
Controlled Enterprise	2000	3.62	2017	2022	

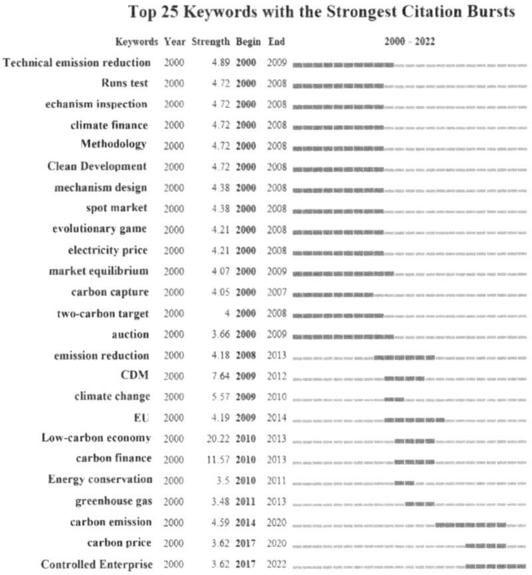

Figure 6. Carbon market research keyword emergence chart.

4 CONCLUSION

In this paper, CiteSpace visual analysis method is adopted to analyze the sample literature in detail from the aspects of publishing institutions, author distribution, research hotspots, and development trends. The main conclusions can be summarized as follows: (1) In terms of the time distribution of carbon market research literature, the carbon market has always been a research hotspot, especially in the past two years. (2) In terms of issuing institutions, there are many types of issuing institutions, large differences in institutional levels, and fewer connections between institutions. Institutions in carbon pilot regions are more concerned with carbon market research than in other regions. (3) In terms of authors, there are fewer prolific authors and less collaboration between authors. (4) In terms of research hotspots, the carbon market mainly includes four major themes: the development process of the carbon market, the trading mechanism of the carbon market, the impact of the carbon market, and the development results and prospects of the carbon market. (5) In terms of the evolution path of the research theme, the development of the carbon market is divided into three stages. From 2000 to 2012, as the initial stage, the concept of the carbon market was introduced and the mechanism of the carbon market was initially explored. From 2012 to 2017, it was a stage of rapid development, and the state paid more and more attention to energy conservation and emission reduction. The period from 2017 to 2022 is the precipitation stage, and most of the research on the carbon market is a continuation of the previous research. (6) In terms of the development trend of keywords, it can be seen that the research on the impact of the carbon market on micro-enterprises is the main development trend.

In terms of future work, Chinese scholars should strengthen mutual cooperation and cooperation with foreign scholars. The research trends of the carbon market in the future have the following aspects: (1) The improvement and development of the carbon market mechanism. Scholars will continue to study the inherent laws of the carbon market, improve the establishment of supervisory agencies, the integrity of the institutional system, the definition and operation of market operators, and other specific issues. (2) The relationship between the carbon market and society, economy, and governance includes the evaluation of the effectiveness of the carbon market and the measurement and avoidance of carbon market risks. (3) Future research can explore the impact of the carbon market on enterprises, energy markets, and financial markets from a micro perspective. For example, discussing the efficiency and benefits of various organizational groups in the context of low carbon, prompting each organization and individual to consciously assume social responsibilities.

REFERENCES

Chaomei Chen, Zhigang Hu, Shengbo Liu, et al. E-merging trends in regenerative medicine: a scientometric analysis in CiteSpace[J]. *Expert Opinion on Biological Therapy*, 2012, 12(5):593–608.

Fei Teng, Xiangzhao Feng, Jie Li. Strengthening the construction of greenhouse gas management system and actively responding to climate change[J]. *Environmental Protection*, 2011(11): 12–15. DOI: 10.14026/j.cnki.0253-9705.2011.11.011.

Jun Yao, Jiao He, Yongfei Wu, Caixia Yan. Energy optimization of wholesale electricity market considering carbon trading and green certificate trading system[J/OL]. *China Electric Power*:1-9 [2022-04-06]. http://kns.cnki.net/kcms/detail/11.3265.TM.20201221.1429.012.html

Xiangzhao Feng, Ji Zou. Economic Analysis of China's CO2 Emission Trend[J]. China Population, *Resources and Environment*, 2008(03):43–47.

Yu Liu, Danhui Wen, Yi Wang, Zhenqing Sun. Assessment of impacts of Tianjin pilot emission trading schemes in China—A CGE-Analysis using term CO2 model[J].*Progress in Climate Change Research*, 2016, 12(06):561–570.

Ze Ye, Jiao He, Xin Zhou, Siqiang Liu. A two-tier planning model for initial allocation of carbon emission rights in the power generation industry [J]. *Systems Engineering*, 2018, 36(11): 140–146.

Zhenqing Sun, Nan Zhang, Xu Jia, Jiajia Liu, Research on China's Regional Carbon Emissions Allowance Allocation Mechanism[J]. *environmental protection*, 2014, 42(01):44–46. DOI:10.14026/j.cnki.0253-9705.2014.01.013.

Zhenqing Sun, Wenshan Gu, Xiaofei Cheng. Research on the Influence Mechanism of Carbon Trading on Green Total Factor Productivity[J/OL]. *Economic Management in East China*:1-8[2022-04-06].DOI:10.19629/j.cnki.34-1014/f.210911013.

Zhenqing Sun, Zirui Lan, Nan Tang. Research on Information Disclosure of my country's Carbon Market[J]. *Economic System Reform*, 2018(04):31–36.

Advances in Renewable Energy and Sustainable Development – Liang & Kasmani (Eds)
© 2023 Copyright the Author(s), ISBN: 978-1-032-39407-7

CiteSpace-based visualization analysis on the hotspots and trends in the field of gas power generation

Bing He* & Ting Ni*
College of Environment and Civil Engineering, Chengdu University of Technology, Sichuan, China

ABSTRACT: With the carbon neutrality target proposed, coal power will not be suitable for further development because it is very polluting to the environment, and natural gas power generation is gradually gaining attention, so research in the field related to natural gas power generation has been the focus of scholars' attention. In this paper, we propose a document mining system to visually analyze the research in the core database of Web of Science[TM] using CiteSpace software, by the analysis shows that the number of publications has been increasing year by year, with the United States and China as the main contributors, and the research focus areas are on technology, system, and environmental issues. The hot spots and trends of research are mainly to develop synergistically with renewable energy and build a low-carbon and stable power generation system to meet the development of the times.

1 INTRODUCTION

In recent years, climate change, green energy, and low-carbon have become irrevocable trends in global energy development. Because of environmental problems, countries are paying increasing attention to the development and use of low-carbon energy (Zeng et al. 2021). The power industry is one of the important sources of carbon emissions and pollutant emissions, of which thermal power is a major emitter. By 2020, coal power has the largest share in the global power supply structure, followed by gas-fired power generation. Although wind and solar energy have doubled in five years, they account for only one-tenth of global electricity (Ember 2021). Coal power is serious pollution, and renewable power generation, although good cleanliness, its instability is difficult to ensure the safe operation of the power grid (Pan et al. 2019), combined with the disadvantages of both, natural gas power generation is also gaining increased attention (Shan 2021; Zhang et al. 2022).

To understand the research focus and future development trends of natural gas power generation, this paper visualizes and analyzes the research results related to natural gas power generation, and draws a knowledge map in the research field of natural gas power generation, to provide reference and reference for the subsequent research in the field of natural gas power generation.

2 RESEARCH TOOLS AND METHODS

With the accumulation of numerous research results on natural gas-fired power generation by scholars from various regions of the world over decades, a suitable analytical framework is required to distill the high-quality literature needed on its technological development. The systematic literature analysis approach provides a good solution for this purpose, overcoming the incompleteness and non-objectivity of the traditional review literature by summarizing and objectively evaluating the

---

*Corresponding Authors: hb2020020296@stu.cdut.edu.cn and niting17@cdut.edu.cn

DOI 10.1201/9781003349648-33

literature as comprehensively and completely as possible, and obtaining more accurate conclusions (Andrea et al. 2008; Margaret et al. 2008).

This paper draws on the literature mining system built by Xu et al. The system structure logic diagram of this paper is built with the help of CiteSpace and the Web of Science™ core collection database, as shown in Figure 1. CiteSpace is an information visualization and analysis software developed by Prof. Chaomei Chen based on Java language. CiteSpace can effectively combine the cluster analysis method, social network analysis method, and multidimensional scale analysis method to analyze the content of scientific literature, and finally, visualize the analysis results in the form of the knowledge graph. Thus, research trends and evolution can be understood (Lia et al. 2018; Xie et al. 2019).

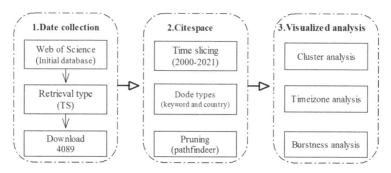

Figure 1. Literature mining system.

In the collection of literature related to natural gas power generation, a combined search was conducted in the Web of Science™ core collection database using the subject terms "natural gas power generation" and "natural gas power generation technology" for the period 2021. A total of 4,089 records were retrieved for the period 2000 to 2021. All the retrieved literature data were exported in plain text format, including title, author, institution, keywords, abstract, and references, in both English and Chinese. The output data were then imported into the software CiteSpace for de-duplication, data conversion, and analysis parameter setting. The research in natural gas power generation in the 21st century is visualized and analyzed in terms of the number of articles published, research institutions, research keywords, and research hotspots.

3 VISUAL ANALYSIS

3.1 *Analysis of annual publication volume*

The number of publications in a research field at a given time reflects the degree of hot research in the field. The higher the number of publications, the more scholars in the research field at that time, and the more the field is booming. In this paper, a total of 4,089 papers were retrieved. The data of 4,089 articles can be imported into Excel to quickly locate the number of articles published each year, and the graphs are shown in Figure 2(a). Before 2007, the number of relevant studies was relatively small, and the number of articles per year remained within 20–100, which can be regarded as a preliminary stage. In 2007, the number of studies steadily increased and reached 300 by 2016. From 2017 to 2021, the number of articles was 300–600, which can be regarded as the stage of the natural gas power generation booming phase.

3.2 *Country cooperation network analysis*

In the operation interface of CiteSpace software, we select the node type as the country for visual analysis and obtain the country region visualization map, as shown in Figure 2(b). A total of 149 nodes and 639 connecting lines are generated. Each node represents a country. The larger the area of the node, the more research in the field in that country. The connecting lines between the nodes represent the cooperation and communication between two countries.

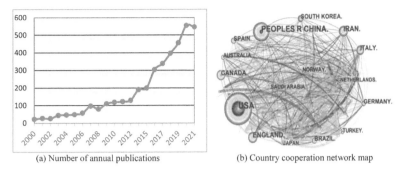

(a) Number of annual publications   (b) Country cooperation network map

Figure 2. Statistical map of published articles.

As can be seen from Figure 2(b), the top six countries with the highest number of publications among the 149 countries involved in published research are the US, China, Iran, the U.K, Italy, and Canada. The US has contributed the most to research related to natural gas power generation, followed by China. This indicates that developed countries are at the forefront of research, with relatively dense nodal links and scholars from different countries collaborating, initially forming an extensive research cooperation network in the field of natural gas power generation.

3.3 *Cluster analysis*

The keyword co-occurrence analysis operation used in this paper is based on the 4,089-literature downloaded earlier, and the period of 2000–2021 is selected in the software operation interface, the network is constructed using the pruning method of minimum spanning tree, the node type is selected as the keyword, and the number of keyword nodes is 643 after running the process, and there are 1,934 connection lines between nodes. The keyword map is shown in Figure 3(a). Keyword analysis, as a co-occurrence analysis, plays an important role in understanding the dynamics of knowledge development (Kamil 2004).

Table 1. Clustering information.

Clusters	Size	S Value	Year	Keywords
0	110	0.69	2011	Saudi gas initiative; natural gas power plant; energy analysis; life cycle assessment; energy storage; carbon dioxide utilization
1	94	0.804	2009	Performance; design; wind energy; storage system; CO_2 capture; electricity generation; techno-economic analysis; plant; storage
2	90	0.795	2015	Power generation; exergy recovery; exergy; optimization; organic Rankine cycle; exergy analysis; waste heat recovery; energy analysis
3	86	0.776	2014	Energy storage; integrated energy system; wind power generation; energy hub; renewable energy; mixed integer; linear program
4	61	0.733	2014	Contingent valuation; electric-thermal ratio; techno-economic analysis; solid oxide fuel cell; hybrid system; exergy analysis
5	52	0.764	2013	In-piston chamber; compression ignition engine; ignition engines; organic Rankine cycle; power generation; cold utilization
6	43	0.852	2008	Renewable energy; electric vehicle; system integration; alternative fuels; hydrogen interim storage; dynamic operation; cost reduction
7	43	0.826	2009	Renewable energy; power generation; wood biomass; supply curve; energy transition; path creation; solar energy; hydrogen application

The keywords are presented in clustering mode, as shown in Figure 3(b). The network density is 0.0128, the modularity degree Q is 0.46, and the average S is 0.7759. There are 24 effective clusters. The value of Q indicates the modularity degree of the network; when Q > 0.3, the clustering effect is significant. The value of S is used to measure the network homogeneity index; when S > 0.7, the clustering has a strong persuasive power (Chen et al. 2015). The clustering results of this study are more desirable. The clustering mapping focuses on reflecting the structural features among clusters, highlighting key nodes and important research areas (Chen & Chen 2006).

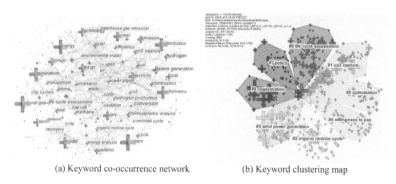

(a) Keyword co-occurrence network   (b) Keyword clustering map

Figure 3. Keyword map.

In this paper, the data related to the top 8 clusters in CiteSpace by cluster size are summarized in Table 1. These keywords help to locate the core research area of natural gas power generation.

By integrating and analyzing the keywords in Table 1 and combining keyword co-occurrence mapping and clustering mapping, the key research areas of 4,089 articles can be obtained. The focus of the research is on optimizing natural gas power generation systems in terms of efficiency, performance, and technology, such as Cluster 1, Cluster 2, Cluster 5, and Cluster 6. We are also considering the study of hybrid power generation systems, such as Cluster 3, Cluster 6, and Cluster 7, as well as the associated environmental issues and the development of sustainable energy, such as Cluster 0, Cluster 1, and Cluster 4.

3.4 *Time zone map analysis*

Converting the keyword co-occurrence map of Figure 3(a) into a keyword time zone view is shown in Figure 4, and the keyword emergence table summarized by CiteSpace software from 2007–2021 was organized as shown in Table 2. Because the number of studies from 2000–2006 was so small that emergent words were not counted in this table, the combined analysis of the time zone diagram and keyword emergence table helped to discover the research hotspots at different stages Combining the time zone mapping and the keyword emergence table, we can see that the period from 2000 to 2006 was the beginning of the development of natural gas-fired power generation because the number of studies was small, the research hotspots were limited, indicating that natural gas-fired combined heat and power systems were the main mode of natural gas-fired power generation at that time, and the research hotspots revolved around the research, design, and optimization of CHP systems.

The keywords that emerged after 2007 include wind energy, CO_2 capture, and energy value analysis, combined with the emergence of "mixture" and "environmental impact", we can conclude that the environmental issues of natural gas power generation have begun to be focused during this period. Carbon dioxide capture technology is the core of carbon capture, utilization, and storage (CCUS) technology, which collects gaseous carbon dioxide through absorption and adsorption technologies. The CCUS technology separates carbon dioxide from the emission source and then directly utilizes or sequesters it to achieve carbon dioxide emission reduction. It is an important technical tool to achieve clean utilization of fossil energy generation.

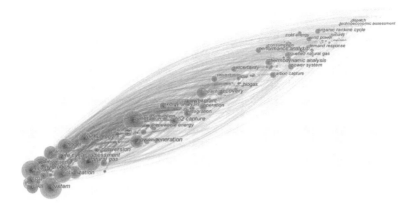

Figure 4. Keyword time zone.

Table 2. Top 15 keywords with the strongest citation bursts.

Keywords	Strength	Begin	End	2007–2021
Coal	4.74	2007	2015	
Hydrogen	4.08	2007	2013	
Combustion	3.76	2008	2011	
Efficiency	8.03	2009	2014	
Environmental impact	5.71	2009	2016	
Gasification	5.07	2010	2015	
Mixture	6.03	2010	2015	
Performance	4.14	2010	2011	
Cycle	4.46	2012	2015	
Greenhouse gas Emission	5.1	2013	2016	
Sustainability	4.65	2016	2018	
Oxygen carrier	3.93	2016	2019	
Energy storage	5.48	2017	2018	
Optimal design	3.94	2017	2019	
Optimal power flow	4.18	2018	2021	

The introduction of energy value indicators to evaluate the sustainability level of natural gas power generation systems, and to a certain extent, to promote the improvement of natural gas power generation efficiency. The emergence of the keywords renewable energy and emergent keywords marks the beginning of scholars' research on hybrid power generation systems of natural gas and renewable energy. Thus, 2007 is recognized as a momentous landmark in the development of natural gas-fired power generation. The period 2007-2015 can be considered the second phase of the natural gas-fired power sector.

After 2016, natural gas power generation technology has matured and natural gas power generation is divided into centralized and distributed. The centralized type is the most extensive and consumes the most gas, with the main purpose being power and heat generation, whereas the distributed type is on the rise, with its characteristics of lower investment and flexible start-stop, which allows for full utilization of resources and can manage peak electricity consumption, thus having a great deal of potential. The keywords energy storage, pollution optimal design, and optimal power flow indicate that the focus of research at this stage is low carbon, sustainable, and affordable power generation, especially in conjunction with renewable energy. Gas power generation plays an important role in peak regulation, peak shaving, and valley filling to relieve the pressure of peak supply.

4 CONCLUSION

In this study, using the period of 2000-2021, we conducted a statistical and visual analysis of 4,089 articles on natural gas power generation retrieved from the WOS database using the visual analysis software CiteSpace and came to the following conclusions.

The number of research publications in the field of natural gas power generation has shown a year-on-year increase. Since 2017, the number of relevant studies has exploded, indicating that the field of natural gas power generation has been experiencing rapid growth. The research is mainly concentrated in developed countries and developing countries with abundant reserves, among which the United States and China are the countries that have contributed the most to make the most. Cooperation between countries is relatively close, and an extensive network of research cooperation in the field of natural gas-fired power generation has initially been formed. The emergence of global consensus on carbon neutrality and energy transition has led to the natural gas sector's bridging role and its position as an alternative energy resource. The integration of natural gas with renewable energy sources has also become widespread. Natural gas power generation has become a frontier subject in world research because of its low-carbon sustainable development and synergistic utilization of renewable energy sources.

ACKNOWLEDGEMENT

This work is supported by the Key Research of Social Sciences Base, Sichuan Oil and Natural Gas Development Research Center (Grant No. SKB17-04).

REFERENCES

Andrea C T, Jennifer T, Margaret S, Dean F, Elise C, Tanya H, David. (2008). Few systematic reviews exist documenting the extent of bias: a systematic review. *Journal of Clinical Epidemiology*, 61(5): 422–434.

Chen CM, Chen Y. (2006) CiteSpace II: Detecting and visualizing emerging trends and transient patterns in scientific literature. *Journal of the American Society for Information Science and Technology*. 57(3):359–377.

Chen Y, Chen C M, Liu Z Y, et al.(2015) Methodological features of CiteSpace Knowledge Graph. *Studies in Science of Science*. 02-0242-12.

Ember. (2021). Global Electricity Review 2021. www.ember-climate.org/global-electricity-review-2021

Kamil K.(2004) Hydropower and the world's energy future. Energy Sources, Gordon R. *Empire energy*. LLC. GHC Bull. 11–3.

Lia M H, Xu J P, Xie H B, Wang Y H. (2018).Transport biofuels technological paradigm-based conversion approaches towards a bio-electric energy framework. *Energy Conversion and Management*, (172):554–566.

Margaret S, Kaveh G, Chantelle G, Tanya H, Mary O, David. (2008). Moher. Systematic reviews can be produced and published faster. *Journal of clinical epidemiology*. 61(6): 531–536.

Pan W H, Wang C, Che XB. (2019). The impact of the China-Russia gas pipeline on the target market impact analysis. *J. International Petroleum Economics*. 27(06):47–55.

Shan T W.(2021). Positioning and development path suggestions for natural gas power generation in China's energy transition period. *China Offshore Oil & Gas*, 33(2).

Xie B, He Y, Lu WX. (2019).Comparative research and the prospect of rural Mingsu based on CiteSpace analysis. *Anhui Agricultural Sciences*, 47(13):123–128.

Zeng P S, Zhao L, Gong H, et al. (2021). Research on the integrated development of natural gas and new energy in Sichuan and Chongqing under the "dual carbon" goal, J, *Energy saving in petroleum and petrochemicals*.

Zhang J P, Wang F P, Mei Q, et al. (2022). Research on the integrated development path of natural gas and new energy. *International Petroleum Economy*. 30(1).

Advances in Renewable Energy and Sustainable Development – Liang & Kasmani (Eds)
© 2023 Copyright the Author(s), ISBN: 978-1-032-39407-7

Study on prevention strategy of coal mine flood accident

Botao Bi[*]
The Management of Sanpan District Office of Daliuta Coal Mine Shendong Coal Mine Group, Shanxi Province, China

Hong Zhang[*]
Shenhua Engineering Technology Co., Ltd., Beijing, China

ABSTRACT: The coal mine flood accidents not only lead to major safety accidents involving mass deaths and injuries, resulting in property losses and a massive waste of resources but also damage the environment. The coal mine flood accident prevention is still an important task in China's coal mine safety production management. Although the number of major and above flood accidents in coal mines has declined significantly and the safety situation has improved significantly, coal mine flood accidents have continued to occur for ten years. Therefore, it is imperative to intensify efforts to prevent major and above flood accidents. To eliminate the occurrence of the coal mine flood accidents from the origin, we must provide a scientific basis for the prevention strategies and implementation methods of the coal mine flood accidents. The research that uses the behavioral safety 2 to 4 model and comparative analysis method shows that the root cause of the coal mine flood accidents is the unsafe behavior of a person. This paper analyzed the coal mine flood accident cases from 2011 to 2020 in China. Research indicates that preventing coal mine flood accidents requires controlling a person's unsafe behavior. According to further research, by using scientific and technological approaches to train safety behavior of people, such as three-dimensional (3D) video education, virtual reality training, and database training, then it could be possible to correct unsafe behavior. The training also can increase a person's safety knowledge, improve safety awareness, and cultivate good safety habits, thereby improving the ability to the prevention of the coal mine flood accidents.

1 INTRODUCTION

China's energy endowment structure is characterized by "poor oil, less gas, and rich coal". As coal accounts for about 74% of China's energy structure, China is already the largest coal producer and consumer in the world (Zhao 2005). Coal still plays a vital role in the development of the economy both now and in the future in China (Cheng et al. 2016). In 2020, China's coal production was 3.9 billion t still occupying the first position in the energy industry. Although China accounted for about 47% of the world's coal mine production, a million-ton death rate of 8.4% is still higher than the 1.1% about seven times in the United States. A total of 124 coal mine flood accidents occurred from 2011 to 2020 in China, including four in 2019, reaching the minimum value in ten years as shown in Figure 1. Over the past ten years, although coal mine production has grown, coal mine flood accidents have declined overall and the death toll from 140 in 2011 is down to 30 in 2020, coal mine safety has improved steadily overall. However, seven coal mine flood accidents happened in 2020; the number of accidents rebounded again, the coal mine flood accidents still exist and the coal mine flood accidents still occur on occasion, which makes the coal mine safety situation grim (Jing & Qing 2021).

Since the 1980s, with the rapid development of the economy, the demand for coal is increasing year by year, and the resulting coal mine safety accidents are also increasing. Thus, coal mine

---

[*]Corresponding Authors: 474198785@qq.com and 10514577@ceic.com

DOI 10.1201/9781003349648-34

accidents are becoming a growing concern, and increased people are conducting research on coal mine flood accidents. At present many experts and scholars have conducted a lot of research on the coal mine flood accidents and achieved fruitful results.

Meng (2010) proposed that rather than paying attention to the geological structure of coal mines, the coal pillars were reserved before mining, and the water levels during rainfall were not adequately monitored.

Deng (Cao 2018; Deng 2022) proposed that remove the surface water, old pit water, fault water, and pore space water to prevent the coal mine flood accidents happened.

Ye (Miao & Ye 2021) proposed reinforcing the supervision and management to avoid the coal mine flood accident.

Wu (Wu et al. 2019) proposed that the occurrence of accidents can be prevented with the control of three factors: individual, organization, and environment.

Zhang (Zhang et al. 2021)proposed that should pay much attention to work time and workplace which maybe make the accident happen. Safety accidents are unexpected events that go against people's will and beyond people's cognitive ability causing losses (Fu & Li 2004).

Above all, the current research is still on the level of engineering technology, and few people pay an eye to the impact of person unsafe behavior on accidents. According to a foreign safety management experience that more than 80% of safety accidents happened by a person's unsafe behavior. Based on the existing research results, this paper uses the behavior safety 2-4 model to focus on the person's unsafe behaviors to fill the gaps in the previous research about the coal mine flood accident. A person's unsafe behaviors were the root cause of coal mine flood accidents. Then all kinds of unsafe behavior that cause the accident are recorded into the database and the unsafe behaviors are corrected using modern scientific and technological means. At the same time, training is carried out through 3D models and virtual scenes to control unsafe behaviors. Continuous training can improve safety awareness, increase safety knowledge, and eliminate the coal mine flood accidents from the root.

2 THE SITUATION OF FLOOD ACCIDENTS IN COAL MINES

China is the country with many coal mine flood accidents, and it is also one of the country's most seriously affected by flood disasters in the world (Meng 2010). The prevention of the coal mine flood accident is one of the main tasks of safety accident prevention (Du 2013) currently. This paper takes the coal mine flood accident as the research object and finds the most fundamental cause of the flood accident by the analysis of the coal mine flood accident in the past 10 years. According to the data of the online inquiry system of the State Mine Safety Supervision Bureau of the accident case investigation report (*National Mine Safety Supervision Bureau* [EB/OL], 2021), the total number of the coal mine flood accidents in 2011 and 2020 were compared for the number of major and above accidents, as shown in Table 1.

Judging from the overall trends in Figures 1 and 2, both the total number of the coal mine flood accidents and the major and above flood accidents have dropped significantly, indicating that the safety situation of the coal mines has greatly improved in the past ten years in China. From 2011 to 2020, a total of 124 flood accidents occurred in the coal mines, including 18 major flood accidents in China. Table 1 indicates that the number of major and above accidents decreased from 6 to 1 between 2011 and 2020, and the major accident rate declined by 83%. No major or above safety incidents were reported in 2017, 2018 and 2019. It can also be seen from Table 1 that although the number of flood accidents has decreased, the coal mine flood accidents still exist. The coal mine flood accident rebounded in 2020, with a total of 7 coal mine flood accidents happening. In the past ten years, an average of 12.4 coal mine flood accidents have occurred every year, including 1.8 major and above incidents, and an average of 36 deaths per year. Therefore, coal mine flood accidents still need to be prevented. Comparing the changes in the number of accidents in Figures 1 and 2, although the number of major and above the coal mine flood accidents has steadily decreased,

Table 1. Details of flood accidents from 2011-2020.

Year	The total number of flood incidents	The number of major and above flood accidents
2011	19	6
2012	14	5
2013	40	2
2014	10	2
2015	6	1
2016	11	1
2017	7	0
2018	6	0
2019	4	0
2020	7	1

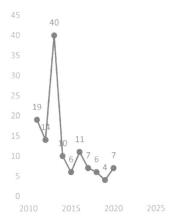

Figure 1. Changes in the total number of flood accidents from 2011 to 2020.

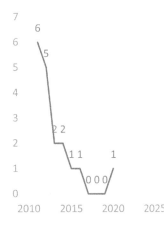

Figure 2. Changes in major and above flood accidents from 2011 to 2020.

the proportion of the reduction is relatively small when compared with the total number of flood accidents; therefore, accident prevention should be re-evaluated.

3 METHODS FOR ACCIDENT PREVENTION

At present, the research on accident prevention strategies is mainly carried out through R & D activities in China. The government funds many R & D activities every year, which are carried out through the National Natural Science Foundation of China and the Ministry of Science and Technology (Chen 2019; Wu et al. 2014). From 2015 to 2019, the distribution of R & D activities in the safety field supported by the National Natural Science Foundation of China is shown in Table 2.

It can be seen from Table 2 that more than 90% of the accident prevention strategies are engineering strategies in China, while less than 10% are behavior control strategies. Behavioral control R & D activities have maintained an upward trend in recent years, but the overall number is still very small, and the proportion has grown slowly. The R & D activities supported by the Ministry of Science and Technology are held every five years and are mainly aimed at major construction projects throughout the country. There are 35 major projects in China's 12th Five-Year Plan, of

Table 2. The distribution of the National Natural Science Foundation in the safety field.

Year	Behavior control project Number	Proportion (%)	Engineering technology project Number	Proportion (%)	Total number
2015	10	6.5	145	93.5	155
2016	13	7.6	157	92.4	170
2017	15	7.9	174	92.1	189
2018	17	7.9	198	92.1	215
2019	27	8.2	302	91.8	329

which only two are related to safety, namely the national public safety emergency technical support project and the food safety key technology project (Lin, Chen, 2014). The relevant proportion of safety projects is shown in Figure 3.

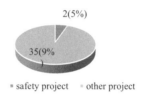

Figure 3. The number and proportion of scientific research projects of the Ministry of Science and Technology.

It is not difficult to see from China's R&D activities that China's current prevention strategy for the accident is based on engineering, supplemented by behavior control. According to Heinrich's accident causation theory (Che 2003) and DuPont's accident statistics rule (Fu 2013) found that more than 80% of accidents are caused by unsafe actions of people. At least one of every major or above flood accident is caused by unsafe actions of people (Yang et al. 2013), which further proved the importance of behavior control in accident prevention. In the case of relatively backward production equipment and technology, the improvement of engineering technology can indeed play a great role in accident prevention and can reduce the occurrence of many accidents, but the accident cannot be eliminated fundamentally. According to the study, when engineering technology improves to a certain level, controlling human behavior will play a decisive role in improving safety status and preventing accidents.

At present, in the prevention of safety accidents, the proportion of person unsafe behavior control is less than 10%, which cannot meet the needs of preventing more than 80% of accidents caused by person unsafe behavior. The proportion of behavior control projects in R & D activities increases year by year, indicating that behavior control strategies are gradually being attention. Behavior control will gradually increase the proportion of accident prevention methods in China.

4 BEHAVIOR CONTROL

4.1 *Theory of behavior control*

In the prevention of the coal mine flood accident, there already have been many discussions on engineering technology. This paper mainly discusses the method of behavior control. As the coal mine flood accidents are still frequent in China, domestic theoretical research on behavior control strategies has been uninterrupted. Fu and his research team, which began their research in 2005 and carried it out for eight years, proposed the behavioral safety 2-4 model (Fu et al. 2014; Gui 2017), as shown in Figure 4.

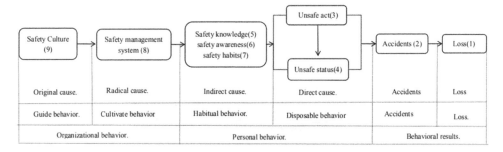

Figure 4. Behavioral safety 24 models.

First, the behavior safety 2-4 model is a modern chain of accident causes. It is based on the accident-prone tendency theory of Greenwood and Woods, the classical accident cause chain of Heinrich, the modern accident cause chain of Bird and Loftus, and a modern accident cause chain proposed by JM Stewart. According to many accidents case analyses and the basic principles of organizational behavior, they proposed more than 80% of accidents by personal behavior. Behavior safety is a common means of behavior control and accident prevention. The application of the behavior safety 2-4 model can analyze the causes of the coal mine flood accidents (Fu et al. 2022), and it is also applicable to accident analysis in construction, aviation, transportation, and other industries. Therefore, the behavior safety 2-4 model has an important role in analyzing the causes of accidents in various industries and preventing the occurrence of similar accidents. The analysis of the behavior safety 2-4 model for the coal mine flood accidents makes up for the neglect of person behavior safety. Using the 2-4 model to analyze the coal mine flood accidents in the past decade clearly shows that eliminating the unsafe behaviors is the most effective measure to prevent accidents.

4.2 *Application of behavior control*

In addition to the behavior safety 2-4 model theory, the behavior control strategy eliminates habitual violations and person habitual unsafe acts through some common safety training methods in practical operation. These training methods are the embodiment of the application of behavior control strategies, mainly including 3D video education, virtual reality, and database training. Behavior control training is a kind of safety training method using scientific and technological means, taking real accidents as samples, taking 3D animation as the carrier, and taking behavior safety 2-4 model as the theoretical basis. First, some coal mine flood accidents are used as a sample for the 3D video education, and then, the unsafe behavior of a person and the unsafe status of objects that lead to these coal mine flood accidents are considered according to the behavior safety 2-4 model. For example, there are obvious signs of flooding in the mine and workers continue to risk work, underground drilling before not exploring the water drill, showing these unsafe acts by 3D animation, and then training individuals. The 3D animation stage should be geared towards highlighting the unsafe act and the unsafe status. To further strengthen the trainees' visual impression, the emphasis should be placed on the unsafe act and the unsafe status. Let the trainees remember that unsafe acts and unsafe status will cause serious accidents, then correct unsafe acts and status by teaching the trainees. The 3D video education aims to address the 2-4 model of behavior safety to address the direct causes of accidents. This can correct the unsafe act that resulted in accidents, to affect their safety knowledge, awareness, and safety habits to conduct behavior control and ultimately achieve the purpose of accident prevention. The scene of 3D video education is shown in Figure 5.

The biggest characteristic of virtual reality is that people of different types of work can simulate the operation of different operating procedures. During prevention training of the coal mine flood accident, in the program set up the different operations to prevent the coal mine flood accidents

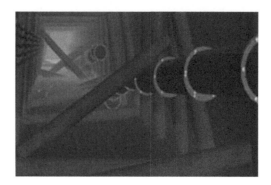

Figure 5. 3D animation scene simulation.

happen, so that the person in the simulation operation process according to different hydrological conditions: to choose different prevention programs and to adjust different parameters. For example, the water inspector can be trained to set the parameters of water exploration drilling, and the shooter can be trained to check the implementation of water drilling measures before firing. The application of virtual reality can not only train employees in accident prevention but also conduct regular assessments of employees to strengthen their safety knowledge and safety awareness of accident prevention. In addition, virtual reality and 3D video education are also different in the sample selection. Through training to correct human safety act to affect personal safety knowledge, safety awareness, and safety habits, to achieve the purpose of behavior control and accident prevention.

Database training is based on the statistical results of the database, which takes the unsafe act and unsafe status of different types of work in the previous accidents, and takes the statistical results of the database to guide the action and state training of different personnel. If the direct cause of the coal mine flood accident is the unsafe act of commanding the workers at the organizational level to cross the line, then the focus is on conducting and correcting the commanding personnel at the organizational level safety. If the shooter does not check the implementation of the measures before firing, then focus on the unsafe movements of the shooter training.

The above three behavior control means all start from the direct cause of accidents in the behavior safety 2-4 model. Through the training on the unsafe act and unsafe status, a person's safety knowledge, safety awareness, and safety habits are affected, and finally achieve the purpose of behavior control and accident prevention.

5 CONCLUSIONS

In this paper, the behavior safety 2-4 model is adopted to study the coal mine flood accident and get the following conclusions:

(1) Current accident prevention technology accounts for more than 90% and behavior control accounts for less than 10%; however, more than 80% of accidents are human accidents, and behavior control strategy is an important means of accident prevention.
(2) The prevention of flood accidents can be completely prevented. Through the control of human unsafe actions, all kinds of accidents can be prevented.
(3) The "2-4" model of behavioral safety is applied to accident analysis and accident prevention and management, and the common training means of behavioral control strategies. 3D video education, virtual reality, and database training are provided to provide new ideas for accident prevention and management.

In terms of future work, the theory of behavior control has been gradually promoted, and eliminating unsafe acts is the most effective measure strategy for accident prevention. With the progress of science and technology and the wide application of digital information, behavior control theory will be more widely used with the help of advanced technology and control methods, and behavior control theory will become an important strategic measure for accident prevention.

REFERENCES

Che H Q. 2003. DuPont executives talk about safety management issues. *J. China's national conditions and national strength.* (2), pp. 57.

Chen X Z. 2019. Application of Behavioral Safety "2-4" Model in Construction Enterprise Safety Training *J Journal of Beijing Polytechnic Institute.*18(1), pp. 102–106.

Cheng Li,Yang C W and G X Jing. 2016. Statistics and law analysis of Coal mine accidents in 2014. *J. Journal of Safety and the Environment.*16(4), pp. 384–389.

Deng Y J. 2022. Study on the causes of mine water damage and water control measures. *J. Shanxi Metallurgy.* 45(195), pp. 327–331.

Du K. 2013. Safety management, cause analysis and prevention of flood accidents in coal mines *J. Technology Information.* (28), pp. 125.

Fu G, Chen Y R and Xu S R. 2022. Connotation Analysis of the "2-4" model and the 6th edition *J The Chinese Journal of Safety Sciences.* 32(1), pp. 12–19.

Fu G, et al. 2014. An expanded version of the Behavioral Safety 24 Model. J. Journal of Coal. 39(6), pp. 994–999.

Fu G, Li X D. 2004. Common causes of accidents and their behavioral science prevention strategies. Proceedings of the first annual meeting of China Occupational Safety and Health Association and Occupational Safety and Health Forum. pp. 211–215.

Fu G.2013 Correction of unsafe behavior [EB/OL]. http://www.chinasafety.gov.cn/zhuantibaodao/2006-05/16/content_167037.htm. 2008-10-26/2013-2-28.

Gui Y D. 2017. "2-4" Model and its Application in Coal Mine Safety Management. *J. Building materials and decoration.* (20), pp. 207–208.

Jing G X and Qing Q R. 2021. Analysis of water actors in China from 2011-2020. *J. The Journal of Safety and the Environment.* pp. 1–9.

Lin Y H, Chen A. 2014. China has made initial achievements in the scientific and technological innovation of natural disasters in safe production and food safety in the field of public safety. *J. Science and technology promotes development,* (1), pp. 79–88.

Meng B L. 2010. Analysis and prevention technical measures of coal mine water damage accident. *J. Mining Technique.* 10(S1): pp. 215–216.

Miao Y W and Ye L. 2021. Analysis of the national coal mine water disaster accident in 2020 and its prevention and control measures. *J. Chinese Coal.* 47(2), pp. 51–54.

National Mine Safety Supervision Bureau [EB/OL]. 2021. https://www.chinamine-safety.gov.cn/zfxxgk/fdzdgknr/sgcc/202202/t20220223-408504.shtml.

Wu J G, Mao J R and Chai P. 2019. From 2000 to 2017, the analysis of the law of major flood accidents in coal mines in China. *J. Coal Mine Safety.* 50(10), pp. 239–242, pp. 247.

Wu Q ,Yang C and Yin W T. 2014. Construction of the coal mine training system based on the "2-4" model of behavior safety. *J. Coal Mine Safety.* 45(9), pp. 238–241.

Yang C, et al. 2013. Unsafe Act Analysis on Major and Particular Major Flood Accident in Coal Mine. The 2nd International Symposium on Mine Safety Science and Engineering C.

Z H Cao. 2018. Explanation of water hazard prevention technology in coal mining process. *J .Technological innovation and application.* (17), pp. 142–143.

Zhang P S, et al. 2021. The statistics and evolution trend analysis of coal mine water disaster accidents from 2018 to 2019 in China. *J. Coal Mine Safety.* 52(8), pp. 194 200, pp. 207.

Zhao J X. 2005. Discussion on China's energy structure and energy strategy composition. *J. Coal Economics Research.* (6), pp. 11–13.

Advances in Renewable Energy and Sustainable
Development – Liang & Kasmani (Eds)
© 2023 Copyright the Author(s), ISBN: 978-1-032-39407-7

Carbon neutral development path under digitization—Based on MATLAB three-party evolutionary game simulation

Kun Xie* & Jingli Wu*

Shanghai Institute of Technology, Shanghai, China

ABSTRACT: Achieving the goal of carbon neutrality is the inherent requirement and the only way for high-quality development. Under the dual background of the vigorous development of digital technology and low-carbon economic transformation, it is of great significance to explore the theoretical mechanism and practical path of digital technology to help carbon neutrality. In this paper, a three-party evolutionary game model is constructed to analyze the interesting linkage among the government, enterprises, and investors, and the results of evolution are simulated by MATLAB, which proves that the development of digital technology can comprehensively promote the transformation and development of low-carbon economy on the premise of alleviating information asymmetry. It is concluded that perfecting the government's carbon emission administrative management system, developing the enterprise's low-carbon technology innovation system, and establishing the carbon emission market trading system are the three main paths to the development of digital carbon neutrality.

1 INTRODUCTION

The global climate environment is deteriorating, which seriously threatens the sustainable development of human society (Zhang 2013). To cope with the worsening climate environment, governments all over the world began to pay attention to low-carbon economic transformation (Bao et al. 2008). At the 75th United Nations General Assembly on September 22nd, 2020, the Chinese government presented the "dual carbon" targets for the first time— "carbon dioxide emissions should reach their peak before 2030, before striving to achieve carbon neutrality by 2060" (Xu et al. 2021). At the same time, the world is also undergoing a new round of scientific and technological revolutions. Digital technologies represented by big data, industrial internet, 5G, artificial intelligence, cloud computing, and blockchain are changing the global economy (Li 2019). The "14th Five-Year Plan" clearly puts forward "accelerating digital development and building digital China" and "driving the change of production mode, lifestyle and governance mode by digital transformation as a whole". Under the dual background of the vigorous development of digital technology and low-carbon economic transformation, it has certain theoretical significance and practical value to explore how digital technology can help achieve the "double carbon" goal. In conjunction with the high and new technology of the digital industrial revolution in China and even worldwide, more and more industries have begun to take advantage of technology, including 5G, big data and cloud computing, and the Internet of things, to improve investment in and use of clean and renewable energy, promote environmental protection and energy saving effects of technological transformation and reduce carbon emissions to address climate change and ultimately achieve carbon neutrality.

    At present, digital technology has been applied in the field of green economy, which plays a positive role in promoting green technological innovation, improving green economic efficiency,

---

*Corresponding Author: xiekun@sit.edu.cn and wjl_christina@163.com

250   DOI 10.1201/9781003349648-35

realizing energy conservation, emission reduction, and low carbon, and promoting green development. It can effectively promote the achievement of energy conservation and emission reduction goals and promote the green transformation of the economy. Research shows that the application of digital technology in various fields can help reduce global carbon emissions by 15-20%. By 2030, digital solutions such as smart manufacturing, smart agriculture, smart buildings, smart mobility, and smart energy can reduce more than 12 billion tons of carbon dioxide equivalent in the global economy, accounting for about one-fifth of the total global emissions. According to the latest version of the 2020 Global Climate Action Summit Index Climate Action Roadmap, digital technology can already help the world reduce carbon emissions by 15% in fields such as energy, manufacturing, agriculture, land, construction, services, transportation, and traffic management (Exponential Climate Action Roadmap 2018).

The global practical experience has proved that the development of digital information technology is an important tool to achieve the goal of carbon neutrality, so it is necessary to explore the practical path and internal mechanism of digital technology's influence on the goal of "double carbon". Therefore, based on the innovative development of digital technology in energy production, consumption, and supervision, this paper analyzes the development path of digital low-carbon economy transformation, focuses on microeconomic entities, builds a game decision model among local governments, production enterprises, and investors, and simulates to analyze the internal mechanism of digital carbon neutrality.

2 DIGITAL CARBON NEUTRALIZATION PATHWAYS

In recent years, with the innovation of digital technology and the rapid development of the digital economy in China, digital technologies such as cloud computing, artificial intelligence, big data, and digital twins have gradually penetrated various fields such as ecology, energy, and finance. According to the situation of digital technology and its application, we will discuss its prospects of promoting the development of a green economy from three perspectives. First, in terms of energy production, digital technology can help the energy sector improve the efficiency of traditional energy and solve the shortcomings of clean energy, thus speeding up the process of energy source structure adjustment. Second, in terms of energy consumption, digital technology can help the three major carbon emission departments of industry, construction, and transportation to accelerate the green transformation of energy conservation, emission reduction, and digitalization. Third, in terms of energy regulation, digital technology can also play an important role in carbon fixation, carbon monitoring, and carbon finance.

The current energy structure and carbon emission situation in China suggests that at least three paths should be explored to achieve carbon neutrality in peak carbon dioxide emissions. First, we should control and reduce carbon emissions, including restricting the use of fossil energy at the energy production end, increasing the use of clean energy, and promoting electrification and energy-saving and efficiency-enhancing technological transformation in key areas at the energy consumption end. Second, we should promote and increase carbon absorption, mainly including technical carbon sequestration and ecological carbon sequestration. Third, we should support the realization of carbon neutrality by establishing a green financial system.

3 SIMULATION OF MICROCOSMIC EVOLUTIONARY GAME MODEL

The government is a proponent of dual carbon targets, a champion of the environmental benefits of carbon emission reduction, a supervisor of enterprises' carbon emissions, and a guide to fostering green development among enterprises. Enterprises are the facilitators of low-carbon production, exchanging short-term R&D technology and other costs for long-term social and economic benefits. Enterprise leaders formulate development strategies according to the development of enterprises and pursue the maximization of comprehensive benefits of enterprises. Investors represent the

Table 1. Application of digital technology in the green economy.

		Typical cases
Energy production perspective	Traditional energy Clean energy	Energy Internet – from "Watts" to "Watts + Bits"
Energy consumption perspective	Industrial sectors	Green intelligent manufacturing—5S manufacturing with both high efficiency and sustainable
	Construction field	Zero carbon building—digital system to achieve sustainable low energy consumption in operation and maintenance
	transportation	New energy vehicles—a combination of digital hardware (automotive chips) and digital software (autonomous driving)
Energy Regulatory perspective	Carbon fixed	Smart irrigation, precision agriculture, and digital pasture
	Carbon monitoring	Carbon monitoring CEMS method, blockchain tracking the carbon footprint
	Carbon finance	Financial information platform, big data modeling, and prediction

providers of green funds. Based on the huge demand for green funds in the future, according to the existing research, the marketization of green finance is the inevitable trend of low-carbon economy development. Digital development is an important tool for efficient energy production and efficient carbon regulation, which can effectively alleviate the problem of information asymmetry. Based on the premise of limited rational people and effective information transmission, the following parameters are set to establish an evolutionary game model.

(1) Government-related parameter Settings

W represents the comprehensive social economic benefits, including economic growth, employment rate, and resource allocation. β $(0 \leq \beta \leq 1)$ is the influence coefficient of the government's encouragement of green production on social and economic benefits. The development of green finance will affect the economic growth rate, employment rate, and resource allocation, and the social and economic benefits will be reduced to βW in the whole process of economic transformation. As the government enforces green production in the early stage, it will increase the cost of enterprises and reduce the enthusiasm of enterprises, which will indirectly harm economic growth. When the environmental benefits of whether the government takes incentive measures are a, -A (A>>b+c+d, which means that the environmental cost of the government not taking measures is far greater than the incentive cost and execution cost of the government); b represents the government's supervision cost for enterprises adopting green financing. At present, China puts the construction of ecological civilization in the first place, so it is assumed that the environmental benefit a of the government's compulsory implementation of green transformation is greater than the cost b of the government's supervision and implementation; c is the government subsidy to green manufacturing enterprises; S refers to the taxes levied on traditional manufacturing enterprises; d refers to the government's preference for investors participating in green finance; T refers to the fees charged for participating in general financial investment.

(2) Investment-related parameter setting

r represents the risk increased probability of investing in enterprises that issue green products, and p represents the cost of financial product exchange; π represents the indirect income from participating in green financial investment. For a financial institution, it includes corporate image and social reputation, while for investors, it includes the satisfaction and honor of fulfilling social responsibility.

(3) The parameter setting of enterprise transformation decision

E represents the profits obtained in the traditional production process; ΔE indicates the profit increased by the enterprise after engaging in green production because of improving energy utilization rate or reducing environmental pollution treatment cost; E indicates the risk probability of loss caused by green financing failing to achieve the expected results. The total cost of enterprise technology upgrading, equipment improvement, and green production transformation is C; the financial cost of enterprises adopting green financing is C_1; the financial cost of traditional financing is C_2. At present, with the support of government policy, the traditional financing method needs to pay higher financing costs than the financing method in the green financial system, so $C_2 > C_1$.

x represents the proportion of enterprises that choose the "green production" strategy among all enterprises, while the proportion that chooses traditional production is 1-x; y represents the proportion of investors who participate in green financial investment, while the proportion of investors who participate in general investment is 1-y; z indicates the scope proportion of government incentives, the proportion of green financial incentives not taken is 1-z. The payoff matrix is shown in the following Table 2. The dynamic replication equation of enterprise A, investor B, and government C is

$$
\begin{aligned}
U(A) &= x(1-x)[(yzA_{111} + z(1-y)A121 + (1-z)yA_{112} + (1-z)(1-y)A_{122}) \\
&\quad -(yzA_{211} + z(1-y)A_{212} + (1-z)yA_{212} + (1-z)(1-y)A_{222})] \\
&= x(1-x)[zc + (1-e)E + eEy] \\
U(B) &= y(1-y)[(xzB_{111} + x(1-z)B_{112} + (1-x)zB_{211} + (1-x)(1-z)B_{212}) \\
&\quad -(xzB_{121} + x(1-z)B_{122} + (1-x)zB_{212} + (1-x)(1-z)B_{222})] \\
&= y(1-y)[\pi x + (1-x)\pi + zd] \\
U(C) &= z(1-z)[(xyC_{111} + x(1-y)C_{121} + (1-x)yC_{211} + (1-x)(1-y)C_{212}) - 1 - z) \\
&\quad \times (xyC_{121} + x(1-y)C_{122} + (1-x)yC_{212} \\
&\quad +(1-x)(1-y)C_{222})] = z(1-z)[-yd - (1-\beta)W + a - b - c + A]
\end{aligned}
$$

Table 2. The payment matrix.

		Investor decision B	
		Green finance investment y	NOT green finance investment 1-y
Green production x (Business decisions A)	Adopt incentive z (Government decision C)	A_{111}=E+ΔE-C-C1+c B_{111}=(1-r)R-P+π+d $C_{111} = \beta$W+a-b+S-c+T-d	A_{121}=(1-e)(E+ΔE)-C-C2+c B_{121}=(1-r)R-P $C_{121} = \beta$W+a-b+S-c+T
	No incentives 1-z (Government decision C)	A_{112}=E+ΔE-C-C1 B_{112}=(1-r)R-P+π C_{112}=W-A+S+T	A_{122}=(1-e)(E+ΔE)-C-C2 B_{122}=(1-r)R-P C_{122}=W-A+S+T
Traditional production 1-x (Business decisions A)	Adopt incentive z (Government decision C)	A_{211}=E-C-C1 B_{211}=R-P+π+d $C_{211} = \beta$W+a-b+S+T-d	A_{221}=(1-e)E-C-C2 B_{221}=R-P $C_{221} = \beta$W+a-b+S-c+T
	No incentives 1-z (Government decision C)	A_{212}=E-C-C1 B_{212}=R-P+π C_{212}=W-A+S+T	A_{222}=(1-e)E-C-C2 B_{222}=R-P C_{222}=W-A+S+T

According to Friedman's research, Jacobi Matrix is constructed according to the above replication dynamic equations, and the value of each replication dynamic equation is zero, and eight stable equilibrium points are obtained. According to evolutionary game theory, ESS obeys Jacobi Matrix eigenvalues are all negative conditions. We checked in turn and found that only points (x=1, y=1, and z=1) conform to the condition that eigenvalues are all negative. It shows that when the government takes incentive and supervision measures, all enterprises adopt green production and investors invest in green financial transactions, the economy, environment, and society can develop in a balanced and stable way. C =1; E = 0.2; Delta E = 4; q=1; d=1; Beta = 0.2; w=10; a=10; b=2; A=100. Visual simulation evolutionary game diagram run with MATLAB is as follows:

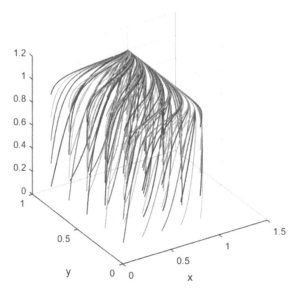

Figure 1. Three-dimensional game simulation diagram.

4 CONCLUSION AND SUGGESTIONS

In this paper, the Evolutionary game method is adopted to study the mechanism of digital technology to help carbon neutrality. The main conclusions can be summarized that Perfecting the government's carbon emission administrative management system, developing the enterprise's low-carbon technology innovation system, and establishing the carbon emission market trading system are the three main paths to the development of digital carbon neutrality. In terms of the future work, digitization should be carried out to enhance the following three aspects

(1) Digital technology to build various financial information platforms
    Use digital technology to build various information platforms to solve information asymmetry problems, including institutions and customers, capital demanders and attackers, regulators, and regulated institutions to improve the efficiency of business, financing, and supervision. In the operation of investment and financing business, a platform for the collection of green financial information statistics and a platform for the evaluative evaluation of green credit provides a method to connect demand and supply for green financing easily, thus solving the problems of difficulty in locating target projects and asymmetric information.
(2) Intelligent analysis, prediction, and modeling based on digital technology
    Intelligent analysis, modeling, forecasting, and other functions based on big data, machine learning, artificial intelligence, and other technologies also play an important role in the supervision and investment of green financial products. For example, in the supervision process, digital technology can be widely used in the audit, daily monitoring, and inspection of green finance business.

(3) Application of Blockchain in Green Finance

With the deepening development of green finance, problems such as asymmetric information, high requirements for coverage scale, more difficult evaluation, and greater regulatory challenges are gradually exposed. The application of blockchain technology can effectively deal with the above problems. Financial institutions can store information about corporate violations, corporate green credit records obtained by credit rating agencies during the credit rating process, analysis of the green degree of fundraising projects, environmental costs, and debt credit rating in the blockchain, and establish a sharing mechanism to provide a basis for financial institutions' loan and investment decisions.

REFERENCES

Bao Jianqiang, Miao Yang, Chen Feng. Low-carbon economy: a new change in the mode of human economic development [J]. *China Industrial Economy*, 2008 (4): 153–160.

Exponential Climate Action Roadmap (Version 1.5), *Future Earth, Global Climate Action Summit*, September 13, 2018.

Li Xiaohua. New features of digital economy and formation mechanism of new kinetic energy of digital economy [J]. *Reform*, 2019 (11): 40–51.

Xu Zheng, Zuo Shengji, Ding Shouhai. peak carbon dioxide emissions, Carbon Neutralization and High-quality Development: Internal Logic and Realization Path [J]. *Economist*, 2021 (11).

Zhang Gaoli. Vigorously promote ecological civilization and strive to build a beautiful China [J]. *Qiushi*, 2013 (24): 3–11.

Advances in Renewable Energy and Sustainable
Development – Liang & Kasmani (Eds)
© 2023 Copyright the Author(s), ISBN: 978-1-032-39407-7

Can ESG investment improve corporate green innovation performance? Evidence from China

Hui Lyu, Yan Sun*, Ruili Zhou & Yue Chu
Department of Economics and Management, China University of Petroleum, Beijing, China

ABSTRACT: Following China's announcement of the carbon-neutral target, ESG investments are expected to help the entity make a green and low-carbon transition. Whether ESG investments will improve corporate green innovation performance is worth studying. Using a data sample comprising Chinese A-share listed firms during the period from 2014 to 2019, we find that corporations invested by ESG funds have significantly better green innovation performance. Further study reveals that ESG investment can reduce financing constraints and improve investment efficiency. The conclusion of this study has certain enlightenment on how to give better play to ESG investment in promoting corporate green sustainable development.

1 INTRODUCTION

Natural disasters and extreme weather caused by climate change, such as forest fires in Australia, high temperatures in the South Pole, floods and hurricanes, have become common challenges facing mankind and have a far-reaching impact on the global economy. Financial institutions in various countries have taken measures to deal with climate change and reduce financial risks, such as assessing climate change risks and encouraging the disclosure of climate-related risk information. In September 2020, China put forward the "net-zero" carbon emissions goal, advocating that corporations should also consider development in environmental, social and governance fields while pursuing economic interests. ESG (environmental, social and governance) investment, as a corporation evaluation system of non-financial performance, conforms with the development trend of the times and has attracted extensive attention from the society. ESG investors consider the impact of their investment on the environment and society when they make investment decisions. Therefore, ESG investment could help corporations achieve green innovation development and long-term development. Green innovation is the core driving force for corporations to realize industrial transformation and upgrading.

Under the specific background of the wide implementation of national green development and innovation in China, this study which focuses on the green innovation of ESG investment, has a profound and significant impact on both theory and practice. Our findings shed light on whether and why ESG investment can affect the green innovation of corporations. In contrast to prior studies, which mainly used ESG rating or ESG scores as the main independent variable, we innovatively focus on ESG investment itself in promoting corporate green innovation with two main reasons listed as follows. First and foremost, different rating agencies have different emphases and priorities when rating, which is likely to lead to different results. Especially for China, which has just introduced the concept of ESG, there is no systematic ESG rating system, leading to a greater difference in ratings. Furthermore, when formulating investment strategies, it is difficult to determine whether and to what extent investment institutions pay attention to the ratings. Therefore, the

---

*Corresponding Author: sylvie1218@126.com

256   DOI 10.1201/9781003349648-36

conclusion obtained by using ESG rating is difficult to be objective and robust. Since discussing the validity and application of ESG rating is not the focus of our study, and in order to avoid the influence of the uncertainty of the results obtained by the ESG rating, this paper innovatively and directly selects the companies that have already obtained ESG investment for analyzing the positive role that ESG played to corporate green innovation.

2 LITERATURE AND HYPOTHESIS DEVELOPMENT

2.1 *The impact of ESG investment on green innovation performance*

ESG investment mainly improves corporate green innovation by three means. The first mean is to fulfill the legitimacy of corporate green transformation. According to the legitimacy theory, corporations seize the green market share through green transformation, obtain the recognition of social participants and foster a good image to obtain richer resources and sustainable competitive advantage.

The second approach is shareholder activism. With the gradual improvement of emerging markets, the scale and quality of Chinese institutional investors have been greatly improved. ESG investors not only have high professional skills, rich capital resources and a high shareholding ratio (Wang et al. 2012), but also pay more attention to the promotion of corporate green innovation than common investors.

The last approach is the redistribution of capital factors. Based on the stakeholder theory, the environmental value of stakeholders can not only increase green innovation, but also helps corporations form a good reputation among stakeholders. Meanwhile, corporations would be required to improve their innovation ability after receiving ESG investment.

Thus, our first hypothesis is:

Hypothesis 1(H1): Corporations with ESG investment have better green innovation performance.

2.2 *Mechanism of the impact of ESG investment on green innovation*

External financing is an important source of funds for corporate green innovation activities. Better environmental performance can gather more long-term external capital investment for corporations, alleviating their financing constraints (Shen & Ma 2014).

First, institutional investors with ESG as their investment philosophy are likely to gradually expand, which would alleviate corporations' financing constraints. Prior research confirmed that it would bring an excess return to investors. Dimson et al. (2015) found that stocks would bring positive returns when ESG investors participated in corporations' ESG governance. The empirical research of Zheng and Lu (2014) showed that the return of social responsibility funds was not worse than the average market level and had a stronger anti-risk performance. Therefore, policy advocacy and investors' preference for the pursuit of excess return are the two main reasons for the promising prospect and sustainability of ESG, which results in more ESG institutional participants. The expanding group of ESG institutional investors would then lead to corporations gaining larger and larger ESG investments in the future. Therefore, ESG investment can effectively alleviate corporations' financing constraints.

Furthermore, herd behavior in the Chinese fund market has been documented in much research (Shi 2001; Tao et al. 2015). When ESG investment continues to flow to certain corporations in large quantities, some other institutional investors and individual investors in the market would then follow the ESG investors to buy stocks, which further reduces the financing constraints of corporations. Sun et al. (2019) and Lu et al. (2019) found that the financing constraints of corporations had an inverted U-shaped relationship with their R & D investment and innovation. It indicates that when the financing constraints of corporations are high, alleviating the financing constraints can significantly promote their R & D and improve innovation.

Thus, we put forward Hypothesis 2.

Hypothesis 2(H2): ESG investment reduces corporations' financing constraints.

On the contrary, if a corporation invests in projects with high pollution and high energy consumption after receiving ESG investment, its investment cost would increase significantly due to more and more strict environmental regulations. This would ultimately lead to the declination of investment efficiency. There are three reasons:

First, in order to comply with the pollutant emissions restrictions, corporations need to invest more in cleaner production machines and equipment. In this way, investment costs are increased, and investment efficiency is reduced. In addition, if the pollutant emissions exceed the regulated amount, corporations are likely to be strictly regulated and fined, increasing costs and reducing investment efficiency. Last but not least, once the media expose the behavior of investment projects with high pollution and high energy consumption, it is more likely to have a negative impact on the social image of corporations, resulting in the decline of stock price, the amount and shareholding ratio of ESG investment. Wong et al. (Jin & Qin 2021) asserted that the negative reports of ESG on corporations from the media had led to a decrease in corporate stock price and corporate value. Kölbel et al. (2017) proved that negative media reports on the ESG of corporations would increase the potential credit risk, improve the cost of debt financing and reduce the efficiency of investment.

Therefore, Hypothesis 3 was created:

Hypothesis 3(H3): ESG investment is significantly improved corporations' investment efficiency.

3 DATA AND METHOD

3.1 *Sample selection and data source*

This study takes all A-share listed companies in China from 2014 to 2019 as samples. Samples were processed as follows. First, we excluded all financial and insurance corporations; Then, we excluded all special treatment (ST) corporations and corporations with an asset-liability ratio less than 0 or larger than 1. Finally, the main continuous variables were winsorized in the 1% tail.

3.1.1 *Corporate green innovation data*

With reference to Wang et al. (2021), the explained variable, which is green innovation, is measured by the number of green patent applications. The number of green utility model patent applications create_num, number of green invention patent applications create_qua and create_sum are used to measure the quantity, quality and total amount of green innovation, respectively.

3.1.2 *ESG investment data*

Referring to the study of Jiang et al. (2021), first, we get information from the "Fund Information List" and "Stock Investment Detailed List" from Fund Market Series in CSMAR Database. If "ESG fund" appears in a corporation's "Stock Investment Detailed List," it indicates that the corporation has ESG investment.

3.1.3 *Other characteristic data of corporations*

Other characteristic data of corporations were gathered from CSMAR.

3.2 *Model design and variables selection*

Based on the regression model with the fixed effect of corporations, this study estimates the following equation to test the impact of ESG investment on Green Innovation:

$$Patent_{i,t} = \alpha + \alpha_1 \times ESGinv_{i,t} + \alpha_{n+1} \times Control_{i,t} + \zeta_{i,t} \tag{1}$$

4 EMPIRICAL RESULTS

4.1 *Main results*

4.1.1 *Descriptive statistics*
The mean value of create_num is 0.697, and the standard deviation is 5.444. The mean value of create_qua is 1.225, and the standard deviation is 13.29. The mean value of create_sum is 1.922, and the standard deviation is 17.31. These indicate great differences in green innovation among sample corporations, and most green patent applications are green invention patent applications. The mean value of ESG investment is 0.426, indicating that 42.6% of the sample corporations have ESG investment.

4.1.2 *Regression: ESG investment in corporate green innovation*
In order to investigate the impact of ESG investment on green innovation, this study uses regression according to model (1), and the results are reported in Table 1. Columns 1, 2 and 3 report the regression results of the quantity, quality and total amount of green innovation, respectively. The estimated coefficient of ESGinv in three columns are all significantly positive, indicating that ESG investment significantly promotes the number, quality and total amount of green innovation of listed companies. Therefore, Hypothesis 1 is preliminarily verified.

Fund shareholding ratio will significantly reduce the number of corporate green innovations, but will not significantly reduce the quality of corporate green innovation. The number and quality of green innovation in corporations will not be significantly affected by the proportion of independent board members, the ratio of book value to market value, asset-liability ratio and ROA. Cash ratio, R&D investment, fixed assets ratio, Tobin Q and duration years will significantly reduce the green innovation of corporations. The ratio of capital expenditure to depreciation and amortization and the size of the workforce will significantly improve the green innovation of corporations.

Table 1. The impact of ESG investment on green innovation.

	(1)	(2)	(3)
VARIABLES	Increate_num	Increate_qua	Increate_sum
ESGinv	0.030***	0.020**	0.036***
	(3.11)	(2.06)	(2.89)
Control variables	YES	YES	YES
Constant	1.815***	1.008***	2.036***
	(11.35)	(6.13)	(9.86)
Observations	13,795	13,795	13,795
Company fixed effect	YES	YES	YES
Adj. R^2	0.0280	0.00855	0.0213

Note: * $p < 0.1$; ** $p < 0.05$; *** $p < 0.01$.

4.1.3 *Robustness test*
1. Excluding Data in 2017

 Considering the Decision of the State Intellectual Property Office on Amending Several Provisions on Regulating Patent Application Behavior issued in 2017, the statistical standard of the patent application has changed. Therefore, we excluded the data in 2017 and then regressed the rest of the data. The results are shown in (1)-(3) of Table 2. ESG investment is significantly positive for the quantity and quality of green innovation at the level of 5% and for the total amount of green innovation at the level of 1%, which is basically consistent with the regression results shown in Table 2. Model (1) is robust.

2. Lagged Explained Variables

Since the green innovation of a corporation increases its probability of receiving ESG investment, endogenous problems like reverse causality would probably occur. In order to mitigate this problem, 1-year lagged explained variables are used respectively in the regression. The regression results are shown in (4)-(6) of Table 2. ESG_inv has a significant positive impact on green innovation at the level of 1%. This again verifies that the model (1) is robust and that Hypothesis 1 is true.

Table 2. Robustness test.

| | Excluding data in 2017 | | | Lag one phase | | |
| | (1) | (2) | (3) | (4) | (5) | (6) |
VARIABLES	Increate_num	Increate_qua	Increate_sum	L.Increate_num	L.Increate_qua	L.Increate_sum
ESGinv	0.027**	0.025**	0.037***	0.134***	0.143***	0.188***
	(2.46)	(2.31)	(2.61)	(7.91)	(7.65)	(7.87)
Controls	YES	YES	YES	YES	YES	YES
Constant	2.062***	1.267***	2.366***	−0.253*	−0.651***	−0.564***
	(11.87)	(7.17)	(10.65)	(−1.71)	(−3.95)	(−2.71)
Observations	11,261	11,261	11,261	11,434	11,434	11,434
Adj. R^2	0.0379	0.0133	0.0302	0.0758	0.0961	0.0995

4.2 Mechanism analysis

4.2.1 Regression: Impact of ESG investment on financing constraints

Then, in order to verify Hypothesis 2 and analyze the impact of ESG investment on financing constraints, the model we designed is as follows:

$$FC_{i,t} = \beta + \beta_1 \times ESGinv_{i,t} + \beta_{n+1} \times Control_{i,t} + \eta_{i,t} \tag{2}$$

FC this paper used is referred to Kaplan and Zingales(Kaplan & Zingales 1997), Wei Zhihua et al. (2014) to measure the financing constraint of corporations with the KZ index. The greater the value of KZ, the greater the financing constraint of the corporation.

Table 5 reports the regression results of model (2). The coefficient of the ESG_inv term is significantly negative at the 5% level, proving that the financing constraints will decrease significantly after corporations receive ESG investment. This supports Hypothesis 2.

4.2.2 Regression: Impact of ESG investment on investment efficiency

In order to verify Hypothesis 3, the following regression model is set to analyze the impact of ESG investment on investment efficiency:

$$Eoi_{i,t} = \gamma + \gamma_1 \times ESGinv_{i,t} + \gamma_{n+1} \times Control_{i,t} + \theta_{i,t} \tag{3}$$

We got the corporate investment efficiency Eoi referring to Richardson's study(Richardson 2006).

Table 3 reports the regression results of model (3). The coefficient of the ESGinv term is positive and significant at the level of 5%, indicating that the investment efficiency will increase significantly after receiving ESG investment. Hypothesis 3 stands.

5 CONCLUSIONS

In this paper, we test whether ESG investment contributes to corporate green innovation performance. The main conclusions can be surmised as follows: (1) Corporations invested by ESG

Table 3. Mechanism analysis: Financing constraints and investment efficiency.

VARIABLES	(1) FC	(2) Eoi
ESGinv	−0.069**	0.004**
	(−2.19)	(2.49)
Control variables	YES	YES
Constant	15.462***	−0.095***
	(14.04)	(−5.41)
Observations	14,681	15,632
Company fixed effect	YES	YES
Adj. R^2	0.208	0.0242

investors have better green innovation performance. (2) ESG investment significantly increases the corporations' investment efficiency. (3) ESG investment reduces the financing constraints of corporations.

In terms of future works, the mechanism of ESG investment on corporate green innovation performance should be further studied, and the moderating effect of Industry and Property Nature should also be carried out to enhance the role of ESG investment, then encourage more investors to invest in ESG and promote ESG investment.

ACKNOWLEDGMENTS

This research was financially supported by the Projects of Beijing Urban Governance Research Base (21CSZL35), the Projects of Beijing Association of Higher Education (YB202104&YB2021162), and the Foundation of China University of Petroleum (Beijing) (ZX20200110).

REFERENCES

Dimson, E., Karakaş, O., Li, X. (2015) Active ownership. *The Review of Financial Studies*, 28: 3225–3268.

Jiang, G., Lu, J., Li, W. (2021) Do Green Investors Play a Role? Empirical Research on Firms' Participation in Green Governance. *J Financ Res*, 491: 117–134.

Jin, B.W., Qin, Z. (2021) Stock market reactions to adverse ESG disclosure via media channels. *The British Accounting Review*.

Kaplan, S.N., Zingales, L. (1997) Do investment-cash flow sensitivities provide useful measures of financing constraints? *The Quarterly Journal of Economics*, 112: 169–215.

Kölbel, J.F., Busch, T., Jancso, L.M. (2017) How media coverage of corporate social irresponsibility increases financial risk. *Strategic Manage J*, 38: 2266–2284.

Lu, C., Lv, H. (2019) Fiscal Subsidies, Financing Constraints and Manufacturing R&D Investment. *Review of Economy and Management*, 35: 17–27.

Richardson, S. (2006) Over-investment of free cash flow. Rev Account Stud, 11: 159–189.

Shen, H., Ma, Z. (2014) Local Economic Development Pressure, Firm Environmental Performance and Debt Financing. *J Financ Res*, 153–166.

Shi, D. (2001) Trading Behavior of Securities Investment Funds and Its market Influence. *The Journal of World Economy*, 26–31.

Sun, B., Liu, S., Jiang, J., Ge, C., Zhou, H. (2019) The Financial Constraints and Firm Innovation: From the Perspective of Human Capital Network. *Chinese Journal of Management Science*, 27: 179–189.

Tao, Y., Liu, Y., Peng, L.(2015) Herding Behavior and Its Influencing Factors of China's Securities Investment Funds. *Journal of Beijing University of Posts and Telecommunications* (Social Sciences Edition), 60–67.

Wang, J., Liu, M., Wang, Z. (2012) An empirical study on the impact of institutional investors' Shareholding on the discount of new share issuance. *Management World*, 172–173.

Wang, X., Wang, Y. (2021) Research on the Green Innovation Promoted by Green Credit Policies. *Management World*, 37: 173–188.

Wei, Z., Zeng, A., Li, B. (2014) Financial Ecological Environment and Corporate Financial Constraints—Evidence from Chinese Listed Firms. *Accounting Research*, 73–80.

Zheng, R., Hu, L. (2014) An Analysis on the Strategy and Performance of Socially Responsible Investment in China. *Business and Management Journal*, 163–174.

Advances in Renewable Energy and Sustainable
Development – Liang & Kasmani (Eds)
© 2023 Copyright the Author(s), ISBN: 978-1-032-39407-7

Integrating technologies into urgent wildlife release

Junlin Shao*

School of Government, Beijing Normal University, Beijing, China

ABSTRACT: When terrestrial wildlife consumption is comprehensively banned, the edible wildlife breeding industry in China faces dramatic challenges, and how to dispose of numerous wildlife stocks becomes a key issue ahead. National Forestry and Grassland Administration specifies releasing into the wild as one of four main disposal methods for the stocks of animals in the farming industry. Traditional administrative means cannot effectively manage the release activities. In this paper, we argue that some emerging technologies could be applied in the process of wildlife release and facilitate data integration, wildlife monitoring, and public participation.

1 INTRODUCTION

The COVID-19 pandemic presses a pause button on the global socio-economic system. At the same time, the threat of mass death posed by the pandemic has raised public health awareness to an unprecedented level. Seventeen years after the outbreak of SARS, the problem of excessive wildlife consumption returns to the political domain in China. On February 24, the National People's Congress (NPC), the highest legislature of China, passed a *Decision* (Xinhuanet 2020). Apart from being one of the swiftest laws ever enacted, the *Decision* has two distinctive features. In one way, unlike previous laws which only banned the consumption of legally protected wildlife, the new law prohibits people from eating all wild terrestrial animals, including those captively bred, aiming to eradicate thousands of years of excessive wildlife consumption and ensure public health; in the other way, the new law bans the whole process of hunting, trading, and transporting of wild terrestrial animals that grow and breed naturally in the wild for food.

The *Decision* has a huge impact on the edible wildlife breeding industry and numerous captive-bred wildlife have to be released into the wild. However, this process faces many challenges, and to address these challenges, we must integrate emerging technologies into the whole process and utilize the power of emerging technologies. This paper is organized as follows. In section 2, we describe the challenges ahead. In section 3, we list how emerging technologies can help in the process of wildlife release. And this paper concludes with some suggestions.

2 CHALLENGES AHEAD

2.1 *Large scale*

With the promulgation of the *Decision*, the edible wildlife breeding industry faces dramatic challenges, and how to dispose of numerous wildlife stocks becomes a key issue ahead. For decades, China has been implementing the strategy of "conservation through utilization" in the field of wildlife conservation (Wang et al. 2019). Due to the high demand for wildlife products and the legislation on commercializing captive breeding animals under *China's Wildlife Protection Law*

---

*Corresponding Author: shaojunlin2009@qq.com

DOI 10.1201/9781003349648-37

(CWPL), China's wildlife farming industry has a rapid growth, which creates employment for 14.09 million people and contributes more than $73.5 billion to the economy (Consulting research project of Chinese Academy of Engineering 2017). Part of the industry that supplies the food market accounts for 24% ($17.6 billion) of the total output and includes nearly 6.3 million workers (Consulting research project of Chinese Academy of Engineering 2017). However, after the enactment of the *Decision*, this industry suffered great economic loss. Many captive-bred wildlife was likely to starve to death on the farm as the breeders could not afford to feed them. Forced by the heavy financial burden, some farmers began to appeal to higher authorities for help.

The urgency of the situation forces the National Forestry and Grassland Administration (NFGA) to dispose of wildlife stocks as soon as possible. Recently, NFGA emphasized that the formulation and implementation of the disposal plan should be completed at the end of September, which put great time pressure on local governments. On May 27, NFGA published *Technical Guidelines for Proper Disposal of Captive-bred Wildlife (the Guidelines)*, specifying four disposal methods for the stocks of animals in the farming industry, including releasing into the wild, conversing to non-edible use, transferring to qualified shelters or rescue organizations, and treating harmlessly, among which release has the greatest impact on the ecosystem and thus becomes the focus of this paper.

The Guidelines point out species that would be mainly disposed of by natural releases, such as flat-breasted turtles, wild geese, and ducks, pheasants, mammals of native species, and birds except for blue peacock; species with a large breeding capacity exceeding the demand of non-edible use, such as cobra, mouse snake, gray mouse snake, and King's snake, can also be partially released to nature. According to the survey conducted in 2016, about 143 million wildlife would be mainly released, including 30 million reptiles, 110 million birds, and 4 million mammals (Consulting research project of Chinese Academy of Engineering 2017). The quantity of releasing should meet the environmental capacity of the continuous natural area where the releasing point is located. Once there is no suitable area or the quantity is too large to be released nearby, the provincial or national forest and grass administrations should coordinate the releasing area.

2.2 *Economic burden and potential social conflicts*

The release process would be complex, involving not only the scientific nature of the disposal plan itself, but also issues such as farmers' compensation, coordination among different regions and departments, and supervision of the implementation process. To facilitate the disposal process, the government needs to compensate farmers. Recently, Hunan province takes the lead to issue a plan on the compensation scope and standard of wildlife disposal among all local governments. Considering the cost of captive breeding of terrestrial wildlife, the input of facilities, and the breeding mode, Hunan province has set compensation standards for the first batch of 14 kinds of wildlife, such as bamboo rat at 75 RMB/kg and civet cat at 600 RMB/unit (Wu 2020). But only subjects with legal license subjects breeding for cooperatives(enterprises) with legal license, or subjects submitting the application for license and meeting essential conditions before February 24, 2020, could be compensated, making the premise to get compensation quite harsh. Due to the lack of income, knowledge, and policy support, many farmers just leave wildlife to their fate. Those who could not get compensation may even secretly release wildlife or trade in the black market. The industry transformation and wildlife disposal are still in a difficult situation.

2.3 *Latent environmental and public health risks*

More importantly, the release of captive-bred wildlife is risky if it is not done properly. On one hand, the foraging ability and wild environment adaptability of the captive-bred wildlife are weaker than those of the wild species, and the recovery of their natural feeding habits needs a certain period, which may affect their survival in the wild. On the other hand, the released wildlife may carry a variety of viruses and bacteria, each of which may endanger the health of the original population (Woodford 2000), especially when it comes to cross-regional release, the potential health risk is even greater. In addition, the release of some dangerous species, such as poisonous snakes, may also bring personal safety hazards.

3 EMERGING TECHNOLOGIES TO OVERCOME CHALLENGES

At present, wildlife disposal is carried out by local governments separately, making it difficult to collect, integrate and analyze related data. Traditional administrative means cannot effectively manage the release activities. In fact, some emerging technologies can be applied in the process of wildlife release.

3.1 Improvement of data sharing

Existing data platforms can promote information integration in the release process. A large number of data records are involved in wildlife release, such as released species and quantity, release time and location, health records, etc. It is conducive to long-term wildlife tracking and scientific research for sharing these data timely among different departments and regions. China has promoted the development of Forestry Big Data and Forestry Cloud for several years. Many National Science & Technology Infrastructures (NSTI) have been established, such as the National Ecological Science Data Center (NESDC) and the National Forestry and Grassland Data Center (NFGDC). These data platforms have rich software & hardware resources and a mature management system in the data standard formulation, data integration, and sharing, which can be easily applied to wildlife release.

3.2 Implementation of regular monitoring of released wildlife

Some technologies can be used to monitor the living condition and reproduction of the released wildlife. In recent years, a lot of wildlife monitoring technologies emerged, such as remote sensing, unmanned aerial vehicle (UAV), GPS telemetry, and camera-trapping (Marvin et al. 2016). Some scientists asserted that we have entered a golden age of animal tracking science (Kays et al. 2015). For example, scientists have combined camera traps with other technologies and innovatively applied them to new environments and species. With the advancement of these technologies, we can not only monitor the distribution and abundance of wildlife populations in a region dynamically but also depict the animal movements in real time. At present, China has also constructed intelligent ecological environment monitoring networks in some areas. Taking the Qilian Mountain area as an example, using satellite remote sensing, UAV, lidar scanner, and soil respiration meter, an omnidirectional ecological observation network system has been established in 8 counties in Gansu Province (National Forestry and Grassland Administration 2020). There are still some legal or technical restrictions on the application of the above monitoring technologies. For instance, UAV flights in some areas are prohibited (Linchant et al. 2015) and the duration of some UAVs need to be further improved (Christie et al. 2016). However, these technologies still have a promising future.

3.3 Enhancement of public participation

New technologies could increase public participation. For one thing, many officers and concerned citizens usually lack essential knowledge to identify wildlife species. Appropriate information tools help to engage people in wildlife conservation and monitoring better. In China, researchers have developed a mobile application called "image and sound recognition of birds" based on the birds' characteristics, containing information on more than 90 species of birds that often inhabit the transmission tower. Transmission line patrols can thus accurately identify and protect birds in situ through this app. We also should allow the public to upload information and realize the public's potential for information collection. For example, one app could help Indian field patrols quickly collect data on tigers and improve the nation's tiger population estimates (EcoWatch 2018). The

Chinese government has provided many public welfare posts for the poor, such as forest rangers, who can also become citizen scientists using appropriate apps. For another, many information tools can transfer early warnings to the public and reduce human-wildlife conflict (Graham et al. 2012). Especially when it comes to poisonous and harmful wildlife release, the release of relevant information is even more important.

In addition, the explosion of data volume and variety has created new challenges and opportunities for information management, integration, and analysis (Kays et al. 2015), the rapid development of artificial intelligence could also help to analyze data more efficiently. In the whole process of wildlife release, many kinds of data, such as pictures and videos, would be generated and collected. The progression of computer vision tools can be used to automatically identify species and speed up the process of data processing.

4 CONCLUSIONS

In this paper, we list the challenges in the process of wildlife release and show how emerging technologies could be integrated to facilitate data integration, wildlife monitoring, and public participation. The main conclusions and implications can be summarized as follows:

(1) Successful wildlife release requires many pivotal conditions: cooperation among different shareholders, constant monitoring, and participatory governance, etc. Emerging technologies provide key tools to meet those requirements through collecting, managing, sharing, and analyzing more and better data, and assisting scientific decisions (Bergermilal & José 2017). China already has a good forestry information infrastructure and is now further promoting "New Infrastructure", which focuses on 5G networks, industrial networks, data centers, and many other advanced fields. However, we have not seen relevant technology in wildlife disposal.
(2) To avoid public health risks, biodiversity loss, and other risks, the formulation, and implementation of the release program should be evaluated and examined more carefully and scientifically.
(3) Since the Chinese government is encouraging the release of some kinds of wildlife in captivity back to nature, a dynamic monitoring system is urgently needed. Technologies should play a more important role in these processes, especially when many wildlife animals must be disposed of in almost a few months.

Many vulnerable protected areas have a low technical capacity and limited infrastructure although these technologies' promise is optimistic. The good news is that many places with low technical capacity and limited infrastructure may gain the chance to deploy appropriate technology in the future. These technologies can not only serve the disposal of captive-bred wildlife but also have long-term scientific research values. For example, in view of so many captive-bred wildlife released into nature, more research on trophic rewilding (Perino et al. 2019) should be performed according to the data collected through these new technologies. The scientific community also needs to assess the impact of rewilding policies on ecosystems. Finally, new technologies alone do not solve problems, we must provide appropriate tools to the right people (Pimm et al. 2015). To promote the change in social norms and gradually eradicate unhealthy habits in traditional foods cultures, new media and social networks should be used to spread awareness of science.

REFERENCES

"Report on the sustainable development strategy of China's wildlife farming industry" (Consulting research project of Chinese Academy of Engineering, 2017. (In Chinese)

A. Perino, H. M. Pereira, L. M. Navarro, et al. *Rewilding complex ecosystems. Science*, Vol.364, eaav5570(2019).

D.C. Marvin, L.P. Koh, A. J. Lynam, S. Wich, A.B. Davies, R. Krishnamurthy, et al., Integrating technologies for scalable ecology and conservation. *Glob. Ecol. Conserv.*, Vol.7, 262 (2016).

EcoWatch "*10 Top Conservation Tech Innovations From 2017*", EcoWatch; https://www.ecowatch.com/conservation-tech-innovations-2522754536.html, (2018).

Graham, M. D., Adams, W. M., & Kahiro, G. N., *Mobile phone communication in effective human elephant–conflict management in Laikipia County*, Kenya. Oryx, Vol.46, 137 (2012).

J. Linchant, J. Lisein, J. Semeki, P. Lejeune, Cédric. Vermeulen, Are unmanned aircraft systems (USAs) the future of wildlife monitoring? A review of accomplishments and challenges. *Mammal Review*, Vol.45, 239 (2015)

J. Y. Wu , "The first provincial-level exit plan for wildlife breeding was announced, with bamboo rats receiving 75 yuan per kilogram", BJnews.com.cn(2020); http://www.bjnews.com.cn/news/2020/05/16/728086.html (in Chinese).

K.S. Christie, S.L. Gilbert, C.L. Brown, et al., Unmanned aircraft systems in wildlife research: current and future applications of a transformative technology. *Front. Ecol. Environ.*, Vol.14, 241 (2016).

M. H. Woodford, *Quarantine and health screening protocols for wildlife prior to transportation and release into the wild*, Wildlife Disease and Zoonotics at Digital Commons @University of Nebraska - Lincoln. Vol.32, (2000).

National Forestry and Grassland Administration, "*Gansu's 'Universe, Earth, and Sky' Methods to Protect Qilian Mountain's Wildlife*", National Forestry and Grassland Administration (2020); http://www.forestry.gov.cn/main/146/20200426/093556995435565.html

O. Bergermilal, José J. Lahozcmonfort, Conservation technology: the next generation. *Conserv. Lett.* Vol.11, 12458(2017).

Pimm, S. L., Alibhai, S., Bergl, R., et al. Emerging technologies to conserve biodiversity. *Trends in ecology & evolution*, Vol.30(11), 685-696 (2015).

R. Kays, M.C. Crofoot, W. Jetz, M. Wikelski, Terrestrial animal tracking as an eye on life and planet. *Science*, Vol.348, aaa2478 (2015).

W. Wang, L. Yang, T. Wronski, S. Chen, Y. Hu, S. Huang. Captive breeding of wildlife resources—China's revised supply-side approach to conservation. *Wildlife Soc. B.*, Vol.43, 425(2019).

Xinhuanet, "Decision of the Standing Committee of the National People's Congress to Comprehensively Prohibit the Illegal Trade of Wildlife, Break the Bad Habit of Excessive Consumption of Wildlife, and Effectively Secure the Life and Health of the People," Xinhua.net; http://www.xinhuanet.com/politics/2020-02/24/c_1125620762.htm (In Chinese) (2020).

Advances in Renewable Energy and Sustainable
Development – Liang & Kasmani (Eds)
© 2023 Copyright the Author(s), ISBN: 978-1-032-39407-7

Spatial planning of eco-industrial park based on reverse logistics

Qing Yang, Chen Wang*, Chenyang Cai & Xingxing Liu
School of Safety Science and Emergency Management, Wuhan University of Technology, Wuhan, China

ABSTRACT: People pay increased attention to environmental protection, and the concept of sustainable development has become increasingly popular. The construction of an eco-industrial park will help to achieve the dual goals of economic benefits and environmental friendliness, and promote the development of a circular economy. Aiming at the problem of disconnection between industrial chain network and spatial planning in the planning of eco-industrial parks in various countries, this paper proposes a spatial planning scheme by combining the methods of reverse logistics, GIS spatial analysis, and goal achievement matrix. Through the spatial planning of the vein industrial park in city A, it shows that this scheme can improve the industrial chain of the park and make the layout of the parking space more in line with the actual land use.

1 INTRODUCTION

People are paying more and more attention to the environmental damage, pollution, and resource depletion caused by economic development. The eco-industrial park is a new type of industrial organization designed based on the theory of circular economy and industrial ecology. It can promote the development of a circular economy and is an effective means to achieve the dual goals of economic benefits and environmental friendliness (Cao & Chen 2015). At present, domestic and foreign research on eco-industrial parks mainly includes areas such as park planning, industrial chain, industrial ecosystem, evaluation system, and operation management. Through a comparative analysis of US fringe cities and Chinese industrial parks, Zhao (2017) presented strategies for eco-industrial park planning from the perspective of cities, regions, and networks. Yedla (2017) analyzed the factors influencing the industrial symbiosis network in terms of economics, technology, information, organization and strategy, society, politics, and institutions. In the case study of Rizhao Eco-Industrial Park in China, Yu F et al. found that strict environmental standards, tax preferences, and financial subsidies are the main economic drivers for enterprises to establish industrial symbiosis (Yu et al. 2015).

The research results of eco-industrial parks at home and abroad are relatively abundant, but there are still some problems: the relationship between the industrial network and the industrial chain in the park is unreasonable, the actual utilization rate of waste is not high, and the benefit sharing effect is not obvious; there are many flexible combination forms in the spatial layout of the industrial symbiotic chain network in the park, at present, there is still a lack of scientific evaluation basis for the selection of planning schemes, and the industrial cycle system is often difficult to be implemented. Based on this, this article attempts to plan the ecological industrial chain network based on the theory of reverse logistics and consider the spatial layout in combination with actual land use. Along the technical route of industrial chain network planning, land use suitability, spatial layout plan, and plan selection, a general method from the selection of ecological industrial park enterprises to the generation of spatial layout plans is proposed to improve the scientificity of the planning process and plan selection reliability.

---

*Corresponding Author: chenyou@whut.edu.cn

2 RESEARCH METHOD

2.1 *Reverse logistics*

The eco-industrial park came into being in the pursuit of people to reduce waste of resources, care for the environment, and achieve the harmonious development of the environment and society. Reverse logistics is to coordinate resources and recycle waste for recycling, it is a powerful concept of circular economy in the eco-industrial park support and specific performance (Yu 2017). Based on reverse logistics, a symbiotic combination of industries is formed, and a circular production approach of "producer-consumer-decomposer" is constructed. Therefore, reverse logistics is a concrete manifestation of the circular economy of the park and a key link of the circular economy of the park. Combining the reverse logistics theory to improve the industrial chain of the park can not only improve the correlation between the industrial network and the industrial chain in the park, reduce the production cost of the park, but also save natural resources and meet social needs.

2.2 *Geographic information system*

Geographic Information System (GIS) can analyze and process spatial information and provide rich spatial data information for spatial decision-making. Based on GIS technology, it can not only tap potential industrial symbiosis opportunities more efficiently, improve resource utilization efficiency and reduce waste generation, but also be applied to the planning and management of eco-industrial parks, such as corporate site selection, spatial layout, and waste in terms of recycling and reuse (Wang & Shi 2017). Therefore, the use of GIS technology in the planning process of the eco-industrial park can make the park's spatial layout more practical, better realize the circular economy, and alleviate environmental pollution problems.

2.3 *Goal achievement matrix*

The goal achievement matrix (GAM) was proposed by Hill in the 1860s. It is a quantitative evaluation method for evaluating whether a planning plan meets the predetermined goals. It is often used to evaluate various planning plans. The specific method is as follows. First, the planning principle is embodied into z predetermined goals $(O_1, O_2, O_3, \ldots, O_Z)$, and the conflict rate between the planning plan under each goal and the predetermined goal is calculated, and then the calculation of each plan is the conflict index $E_{o,i}$ and the comprehensive conflict index E_i under each goal. The calculation formulas for conflict rate and conflict index are:

$$CMA = A \cap B / A \tag{1}$$

where CMA is the conflict rate, A is the planning plan, B is the predetermined goal, and $A \cap B$ is the conflict range between the planning plan and the predetermined goal.

$$E_i = \sum_{o=1}^{z} E_{o,i} = \sum_{o=1}^{z} (W_o \times CMA_{o,i} \times 100\%) \tag{2}$$

where W_O is the weight value of the o-th predetermined target, $CMA_{o,i}$ is the conflict rate of the i-th scheme under the o-th predetermined target, and z is the total number of targets.

3 SCHEME GENERATION

The venous industrial park is to establish an eco-industrial park led by the venous industry. The venous industrial park collects, transports, recycles, and finally safely disposes of all kinds of waste generated from production and consumption. It is the carrier and specific practice form of the venous industry. This paper studies the venous industrial park in city A. Based on the goal of reduction, resource utilization, and harmlessness, the park should establish three major cycles of water, energy, and material to realize a circular economy. Combining the theory of reverse logistics, planning for supply-oriented enterprises, production-oriented enterprises, consumer-oriented enterprises, and decomposing enterprises in the park, forming an ecological industrial chain network.

3.1 *Water cycle*

In the water cycle of the park, the decomposing enterprise in a key position is the wastewater treatment plant, which treats production wastewater, sludge drying wastewater, kitchen garbage treatment wastewater, and landfill percolation water and other wastewater, and passes through the reclaimed water reuse system. The reused water is used in incineration plants of production-oriented enterprises, food waste treatment plants, etc., for continuous recycling. The safe landfill and sludge treatment plant will be used as consumer enterprises to share wastewater to the wastewater treatment plant for recycling. A soil vegetation restoration plant will also be built in the park to use recycled water, and the newly generated waste will be processed by a production-oriented enterprise, which is a supply-oriented enterprise in the water cycle.

3.2 *Energy cycle*

The energy cycle of the park is mainly the flow and transformation of electric energy, chemical energy, water, and steam heat energy in the system. Food waste produces biogas, and the separated waste oil is made into biodiesel, which is further used as fuel or chemical raw materials; the electricity generated by the power plant can maintain the normal production and life of the park, and the hot water can be used for heating in the residential area. Therefore, power plants and food waste treatment plants are decomposing enterprises in the energy cycle. In addition, biogas is used for waste incineration power generation and hazardous waste incineration to support combustion, and boiler flue gas can be used for the sludge drying process, so waste incineration plants and sludge treatment plants are production-oriented enterprises.

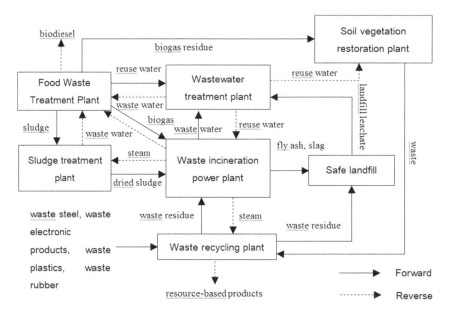

Figure 1. Industrial chain network of the park.

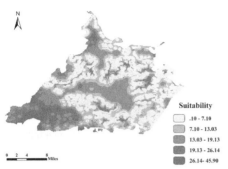 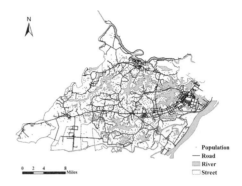

Figure 2. Suitability of the parkland.   Figure 3. City traffic conditions.

3.3 Material cycle

The landfill receives the slag generated from the recycling of various waste resources and construction waste, as well as the fly ash and slag generated from the incineration of various waste materials. As a result of wastewater treatment, sludge and river bottom sludge are further dehydrated and a portion of them are incinerated. A portion of them may also be co-processed with food waste. The residue from the fermentation of food waste is used as a fertilizer for soil remediation. Therefore, waste incineration plants, sludge treatment plants, kitchen waste treatment plants, and soil vegetation restoration are decomposing enterprises responsible for decomposing waste and achieving material recycling. Safe landfills and waste resource treatment plants that handle waste residues, fly ash, fertilizers and other wastes are production-oriented enterprises.

Through the coupling of reverse logistics and industry, the single urban waste treatment and disposal unit in the intravenous industrial park of city A is organically connected, and the circulation of materials and the efficient use of energy are realized. The overall industrial layout of the park, the synergistic relationship of the industrial chain, and the relationship of material circulation are shown in Figure 1.

3.4 Spatial layout plan

3.4.1 Land suitability

In the study, the suitability of the current land use that affects the layout of future residential areas and industrial areas is evaluated, and the areas suitable for construction and forbidden construction are determined. The evaluation of the suitability of the park's land is considered from three factors that affect the residents' quality of life, economic benefits, and ecological benefits. Five factors are selected: the distance from the river, the distance from the existing residential area, the distance from the green plants, the distance from the national and provincial highways, and the slope of the land.

The results of the evaluation of the suitability of the land in the park are shown in Figure 2. From the evaluation of the suitability of the land in the park, the north and southwest areas of the city are more suitable for the construction of industrial parks. Considering that the intravenous industrial park will generate noise and harmful substances, the industrial park should be built far away from residential areas and water sources. However, to reduce transportation costs and facilitate water access, the industrial park cannot be located too far away from the river and has convenient transportation. According to Figure 3, the city is suitable for the construction of an industrial park. Only the northern area meets the above conditions. The general plan of the parkland layout is shown in Figure 4.

3.4.2 *Industrial land layout*

The industrial land layout revolves around the three major cycles in Figure 1, with a waste inciner-ation power plant as the center, forming a radial single-core layout structure. Then the land layout is equivalent to a circular arrangement of n different types of land with a waste incineration power plant as the center. If each type of land is regarded as an object unit, the park plans to include kitchen waste treatment (X1), sludge treatment plant (X2), wastewater treatment plant (X3), soil vegetation restoration plant (X4), safe landfill (X5), waste resource treatment plant (X6) 6 different industrial land. To simplify the arrangement and consider the objects with more reverse logistics together, only the four objects A1 (X1, X2, X3), A2 (X4), A3 (X5), and A4 (X6) need to be arranged and combined. In the end, 24 different combinations were obtained, which corresponded to 24 land use planning schemes.

4 SCHEME SELECTION

4.1 *Setting of target constraints*

The general plan of the parkland layout has considered the suitability of the land in terms of transportation, location, ecology, etc. Then, according to the "3R principle" of the circular econ-omy industrial park, evaluate the generated 24 combination schemes and select the final planning scheme.

The damage to environmental resources caused by improper human development and construc-tion activities occurs from time to time. To reduce the negative environmental impact caused by improper land use in vulnerable areas, ecological sensitivity is selected to limit the environment.

The wastes or by-products generated by the enterprises in the park also become the nutrients for other enterprises, forming interdependent reverse logistics between enterprises. Improving the efficiency of reverse logistics can effectively reduce the time, economic and environmental governance costs of exchanging by-products between enterprises, and maintain the improvement of the ecological industrial chain to a certain extent.

The failure of a node will cause the disappearance of the associated path, which will interfere with the network association effect of the entire industrial park. The importance of each enterprise in the campus network is measured according to the degree of association of the enterprise (the number of paths associated with a node) to obtain the resilience of the campus network.

4.2 *Results and planning*

The index system of ecological sensitivity, reverse logistics efficiency, and park network resilience selected in the previous article is used as the evaluation target of the land use scenario simulation program. The first level represents the largest range of conflicts and the most serious conflicts; the second level represents the second level of conflicts. Otherwise, the conflict is more serious; the third level means that the scope of the conflict is the smallest and the conflict is minimal (Table 1).

Using the spatial overlay function of the GIS thematic map, the 24 land layout schemes and the above-mentioned 3 target constraints are superimposed and analyzed, and the conflict rate of each level under the influence of the 3 single target constraints can be calculated. At the same time, the first conflict level is assumed the weight is assigned a value of 0.6, the second conflict level is assigned a weight of 0.3, and the third conflict level is assigned a weight of 0.1. According to Equations (1) and (2), the comprehensive conflict index under the constraint target for each land layout scenario can be calculated. It can be seen from table 2 that among the top four scores, Scheme 11 has a larger ecological sensitivity conflict index, Scheme 7 has a larger conflict index for reverse logistics efficiency, and Scheme 18 has a larger conflict index for park network resilience, Scheme 16 has significant advantages in terms of ecological sensitivity and reverse logistics efficiency. Therefore, Scheme 16 is selected as the best scheme for this planning for land use layout planning.

Table 1. Target constraints and their conflict levels.

Objective constraints	Conflict level		
	Level 1	Level 2	Level 3
Ecological sensitivity	Within 1 km of rivers and residential areas	Within 1.5 km of rivers and residential areas	Within 2 km of rivers and residential areas
Reverse logistics efficiency	The exchange rate is less than 30%	The exchange rate is between 30%-70%	The exchange rate is more than 70%
Campus network resilience	The number of paths associated with the node is 1 and below	Number of paths associated with the node 2-3	The number of paths associated with the node is 4 and above

Table 2. Environmental conflicts of the top four simulation schemes in total score.

Proposal	Conflict rate of target constraints (%)			Total conflict rate (%)	Overall ranking
	Ecological sensitivity	Reverse logistics efficiency	Campus network resilience		
Scheme 7	15.9	4.2	4.3	24.4	2
Scheme 11	16.8	2.9	11.4	31.1	3
Scheme 16	15.4	1.6	5.2	22.2	1
Scheme 18	15.1	3.5	12.7	31.3	4

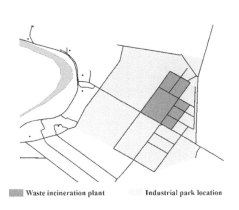

Figure 4. Land use layout plan.

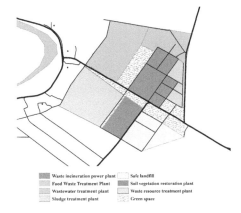

Figure 5. Land use distribution in the park.

Based on the spatial arrangement of industrial land elements in the optimized Scheme 13, the industrial land is spatially implemented, and the final land use layout scheme can be obtained as shown in Figure 5.

5 CONCLUSIONS

In this paper, reverse logistics, GIS spatial analysis, and goal achievement matrix method is adopted to study eco-industrial park space planning. The main conclusions can be summarized as follows:

(1) with the theory of reverse logistics, the industrial chain network of the park can be optimized to meet the environmental protection requirements of society; (2) GIS spatial analysis can be used to optimize the use of land in the park and reduce environmental pollution; (3) For park space planning, the best solution was selected using the goal achievement matrix. In terms of future work, the selection of various types of enterprises and the more complex combination forms need further research.

REFERENCES

Cao Zi; Chen Hongbo (2015) Analysis on the International Experiences of Formation and Development Mechanism of Ecological Industrial Park. *Science & Technology Progress and Policy* 32(9):41–47.

Wang Xue; Shi Xiaoqing (2017) A review of industrial ecology based on GIS. *Acta Ecologica Sinica* 37(04):1346-1357.

Yedla S, Park H (2017) Ecoindustrial networking for sustainable development: review of issues and development strategies. *Clean Technologies and Environmental Policy* 19 (2):391–402.

Yu F, Han F, Cui Z (2015) Evolution of industrial symbiosis in an eco-industrial park in China. *Journal of Cleaner Production*, 2015(87):339–347.

Yu Ying (2017) The construction of my country's reverse logistics system under the condition of circular economy. *Journal of Commercial Economics*, 2017(01):78–80.

Zhao Sidong, Bi Xiaojia, Zhong Yuan, et al. (2017) Chinese industrial park planning strategies informed by American edge cities' development path: a case study of China (Changzhou): Thailand industrial park. *Procedia Engineering* 180: 832–840.

*Advances in Renewable Energy and Sustainable
Development – Liang & Kasmani (Eds)
© 2023 Copyright the Author(s), ISBN: 978-1-032-39407-7*

Possibility analysis of three Gorges Dam to realize sustainable urban development goals—Taking Yichang City as an example

Hsiao-Hsien Lin
School of Physical Education, Jiaying University, Meizhou, China

Su-Fang Zhang
Director, Tourism Management, Athena Institute of Holistic Wellness, Wuyi University, Wuyi Avenue, Wuyishan, Fujian Province, China

Penghui Liu, Heyong Wei & Mei-Ling Chan
School of Physical Education, Jiaying University, Meizhou, China

Chin-Hsien Hsu*
Department of Leisure Industry Management, National Chin-Yi University of Technology. Taichung, Taiwan

Po Hsuan Wu*
Department of Environmental science and Engineering, National Pingtung University of Science and Technology. Pingtung, Taiwan

ABSTRACT: Despite numerous attempts by the government to control floods and boost the economy through water conservancy projects (WCP), a gap between planning and actual results still exists. Therefore, this study explores whether WPC can promote the development of village tourism. Considering the residents and tourists as objects, a mixed research method was adopted, questionnaires were collected, and the impact of village development and natural ecology on re-consumption willingness was analyzed by Pearson performance difference correlation. The results found that the Three Gorges Dam project provides rich tourism resources and vast land for development, further promoting economic development in surrounding rural areas, and improving infrastructure and the community environment. However, if new industries can be developed, diversified tourism activities can be planned, the industrial youth workforce can be maintained, emotional exchanges can be emphasized, water quality can be maintained, garbage can be reduced, and the tourism and living environment can be improved. This will enhance people's desire to relocate and travel as well as promote consumption willingness and behavior.

1 INTRODUCTION

The Yangtze River is one of the main birthplaces of Chinese culture (Chen 2007; Xu et al. 2021; Zhang et al. 2020). During the subtropical monsoon, there is abundant rainfall in spring and summer, severe soil erosion, river silting, the elevation of the riverbed, and loss of dyke function, resulting in frequent floods, affecting village development, and personal safety (Chen 2007; Monkhouse 2017; Mura & Gope 2020; Wang 2021; Xu et al. 2021; Zhang et al. 2020).

The Chinese government proposed the Three Gorges Dam program in 1919, and the construction began in 1993 (Chang et al. 2018; Chen 2007; Gleick 2009; Monkhouse 2017; Mura & Gope 2020; Xu et al. 2021; Zhang et al. 2020; Wang 2021); it took 16 years to complete the construction

---

*Corresponding Authors: hsu6292000@yahoo.com.tw and Peggy.wu.19918012@gmail.com

DOI 10.1201/9781003349648-39

(Rice 2019). The multi-functional reservoir water source area mainly integrates flood control, power generation, transportation, irrigation, shipping, tourism, and water source scheduling (Rice 2019), which has created abundant leisure tourism resources and provided urban development and utilization (Li &Ye 2001).

Yichang City, located toward the end of the Three Gorges Dam, has had an important strategic position since ancient times (Yi 2020). After the completion of the Three Gorges Dam, the government quickly invested in improving public construction (Chang et al. 2018) and developing various leisure and tourism industries (Rice 2019; Li & Ye 2001).

The Dam also promotes the development of the village tourism industry and creates high business opportunities. According to the research statistics, in 2019, the number of tourists was 89.01 million, and the total tourism revenue was USD 98.6 billion, including USD 97.2 billion in domestic tourism revenue and USD 198.05 million in foreign exchange earnings from tourism (Yi 2020; Zhao et al. 2021). The local population has increased to 3.899 million in the surrounding villages (Yi 2019). The construction and application of this dam will promote economic development in Yichang City and improve the livable environment.

Decision-making can use existing natural ecological and cultural resources to promote leisure and tourism activities, improve local predicaments (Yu et al. 2021), and promote economic development (Wu 2021). However, the development and effectiveness of decision-making will vary according to different factors (Wang 2020). Therefore, studying the impact of the economic, social, and environmental planning effects on the surrounding cities, as well as the natural environment and ecological status (Hsu et al. 2020), will contribute to the actual benefits of the Three Gorges Dam on the surrounding cities, natural environment, and ecological development.

Although scholars believe that the quality of decision-making needs to evolve, which is interpreted from the residents' perspective (Lin et al. 2018), and then refers to the views of tourists (Shen et al. 2020). However, we believe that the quality of decision-making is the key to influencing residents' willingness to stay and tourists to travel. Scholars Baker et al. (2020) pointed out that consumption willingness could predict whether actual consumption behaviors meet individual consumption expectations (Baker et al. 2020). Correct decision-making can obtain a stable and safe living environment and meet living conditions (Tang & Hao 2018). When living conditions are safe, sufficient leisure time can be obtained to engage in leisure behaviors and promote consumption, such as buying a house (Lehto & Lehto 2019).

Therefore, the safer the living and traveling environment and the more perfect life functions, the more popular and recognized people will be (Privitera et al. 2019). When the people's willingness to live and travel is strong, it means that the decision-making is trusted by the people (Li &ye 2001; Yi 2020). Therefore, we believe that objective answers can be obtained by integrating residents' and tourists' perceptions of decision-making effectiveness and residence and travel intentions and then evaluating them (Hsu et al. 2020; Wang 2020).

In addition, most of the current related studies discuss development (Wu 2021), freshwater sources, and water quality (Koh & Fakfare 2019), industrial carbon emissions (Yeleliere et al. 2018), hydrology, water pollution, and marine and freshwater ecosystems (Lehner et al. 2021). Many research studies on the Yangtze River are only focused on air pollution and water pollution (Yuan et al. 2020), energy utilization, and land development and utilization (Wu et al. 2021), cruise tourism development, and residents' environmental literacy (Niu et al. 2021). Moreover, most of them are discussed by a single party, such as residents and tourists (Zhu et al. 2021), and they are rarely discussed together. Therefore, we believe that using a variety of research methods to obtain multi-faceted information, through the process of classification, induction, and sorting, after summarizing the research information, cross-comparison is carried out (Shi et al. 2022). Then, the significance of the research information will be discussed by the multivariate inspection analysis method. The influence of river dam facilities on the development of surrounding villages and the natural environment can be analyzed, and more in-depth answers can be obtained (Fischer et al. 2018; Janesick 2000; Strauss & Corbin 1990).

Therefore, we expect to reflect the authenticity of the effect through the willingness of residents and tourists to live, revisit, and provide decision-making suggestions for river water conservancy projects and urban development. This is the value and contribution of this study.

2 METHODS

2.1 Research framework

We use the cognition of village development (economy, society, and environment) and natural ecological development, as well as topics such as willingness to consume (willingness to live and revisit) to obtain information from the perspectives of residents and tourists and use a multi-dimensional examination method for discussion. We propose four hypotheses based on the derivation of the literature. Figure 1 illustrates the main research framework.

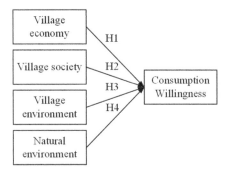

Figure 1. Research framework.

2.2 Research processes and tools

A total of 100 samples were collected in May 2022, and SPSS 26.0 statistical software was used to test by basic statistical verification method. Scholars believe that when the Kaiser-Meyer-Olkin (KMO) > 0.06 and the p-value in the Bartlett test is less than 0.01 ($p < 0.01$), it indicates that the scale is suitable for continuous factor analysis (Wang et al. 2019). When the coefficient α is greater than 0.60, it indicates that the questionnaire has good reliability (Devellis 1991), and the continuous analysis can be reserved. All question alpha values above 0.6 turned out to be acceptable.

3 ANALYSIS

3.1 Correlation analysis of village economic status on cognition of consumption willingness

The analysis results show that the economic development of the village has a significant effect on the willingness to consume ($p < 0.01$), and the results support Hypothesis 1. For the idea of revisiting or living, industry and construction (0.660) have the highest influence, and public facility maintenance (0.415) has the lowest influence. For the idea of recommending relatives and friends, industry and construction (0.627) have the highest influence, and public facility maintenance (0.540) has the lowest influence. Public facilities (0.686) score highest in the idea of sharing consumption experience, and entrepreneurship and employment opportunities (0.508) score the lowest.

3.2 Correlation analysis on the cognition of consumption willingness by the current social conditions of villages

The analysis results show that the social development of villages has a significant effect on the willingness to consume ($p < 0.01$), and the results support Hypothesis 2. Among them, people's communication (0.584) scores the highest in the idea of revisiting or living, and leisure opportunity (0.353) scores the lowest. The idea of recommending relatives and friends is the highest in characteristic culture (0.569) and the lowest in leisure opportunities (0.388). For the idea of sharing consumption experience, tourism service quality (0.640) is the highest, whereas public communication (0.479) is the lowest.

3.3 Correlation analysis of the current situation of the village environment on the cognition of consumption willingness

The analysis results show that the development of the village environment has a significant effect on the willingness to consume ($p < 0.01$), and the results support Hypothesis 2. Among them, the idea of revisiting or living is that the environment is affected by tourists (0.478), and the pollution increase (0.220) is the lowest. The idea of recommending relatives and friends is the highest with sufficient trashcans (0.540) and the lowest by the environment affected by tourists (0.407). The idea of sharing consumption experience scores the highest with adequate public toilets (0.606) and the lowest with increased pollution (0.435).

3.4 Correlation analysis of the cognition of consumption willingness by the status of natural ecology and environmental development of villages

The analysis results show that the development of the village's natural ecology and environmental development t has a significant effect on the willingness to consume ($p < 0.01$), and the results support Hypothesis 2. Among them, the turbidity of river water (0.273) is the highest in the idea of revisiting or living, and the soil erosion along the river course (0.203) is the lowest. For the idea of recommending relatives and friends, the river turbidity and tourism waste (0.337) score the highest, and the original ecological habitat change (0.283) scores the lowest. The increase of tourism waste (0.446) scores the highest in the idea of sharing consumption experience and the lowest in the turbidity of river water (0.377).

4 CONCLUSIONS AND RECOMMENDATIONS

This study has adopted quantitative research, taking residents and tourists in Yichang City as the object and using the Pearson performance difference correlation coefficient to analyze the questionnaire. The main conclusions can be summarized as follows.

(1) The Three Gorges Dam project provides rich tourism resources and vast land for development, promoting the economic development of surrounding rural areas and improving infrastructure and the community environment.
(2) Due to the high similarity of industrial types and low technology, the outflow of young people, lack of labor, neglect of emotional communication, and loss of the simple humanistic characteristics of the village, water pollution and tourism waste increase, and the quality of tourism and living environment sanitation decline.

Thus, the idea of migrants and residents of that area migrating and settling there is hampered, the intention of tourists to consume is diminished, and the willingness of consumers to consume is influenced.

REFERENCES

B Lehner, L Katiyo, F Chivava, HM Sichingabula, E Nyirenda, NA Rivers-Moore, BR Paxton, G Grill, F Nyoni, B Shamboko-Mbale, K Banda, ML Thieme, OM Silembo, A Musutu, R Filgueiras. Identifying priority areas for surface water protection in data scarce regions: An integrated spatial analysis for Zambia. Aquatic Conservation: Marine and Freshwater Ecosystems, 2021, 31(8), 1998–2016. https://doi.org/10.1002/aqc.3606

B Niu, D Ge, R Yan, Y Ma, D Sun, M Lu, Y Lu. *The evolution of the interactive relationship between urbanization and land-use transition: A case study of the Yangtze River Delta. Land*, 2021, 10(8), 804. https://doi.org/10.3390/land10080804

CC Shen, CF Liang, CH Hsu, JH Chien, HH Lin. Research on the Impact of Tourism Development on the Sustainable Development of Reservoir Headwater Area Using China's Tingxi Reservoir as an Example. *Water*, 2020, 12(12), 3311. Doi:10.3390/w12123311

CH Wu. Research on the Coastal Marine Environment and Rural Sustainable Development Strategy of Island Countries-Taking the Penghu Islands as an Example. Water 2021, 13(10), 1434. https://doi.org/10.3390/w13101434

Chin-Hsien Hsu, Hsiao-Hsien Lin, Shang-Wun Jhang, Tzu-Yun Lin. Does environmental engineering help rural industry development? Discussion on the impact of Taiwan's "special act for forward-looking infrastructure" on rural industry development. *Environmental Science and Pollution Research*, 2020, 28, 40137–40150. Doi:10.1007/s11356-020-11059-6

CYA Chang, Z Gao, A Kaminsky, TG Reames. Michigan sustainability case: Revisiting the Three Gorges Dam: Should China continue to build dams on the Yangtze River? Sustainability: *The Journal of Record*, 2018, 11(5), 204–215. http://doi.org/10.1089/sus.2018.29141.cyac

D Fischer, R Mauer, M Brettel. Regulatory focus theory and sustainable entrepreneurship. *International Journal of Entrepreneurial Behaviour & Research*, 2018, 24(2), 1–15. DOI: 10.1108/IJEBR-12-2015-0269

D Privitera, A Nedelcu, V Nicula. Gastronomic and food tourism as an economic local resource: Case studies from Romania and Italy. *GeoJournal of Tourism and Geosites*, 2019, 21(1), 144 33–11557. Retrieved from: http://gtg.webhost.uoradea.ro/

D Wang. Artificial intelligence–based mountain soil erosion and the impact of climate conditions on marathon competitions. *Arabian Journal of Geosciences*, 2021, 14, 948. https://doi.org/10.1007/s12517-021-07245-6

D Zhang, X Shi, H Xu, Q Jing, X Pan, T Liu, H Wang, H Hou. A GIS-based spatial multi-index model for flood risk assessment in the Yangtze River Basin, China. *Environmental Impact Assessment Review*, 2020, 83, 106397. https://doi.org/10.1016/j.eiar.2020.106397

D Zhao, M Xiao, C Huang, Y Liang, Z An. Landscape Dynamics Improved Recreation Service of the Three Gorges Reservoir Area, *China. Int. J. Environ. Res. Public Health* 2021, 18(16), 8356. https://doi.org/10.3390/ijerph18168356

Devellis, R.F. Scale Development: Theory and Applications; Sage: Newbury Park, CA, USA, 1991.

E Koh, P Fakfare. Overcoming "over-tourism": The closure of Maya Bay. *International Journal of Tourism Cities*, 2019, 6(2), 279-296. https://doi.org/10.1108/IJTC-02-2019-0023

E Yeleliere, SJ Cobbina, AB Duwiejuah. Review of Ghana's water resources: the quality and management with particular focus on freshwater resources. *Applied Water Science*, 2018, 8, 93. https://doi.org/10.1007/s13201-018-0736-4

FJ Monkhouse. *A dictionary of geography, 2nd Edition. Routledge*, Taylor & Francis, New York, 2017. https://doi.org/10.4324/9781315083629

G Li, W Ye. Influence of the Three Gorges Project on the evolution of the Yangtze Three Gorges tourism layout and countermeasures. *Geography and Territorial Research*, 2001, 4. Retrieved from: http://en.cnki.com.cn/Article_en/CJFDTotal-DLGT200104007.htm

HH Lin, SS Lee, YS Perng and ST Yu. Investigation about the Impact of Tourism Development on a Water Conservation Area in Taiwan. *Sustainability*, 2018, 10(7), 2328. Doi: 10.3390/su10072328

J Shi, K Xu, K Duan. Investigating the intention to participate in environmental governance during urban-rural integrated development process in the Yangtze River Delta Region. Environmental *Science & Policy*, 2022, 128, 132–141. https://doi.org/10.1016/j.envsci.2021.11.008

J Zhu, H Wang, B Xu. Using Fuzzy AHP-PROMETHEE for Market Risk Assessment of New-Build River Cruises on the Yangtze River. *Sustainability*, 2021, 13(22), 12932. https://doi.org/10.3390/su132212932

Janesick, V. J. (2000). *The choreography of qualitative research design: Minuets, improvisations, and crystallization*. In N. K. Denzin & Y. S. Lincoln (Eds.), Handbook of qualitative research, 379-399. Thousand Oak, CA: Sage.

JE Rice. Three Gorges Dam. Muitchell lane Publishers, USA, 2019. Retrieved from: https://books.google.com. tw/books?hl=zh-TW&lr=&id=qJKgDwAAQBAJ&oi=fnd&pg=PT7&dq=The+Yangtze+River+proposed+ the+construction+of+the+Three+Gorges+Dam+in+1919,+and+finally+started+construction+in+1993.+& ots=O1c_texjZw&sig=sYXitGHjwtT6fV6FhibAfHub6Zg&redir_esc=y#v=onepage&q&f=false

JH Yu, HH Lin, YC Lo, CH Hsu, Y Liang, Z An. Is the travel bubble under COVID-19 a feasible idea or not? *Int. J. Environ. Res. Public Health*, 2021, 18(11), 5717. https://doi.org/10.3390/ijerph18115717

PH Gleick. Three Gorges Dam Project, Yangtze River, China. The World's Water 2008-2009: *The Biennial Report on Freshwater Resources*, 2009.Retrieved from: https://books.google.com.tw/books?id=uIGRAsSAv tEC&printsec=frontcover&hl=zh-TW#v=onepage&q&f=false

Q Yuan, B Qi, D Hu, J Wang, J Zhang, H Yang, S Zhang, L Liu, L Xu, W Li. Spatiotemporal variations and reduction of air pollutants during the COVID-19 pandemic in a megacity of Yangtze River Delta in China. Science of The Total Environment, 751, 141820. https://doi.org/10.1016/j.scitotenv.2020.141820

S Baker, J Waycott, E Robertson, R Carrasco, BB Neves, R Hampson, F Veterea. Evaluating the use of interactive virtual reality technology with older adults living in residential aged care. Information Processing & Management, 2020, 57(3), 102105. https://doi.org/10.1016/j.ipm.2019.102105

S Tang, P Hao. Floaters, settlers, and returnees: Settlement intention and hukou conversion of China's rural migrants. China Review, 2018, 8(1), 11-34. Retrieved from: https://www.jstor.org/stable/26435632.

SNS Mura, A Gope. Anthropogenic Impact on Forms and Processes of the Kangsabati River Basin. Anthropogeomorphology of Bhagirathi-Hooghly River System in India, 2020. Retrieved from: https://www.taylorfrancis. com/chapters/edit/10.1201/9781003032373-8/anthropogenic-impact-forms-processes-kangsabati-river-basin-shambhu-nath-sing-mura-ananta-gope

Strauss & Corbin, (1990). Basics of qualitative research: Grounded theory procedures and techniques. Newbury Park, CA: Sage.

W Wu, T Zhang, X Xie, Z Huang. Regional low carbon development pathways for the Yangtze River Delta region in China. Energy Policy, 2021, 151, 112172. https://doi.org/10.1016/j.enpol.2021.112172

X Chen. A tale of two regions in China: Rapid economic development and slow industrial upgrading in the Pearl River and the Yangtze River Deltas. *International Journal of Comparative Sociology*, 2007, 48(2–3), 167–201. https://doi.org/10.1177/0020715207075399

XY Lehto, MR Lehto. Vacation as a public health resource: toward a wellness-centered tourism design approach. *Journal of Hospitality & Tourism Research*, 2019, 43(7), 935–960. DOI: 10.1177/1096348019849684.

Y Xu, Z Jin, LF Gou, A Galy, C Jin, C Chen, C Li, L Deng. Carbonate weathering dominates magnesium isotopes in large rivers: Clues from the Yangtze River. *Chemical Geology*, 2022, 588, 120677. https://doi.org/10.1016/j.chemgeo.2021.120677

Yichang Municipal Bureau of Statistics. *Yichang Statistical Yearbook-2020* (PDF) First Edition. Beijing: China Statistics Press.

Yichang Municipal People's Government. Yichang City Situation - Population Statistics. Retrieved from: yichang.gov.cn (2019-09-09).

Yi-Nan Wang. Have the "prophecies" that opposed the construction of the Three Gorges been fulfilled? Why do we need the Three Gorges? *China Economic Weekly*, 2020. Retrieved from: http://www.ceweekly.cn/2020/0717/305465.shtml

Z Wang, L Yin, X Qin, S Wang. Integrated assessment of sediment quality in a coastal lagoon (Maluan Bay, China) based on AVS-SEM and multivariate statistical analysis. *Marine pollution bulletin*, 2019, 146, 476-487. https://doi.org/10.1016/j.marpolbul.2019.07.005

Author index

Bai, L. 137
Bi, B. 243

Cai, C. 268
Cao, J. 35
Cao, Y. 230
Chan, M.-L. 275
Chen, F. 137
Chen, L. 19, 102, 195, 217
Chen, M. 59
Chen, Q. 195, 217
Chen, R. 179
Chen, S. 67
Chen, T. 52
Chen, Z.-H. 190
Chu, Y. 256

Deng, Z. 59
Ding, Y. 19, 52
Duan, Y. 230

Fan, X. 205
Fan, Z. 102
Fang, R. 205
Feng, P. 52
Feng, Y. 145

Gao, J. 28, 73
Guan, P. 12
Guo, M. 102

He, B. 237
He, L. 205
Hou, J. 59
Hou, T. 205
Hsu, C.-H. 275
Hu, L. 160
Hua, W. 89
Huang, L. 166
Huang, X. 12
Hung, C. 67

Ji, W. 145
Jin, M. 195, 217

Kang, W. 59

Li, F. 35, 89
Li, L. 89
Li, M. 172
Li, N. 84
Li, Q. 67
Li, X.-Y. 190
Li, Y. 52
Li, Z. 28
Liang, L. 59
Liang, Z.-F. 190
Liao, C. 67
Liao, W. 205
Lin, H. 67
Lin, H.-H. 275
Lin, X. 113
Liu, P. 275
Liu, Q. 12, 73, 79
Liu, X. 12, 268
Liu, Y. 28
Liu, Z. 89, 95, 155
Lu, H. 130
Lu, Q. 113
Lu, Z. 89
Luo, C. 205
Lv, X. 3
Lyu, H. 256

Ma, M. 59
Men, G. 145
Meng, X. 113
Mou, H. 102

Ni, T. 237

Pan, J. 190

Ren, J. 102

Shao, J. 263
Shi, C. 35

Shiu, J.-Y. 121
Sun, C. 35
Sun, W. 40
Sun, Y. 256

Tan, L. 102
Tang, S.-Y. 190

Wang, C. 268
Wang, H. 45
Wang, J. 137
Wang, K. 52
Wang, X. 121
Wang, Y. 212
Wang, Z. 205
Wei, D. 12
Wei, H. 275
Wu, C. 67
Wu, C.-H. 121
Wu, H. 195
Wu, J. 250
Wu, P.H. 275
Wu, S. 79
Wu, Y. 230

Xie, K. 250
Xie, Y. 179
Xie, Z. 84
Xin, Y. 28
Xing, Y. 195, 217
Xu, J. 179

Yang, B. 3, 137
Yang, H. 35
Yang, M. 73
Yang, Q. 268
Yang, S. 212
Yang, X. 195, 217
Yang, X.-Q. 190
Yang, Y. 102
Yang, Z. 217
Yang, Z.-L. 190

Yu, G. 19
Yu, H. 102
Yuan, Y. 79

Zhang, B. 145
Zhang, H. 243
Zhang, K. 137
Zhang, L. 3

Zhang, M. 102, 195, 217
Zhang, N. 59
Zhang, P. 40
Zhang, S. 137, 179
Zhang, S.-F. 275
Zhang, X. 19, 212
Zhang, Y. 12
Zhang, Z. 3, 145

Zhao, L. 84
Zhao, Z. 113, 212
Zheng, Z. 102
Zhou, R. 256
Zhou, Y. 195, 217
Zhu, G. 28
Zhu, L. 52
Zhu, Z. 84